Biotechnology Intelligence Unit

Biotechnological Applications of Photosynthetic Proteins: Biochips, Biosensors and Biodevices

Maria Teresa Giardi, Ph.D.
Group on Photosynthetic-Based Biosensors
National Council of Research-IC
Institute of Crystallography, CNR
Monterotondo Scalo, Rome, Italy

Elena V. Piletska, Ph.D.
Institute of Bioscience and Technology
Cranfield University
Silsoe, Bedfordshire, U.K.

Landes Bioscience / Eurekah.com
Georgetown, Texas
U.S.A.

Springer Science+Business Media
New York, New York
U.S.A.

Biotechnological Applications of Photosynthetic Proteins:
Biochips, Biosensors and Biodevices

Biotechnology Intelligence Unit

Landes Bioscience / Eurekah.com
Springer Science+Business Media, LLC

ISBN: 0-387-33009-7 Printed on acid-free paper.

Springer Science+Business Media, LLC, 233 Spring Street, New York, New York 10013, U.S.A.
http://www.springer.com

Please address all inquiries to the Publishers:
Landes Bioscience / Eurekah.com, 810 South Church Street, Georgetown, Texas 78626, U.S.A.
Phone: 512/ 863 7762; FAX: 512/ 863 0081
http://www.eurekah.com
http://www.landesbioscience.com

Printed in the United States of America.

9 8 7 6 5 4 3 2 1

Library of Congress Cataloging-in-Publication Data

Biotechnological applications of photosynthetic proteins : biochips, biosensors, and biodevices / [edited by] Maria Teresa Giardi, Elena V. Piletska.
p. ; cm. -- (Biotechnology intelligence unit)
Includes bibliographical references and index.
ISBN 0-387-33009-7 (alk. paper)
1. Proteins--Biotechnology. 2. Biochips. 3. Biosensors. I. Giardi, Maria Teresa. II. Piletska, Elena V. III. Series: Biotechnology intelligence unit (Unnumbered)
[DNLM: 1. Photosynthetic Reaction Center Complex Proteins. 2. Biosensing Techniques. 3. Biotechnology. 4. Oligonucleotide Array Sequence Analysis. 5. Photosystem II Protein Complex. QU 55 B6156 2006]
TP248.65.P76B56 2006
660.6'3--dc22

2006004699

About the Editors...

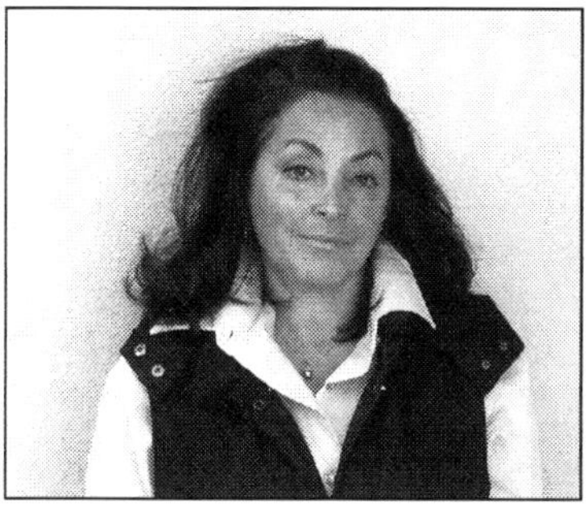

MARIA TERESA GIARDI is a coordinator of multi-disciplinary studies on the realization of biosensors for the European Community and the European Space Agency. She is research leader of a group of young post docs with a special role in direct participation in several of the results presented in this book. She has a background in organic chemistry; her scientific work includes studies on biochemical mechanisms of photosynthesis, stress biochemistry and photosynthetic biosensors. She supports a sustainable technological development.

ELENA V. PILETSKA graduated from Moscow State University in 1985 and gained her Ph.D. in biochemistry from A.N. Bach Institute of Biochemistry (Moscow) working on genome of chloroplasts (1989). Dr. Piletska joined the Institute of BioScience and Technology, Cranfield University, U.K. in 1998. Her current research interests include molecular recognition using synthetic and natural receptors, polymer and analytical chemistry and computational modelling.

This book is dedicated to Gianni, Raffaele,
Ginko, Kyria, Gea, Spina, Ruya and Eliott.

—*Maria Teresa Giardi, Ph.D.*

CONTENTS

EDITORS

Maria Teresa Giardi
Group on Photosynthetic-Based Biosensors
National Council of Research-IC
Institute of Crystallography, CNR
Monterotondo Scalo, Rome, Italy
Chapters 1, 8, 10, 13, 17, 18

Elena V. Piletska
Institute of Bioscience and Technology
Cranfield University
Silsoe, Bedfordshire, U.K.
Chapters 7, 12, 14-16

CONTRIBUTORS

Angela Agostiano
Dipartimento di Chimica
Università di Bari
and
CNR-IPCF sez
Bari, Italy
Chapter 9

Giovanni Basile
Biosensor Srl
Palombara Sabina, Italy
Chapter 18

Ivo Bertalan
Institut für Pflanzenphysiologie
Martin-Luther Universität
Halle-Wittenberg
Halle, Germany
Chapter 5

Florent Breton
Université de Perpignan
Centre de Phytopharmacie
Perpignan, France
Chapters 14, 15

Robert Carpentier
Groupe de Recherche en Énergie
et Information Biomoléculaires
Université du Québec à Trois-Rivières,
Trois-Rivières, Québec, Canada
Chapters 7, 15

Pinalysa Cosma
Dipartimento di Chimica
Università di Bari
and
CNR-IPCF sez
Bari, Italy
Chapter 9

Flavia di Costa
Institute of Crystallography, CNR
National Council of Research-IC
Monterotondo Scalo, Rome, Italy
Chapter 18

Dania Esposito
Institute of Crystallography, CNR
Monterotondo Scalo, Rome, Italy
Chapters 8, 17

Pierre Euzet
Université de Perpignan
Centre de Phytopharmacie
Perpignan, France
Chapter 7

Cecilia Faraloni
Istituto per lo Studio degli Ecosistemi, CNR
Sezione di Firenze
Florence, Italy
Chapter 17

Floriana Fasolo
Istituto Nazionale Fisica Nucleare
Turin, Italy
Chapter 17

Mayank Garg
Institute of Crystallography, CNR
Monterotondo Scalo, Rome, Italy
Chapter 10

Demetrios F. Ghanotakis
Department of Chemistry
University of Crete
Heraklion, Greece
Chapter 3

Donato Giannino
Institute of Biology and Agricultural Biotechnology, CNR
Monterotondo Scalo, Rome, Italy
Chapter 10

Louisa Giannoudi
Institute of Bioscience and Technology
Cranfield University
Silsoe, Bedfordshire, U.K.
Chapter 16

Licia Guzzella
IRSA, CNR
Brugherio, Milan, Italy
Chapter 11

Lydia Hilbig
Institut für Pflanzenphysiologie
Martin-Luther Universität Halle-Wittenberg
Halle, Germany
Chapter 5

Udo Johanningmeier
Institut für Pflanzenphysiologie
Martin-Luther Universität Halle-Wittenberg
Halle, Germany
Chapter 5

Khalku (Kal) Karim
Institute of Bioscience and Technology
Cranfield University
Silsoe, Bedforshire, U.K.
Chapter 14

Prashant Katiyar
Institute of Crystallography, CNR
Monterotondo Scalo, Rome, Italy
Chapter 10

Josef Komenda
Institute of Microbiology Academy of Sciences
Trebon, Czech Republic
and
Institute of Physical Biology
University of South Bohemia
Nové Hrady, Czech Republic
Chapter 4

Stanisiava Kuviková
Institute of Microbiology Academy of Sciences
Trebon, Czech Republic
and
Institute of Physical Biology
University of South Bohemia
Nové Hrady, Czech Republic
Chapter 4

Francesco Longobardi
Dipartimento di Chimica
Università di Bari
and
CNR-IPCF sez
Bari, Italy
Chapter 9

Lenka Lupínková
Institute of Microbiology
Academy of Sciences
Trebon, Czech Republic
and
Institute of Physical Biology
University of South Bohemia
Nové Hrady, Czech Republic
Chapter 4

Andrea Margonelli
Institute of Crystallography, CNR
Monterotondo Scalo, Rome, Italy
Chapter 17

Domenico Mariotti
Institute of Biology and Agricultural
Biotechnology, CNR
Monterotondo Scalo, Rome, Italy
Chapter 10

Marco Mascini
Biosensors Laboratory
Department of Chemistry
University of Florence
Florence, Italy
Chapter 2

Jiri Masojídek
Institute of Microbiology
Academy of Sciences
Trebon, Czech Republic
and
Institute of Physical Biology
University of South Bohemia
Nové Hrady, Czech Republic
Chapter 4

Margarita S. Odintsova
A.N. Bach Institute of Biochemistry
Russian Academy of Sciences
Moscow, Russia
Chapter 6

Walter Oettmeier
Biochemie der Pflanzen
Ruhr-Universität Bochum
Bochum, Germany
Chapter 5

Emanuela Pace
Institute of Crystallography, CNR
National Council of Research-IC
Monterotondo Scalo, Rome, Italy
Chapters 10, 13, 18

Sergey A. Piletsky
Institute of Bioscience and Technology
Cranfield University
Silsoe, Bedfordshire, U.K.
Chapters 7, 12, 14-16

Fiorenzo Pozzoni
IRSA, CNR
Brugherio, Milan, Italy
Chapter 11

Emmanuel Psylinakis
Department of Human Nutrition
and Dietetics
School of Food Technology
and Dietetics
Technological Educational Institute
of Crete
Crete, Greece
Chapter 3

Régis Rouillon
Université de Perpignan
Centre de Phytopharmacie
Perpignan, France
Chapters 7, 12, 14, 15

Jana Schulze
Institut für Pflanzenphysiologie
Martin-Luther Universität
Halle-Wittenberg
Halle, Germany
Chapter 5

Aspasia Spyridaki
Department of Human Nutrition
and Dietetics
School of Food Technology
and Dietetics
Technological Educational Institute
of Crete
Crete, Greece
Chapter 3

Giulio Testone
Institute of Biology and Agricultural
Biotechnology, CNR
Monterotondo Scalo, Rome, Italy
Chapter 10

Giuseppe Torzillo
Istituto per lo Studio degli Ecosistemi,
CNR
Sezione di Firenze
Florence, Italy
Chapters 8, 17

Eleftherios Touloupakis
Department of Chemistry
University of Crete
Crete, Greece
Chapter 18

Stefan Wilski
Biochemie der Pflanzen
Ruhr-Universität Bochum
Bochum, Germany
Chapter 5

Nadezhda P. Yurina
A.N. Bach Institute of Biochemistry
Russian Academy of Sciences
Moscow, Russia
Chapter 6

Alba Zanini
Istituto Nazionale Fisica Nucleare
Turin, Italy
Chapter 17

Edda Zeidler
Institut für Pflanzenphysiologie
Martin-Luther Universität
Halle-Wittenberg
Halle, Germany
Chapter 5

Acknowledgements

The authors thank the European Union
(contract QLK3-CT-2001-01629),
MIUR (prot. 1633/ric FISR)
and ASI-ESA (Photo I/004/05/0) for their support.

CHAPTER 1

Introduction:
The Emergence of a New Technology

Maria Teresa Giardi*

The possibility of producing a new generation of technological devices that integrate the knowledge coming from various fields (chemistry, biology, computer science, electronics, engineering) is attracting increasing attention. This trend has introduced a new technological science called "molecular electronics" or "nanotechnology". It is a technology based on the use of molecular scale components such as a single or a few molecules, carbon nanotubes, nanoscale metallic and/or semiconductor wires, etc. that function as electronic components.

RC-biotechnology refers to the use of Reaction Centres (RC) and more in general of photosynthetic proteins, for technological purposes. It regards the construction of photo optical-electrical devices based on photosynthetic proteins. Photosynthetic RC proteins are suitable biological material for the construction of devices because they exhibit light-induced electron transfer across lipid membranes. Many chromophore molecules, such as bacteriochlorophylls, bacteriopheophytins and quinones, are arranged in RCs with relevant interchromophore distances and relevant gaps in the energy levels of each chromophore to ensure unidirectional electron transfer.

The development of biosensors represents a valuable step towards the advancement of pollutant monitoring in ecosystems. Biosensors are analytical devices that consist of a biosensing element (enzyme, tissue, living cell) that provides selectivity and a transducer that transfers the chemical signal to an electrical signal for further processing. Therefore even a single protein molecule of an RC is a sophisticated molecular device. They are able to generate supramolecular and self-assembling structure and, hence, are natural nanostructures.

In recent years, progress on isolation of RC and of photosystem II (PSII) particles has been obtained, and it is now possible to isolate quite stable and pure preparations from plant thylakoid and cyanobacterial membranes by detergent solubilization. These preparations are capable of light-induced oxygen evolution, at high rates, and/or electron transfer in the absence or presence of benzoquinones as artificial electron acceptors. The RC isolated from photosynthetic bacteria is particularly stable against denaturation. Moreover, recent advances in RCII biochemistry and molecular biology (site-directed mutagenesis) have produced a number of mutants resistant to extreme conditions, showing altered amino acid composition of the D1 protein.

RC-biotechnology exploits the characteristics of the pigment-protein complexes located within the membrane of plants, algae, cyanobacteria and bacteria. However, the structures, functions and potentials of the photosynthetic complexes are different in the various photosynthetic organisms. We can distinguish the technological applications obtained from the three types of photosynthetic proteins from bacteria (RC), from cyanobacteria, algae and higher plants (RCII) and rhodopsin from halobacteria (bR). RC from bacteria was utilised for building several biochip types; RCII-technology includes applications such as photonic-crystal bandgap materials, biosensors and

*Corresponding Author: Maria Teresa Giardi—Group on Photosynthetic-Based Biosensors National Council of Research-IC, Via Salaria km 29.3, Area of Research of Rome, Rome, Italy. Email: giardi@mlib.cnr.it

Biotechnological Applications of Photosynthetic Proteins: Biochips, Biosensors and Biodevices, edited by Maria Teresa Giardi and Elena V. Piletska.

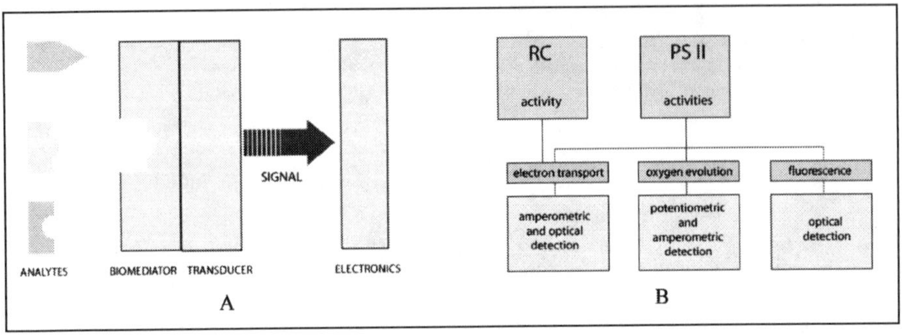

Figure 1. Schematic representation of a biosensor (A); Photosynthetic Photosystem II and Reaction Centre activities (B).

biodevices; finally, bR-technology includes a reversible holographic memory, an ultrafast random-access memory, pattern-recognition systems and photoelectrical cells.

The advantage of using RC biodevices mainly depends on the specificity of the enzyme to recognize certain analytes or particular physicochemical conditions. Moreover, RC and PSII are especially suitable because of their physiological activities that can be easily monitored by amperometric, potentiometric and optical systems (Fig. 1).

Why and How to Make a Photosynthetic-Based Biosensor

Despite initial enthusiasm, biosensors have not yet achieved the prominent commercial application that was initially predicted. This slow progress is not surprising since biosensor development requires the combined expertise of biologists, chemists, biotechnologists, biochemists, physicists, and mechanical and electrical engineers. It is rare to have so many disciplines in a commercial company.

The construction of RC-based devices requires a multidisciplinary approach where the device is obtained in various stages. First of all the physiology of the photosynthetic organism should be considered since it is essential before designing the biodevice. For instance, the choice of a thermophilic cyanobacterium as a biomediator guarantees the stability of biosensors for monitoring herbicides. The second stage is the isolation of the biomediator, its stabilization and immobilization by biochemical and chemical techniques. Using molecular biology, mutations on PSII complexes that produce specific properties can be carried out; e.g., a modification of a single amino acid on D1 protein generates a resistance towards herbicide subclasses. Bioinformatics is then applied to optimise the molecular modifications on the biomediator. The next stage is the study of suitable transduction system for the detection of the chosen PSII activity and the analyses of the data. Finally, biosensor prototypes for specific applications are designed (e.g., field portable, device for laboratory, miniature size prototype etc.).

Although it utilises the same basic biomediator, the tech applications can be divided into different classes, based on different concepts of the RC properties.

Potential and Prospective of the RC-Biotech for Basic Research and Applications

The production of mono/multi-molecular layers is now a reality. The layers are made up integrating natural or engineered photosynthetic systems with synthetically derived molecules that act as a means of transduction and immobilisation.

A common feature of the various RCs is the trigger of the photochemistry by solar energy, but there are differences in the way they convert energy and consequently on their potential application for the building of biochips, biosensors, photovoltaic and photoelectrochemical cells, holographic memory etc.

The present RC-technology is geared around the concept that the engineered RC can be the core of numerous innovative devices. The technological applications described above can be divided into distinct classes of innovative products based on the different RC-biochip properties. However, the skills and knowledge required to manipulate and settle the RC complex properties are common. Single changes in amino acid sequences can create more stable biomediators and increase the efficiency of selected photochemical processes. Moreover, the required functionality for new devices can be found in molecules that are in natural abundance.

The emergence of RC-technology would perhaps be a benefit to those looking to establish commercial devices. Certainly the number of patents based on photosynthetic proteins that are applied for and granted every year is increasing, and that is a clear indication of its future commercial success.

The data summarised here can serve as a basis for the development of a commercial biosensor for use in rapid prescreening analyses of PSII pollutants, minimising costly and time-consuming laboratory analyses.

The aim of this book is to give a general description of the basic and technical research in this sector.

CHAPTER 2

A Brief Story of Biosensor Technology

Marco Mascini*

Introduction

The vast literature in the last 40 years related to the keyword **Biosensor** reveals without doubt that the scientific field is **attractive**! We realized at once that several researchers with different background are involved in this field of research, from chemistry to physics, to microbiology and of course to electrical engineering, all are deeply involved in several facets of the assembly of the object "Biosensor".

Looking at the past we realize also that the concept of Biosensor has evolved!

For some authors, especially at the beginning of this research activity, i.e., about 40 years ago, Biosensor is a self contained analytical device that responds to the concentration of chemical species in biological samples! This is clearly wrong, but it has been very difficult to clarify this point! No mention of a biological active material involved in the device! Thus any physical (thermometer) or chemical sensor (microelectrode implanted in animal tissue) operating in **biological samples** could be considered a Biosensor. We agree that a biosensor can be defined as a device that couples a biological sensing material (we can call it a molecular biological recognition element) associated with a transducer.

In 1956 Professor Leland C. Clark publishes his paper on the development of an oxygen probe and based on this research activity he expanded the range of analytes that could be measured in 1962 in a Conference at a Symposium in the New York Academy of Sciences where he described how to make electrochemical sensors (pH, polarographic, potentiometric or conductometric) more intelligent by adding "enzyme transducers as membrane enclosed sandwiches".[1] The first example was illustrated by entrapping the enzyme Glucose Oxidase in a dialysis membrane over an oxygen probe. The addition of glucose determined the decrease of oxygen concentration in proportional relation! The first biosensor was described in the published paper coining the term "enzyme electrode".[2] Then subsequently in 1967 Updike and Hicks use the same term "enzyme electrode" to describe a similar device where again the enzyme glucose oxidase was immobilized in a polyacrylamide gel onto a surface of an oxygen electrode for the rapid and quantitative determination of glucose.[3]

Besides amperometry Guilbault and Montalvo in 1969 use glass electrodes coupled with urease to measure urea concentration by potentiometric measurement.[4]

Starting from 1970, several others authors start to prove the concept of Biosensors, the coupling of an enzyme and electrochemical sensors. This was at the beginning a Biosensor, a strange research where biological elements were combined with electrochemical sensors.

In the electrochemical community at that period the research on ion selective electrodes (ISE) was very active and the idea to extend the range of sensors to non electrochemical active compounds, and even to non ionic compounds, like glucose, has been very well accepted. We saw at that time the possibility to extend much more the research activity. The groups active in ISE development have been definitively the first to shift to the development of electroanalytical biosensors.

*Corresponding Author: Marco Mascini—Biosensors Laboratory, Department of Chemistry, University of Florence, Florence, Italy. Email: mascini@unifi.it

Biotechnological Applications of Photosynthetic Proteins: Biochips, Biosensors and Biodevices, edited by Maria Teresa Giardi and Elena V. Piletska. ©2006 Landes Bioscience.

Table 1. Biosensors

Receptors	Transducers
• Tissues • Microorganisms • Organelles • Cell receptors • Enzymes • Antibodies • Nucleic acids • Biomimetic receptors	• Electochemical • Optical • Thermometric • Piezoelectric • Magnetic

I recollect very elegant research starting from Prof. G. Rechnitz involving the development of an "amygdaline" sensor based on the coupling of an Ion Selective Electrode (cyanide ISE) with betaglucosidase to give benzaldeyde and cyanide.[5]

But this was just the beginning of a large activity where the obtained couplings have been multiplied by changing the "biological element" and the kind of transducer! Enzyme, multiple enzymes, organelles, bacteria, specialized biological tissue, containing specific enzymes were coupled to potentiometric or amperometric devices, then optical, thermometric, piezoelectric, etc. We continue also today to enlarge the list of physical sensors with the last entry of "magnetic devices". Recently the concept evolved again in the tentative to replace or mimic the biological material with synthetic chemical compounds! Table 1 demonstrates all kinds of couplings have been used in order to obtain Biosensors.

Enzymes (and all biological elements based on the enzymes contained in it) represent the class of what is now called "catalytic elements". The other important class is represented by the "affinity elements", namely antibodies, lectins, nucleic acids (DNA and RNA) and recently synthetic ligands.

Biomolecular sensing can be then defined as the possibility to detect analytes of biological interest, like metabolites, but also of environmental concerns or of any other technological field where the concentration of a specific compound is important to be quantified in a complex sample.

The exploitation of the selectivity of the biological element is the "driving force" of the Biosensor.

The Problem of Amplification

Catalytic events or affinity events have not the same scheme of transduction. If the biological recognition element present in the sensing layer is an enzyme or generally a biocatalyst, a reaction takes place in the presence of the specific target analyte and an increasing amount of coreactant or product is consumed or formed, respectively, in a short time depending on the turnover. In this scheme the amplification step is inherent and a large chemical amount can be obtained from the sensing layer.

In contrast the use of the antibodies for the detection of antigens has not an amplification stage involved and then the "affinity" reaction should be amplified in order to have a clear transduction. We have two possibilities, one is the use of a bioconjugate involving a bound enzyme, like in the classical ELISA test; the second is the inherent amplification given by the mass of the biological element involved, a piezoelectric device (sensitive to mass) can detect minute amount of large proteins (like antibodies) if they are attracted on the surface of the sensor.

With the same scheme surface plasmon resonance can be sensitive to minute amount of large molecule reacting at the surface of the electrode.

The Biological System

The main problem of the biological system, catalytic or affinity, is the associated fragility and the operational activity. Most proteins have an optimal pH range in which their activity is maximal; this pH range should be compatible with transducer. Moreover the most of the biological systems have a very narrow range of temperature (15-40°C).

The most important problem and main drawback for industrial exploitation is the short lifetime associated with the biological elements. During last years several Meeting and Joined Actions were specifically dedicated to this point which is still object of research work. Lifetime or at least shelf lifetime of months or few years are the prerequisite for a suitable market and the fragility of the assembled systems has always limited the diffusion of biosensors in the market.

Immobilization of the Biological System

The technique of the immobilization of the biological elements has changed according with the different events, catalytic or affinity. The simplest way to retain enzymes on the tip of a transducer is to trap them behind a perm-selective membrane. This method has been mainly used in addition to embedding procedures in polyacrylamide gels. Then, mainly in the 80th, the trend shifted to use disposable membranes with bound bioactive material. Several companies put on the market preactivated membranes suitable for the immediate preparation of any bioactive membrane and this appeared as a real improvement at least for the easy use of enzyme sensors.

The removal of intereference has been also the other important aspect for the wide use of biosensors for industrial processes. The two problems have been solved by using multilayer membranes, such as those developed by Yellow Springs Instrument Co. (for glucose or lactate electrodes), with the enzyme sandwiched between a special cellulose acetate membrane and a polycarbonate nucleopore membrane. The main role of the membrane is to prevent proteins and other macromolecules from passing into the bioactive layer. Cellulose acetate membrane allows only molecule of the size of hydrogen peroxide to cross and contact the platinum anode, thus preventing intereference from ascorbic acid or uric acid, for example, at the fixed potential. Such configuration has been used by several researchers in their biosensor assembly. But at the same time several recipes of immobilization of enzymes were published and several laboratories developed their own procedure for immobilizing the biological element, sometimes also patented!

One approach was also the development of disposable sensor, based on combination of screen printed electrochemical sensors with enzyme adsorbed on the electrode surface (in this case mainly carbon). The use of the sensor just for one measurement limited the use of complicated immobilization procedures to simplest as possible, like only based on adsorption on carbon surface. This electrode surface acted as a sponge, and the large protein was easily immobilized even if the bond was weak. This approach was useful only for a quick and rapid measurement.

The immobilization of antibodies soon revealed that random immobilization of proteins was not effective and a new research in this direction started. Several researchers start to think how to immobilize proteins using an exact deposition. Technology, like self-assembling, based on gold surface and thiol groups prove to have a high potential. Proteins immobilized on the surface of the transducers were now aligned and their ligands group were directed toward the exterior ready to fix the metabolite.

This technique become important in the antigen-antibody reactions and even more in immobilization of nucleic acids such as 20-30 bases' oligonucleotides, which single strand conformations were looking for their complementary sequence of bases in the sample material.

Important Steps in the Biosensor Research

I want just concentrate my talk on three points which I consider very relevant in the Biosensors development in the last 30 years:

The Case of Glucose Pen

In 1984 Cass and coworkers publish a scientific paper where the team prove the use of ferrocene and its derivatives as mediators for amperometric biosensors.[6] Few years later the Medisense Exatech Glucose Meter was launched in the market and become the world's best selling biosensor product. The initial product was a pen-shaped meter with a disposable screen printed electrodes.

There were several advances in this product; first of all, the miniaturized instrument, just a pen with a small screen, where the current, recalculated in mg/dL of glucose was directly displayed. Then the concept of disposable screen printed electrodes, which allowed discarding the sensor after the use, and more important allowed the elimination of the calibration step. This was an incredible step in the sensor community. All sensors known, from pH glass electrodes to all kind of ISE, etc. should pass the "calibration step", where the sensor must be calibrated every day and sometimes before every measurement. The disposable screen printed electrodes do not need the calibration which simplifies enormously the use of it. The sensors became simple and user-friendly objects and diabetic people started to use it for individual monitoring and a large market was created. In 1996 Exatech was sold to Abbott for 867 million of US dollars.

The performance and design of several strip analyzers based on different electrochemical principles and meters has been published.[7] The book described around 10 different home glucose meters but it is not the final number because the new instruments continue to appear on the market.

The Wearable Artificial Pancreas

One major interesting application of Biosensor has been the development of a wearable artificial pancreas and the studies associated with development. This devise has never reached the market stage even if several scientists addressed the problem and demonstrated the possibility to resolve it. In 1976 Clemens et al. incorporated an electrochemical glucose biosensor in a "bedside artificial pancreas".[8] It was later marketed by Miles (Elkhart) as the Biostator Glucose-Controlled Insulin Infusion System (60 Kg, 42 x 46 x 46 cm) (Fig. 1).

This instrument became very well known in the medical endocrinology community. It regulated the glucose value injecting insulin or glucose into the bloodstream of the patient.

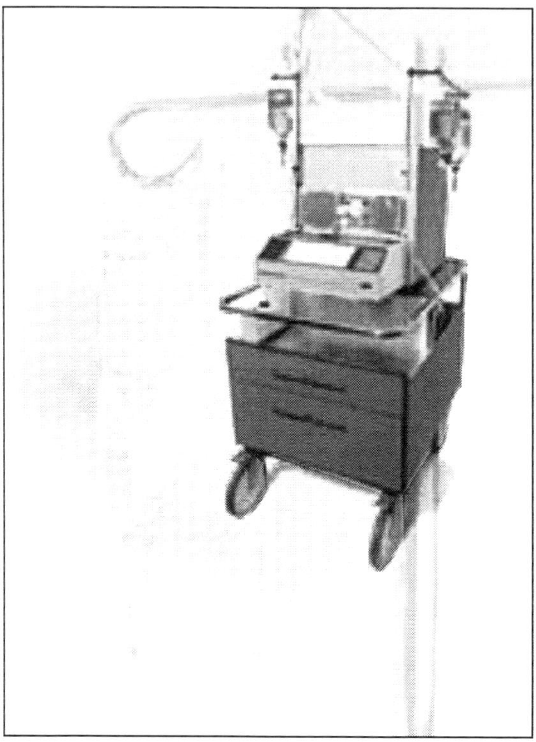

Figure 1. A general view of Biostator.

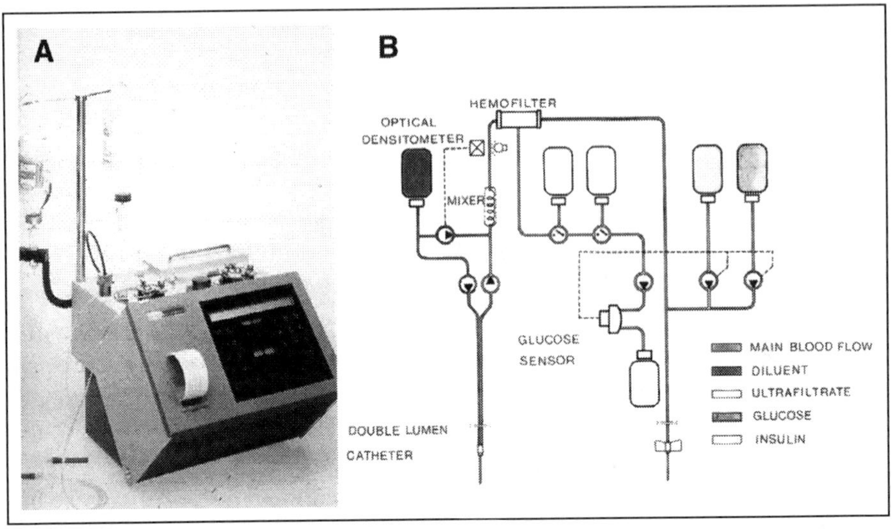

Figure 2. A general view of the Betalike (A) the artificial pancreas developed and marketed by Esacontrol, and a scheme of the system (B). Principle of operation: The device draws a small amount of blood from a peripheral vein and then infuse it into another one. In order to prevent coagulation inside the tubes the blood is diluted and heparinized at the tip of the drawing needle. The glucose measurements are carried out on the ultrafiltrate liquid obtained from a micro hemofilter cartridge. On a minute basis the Betalike infuses into the blood stream the amounts of insulin and glucose, calculated by means of a mathemathic algorithm, which are needed to reach and maintain the selected glucose level.

Although Biostator production was soon discontinued, it was later substituted by a similar instrument called Betalike produced and distributed by a small Italian company in 1990.

The Betalike had some innovations and improvement over Biostator: blood was taken from the patient via a double lumen catheter (6 ml/h) diluted with a buffer solution (1:9) with the addition of 3 units/ml of heparin. The diluted blood was then dialyzed in a miniaturized hollow fibre haemofiltering cartridge (filtration surface was 50 cm^2; membrane cut off about 35 000 daltons) which allows only the haemofiltrate to reach the sensors while the blood cells and proteins are reinfused into the patient (Fig. 2). The value of glucose could then give a signal for feeding in another needle positioned in the bloodstream, insulin or glucose according to the glucose profile (value and trend). This instrument opened new opportunities to study the diabetes and the glucose variation during the day.

Figure 3, shows one representative case of three days continuous record of an insulin-dependent diabetic treated with continuous subcutaneous insulin infusion. The continuous monitoring of glucose concentration disclosed a day-by-day variation of glycemia in diabetics. Then a large research activity started to miniaturize the system in order to obtain a real wearable artificial pancreas. The first step was miniaturizing the sensor. Today we have on the market several small instruments able to monitor glucose continuously up to one week.[9,10]

The Appearance of BIAcore on the Market

In 1982 researchers from Pharmacia started to work jointly with physics and biochemistry professors at Linkoping University in order to develop a new bioanalytical instrument able to monitor the interactions between biomolecules. In 1984 a new company Pharmacia Biosensor was created.

The company introduced a new instrument, BIAcore, in 1990 (Fig. 4). The instrument had a very high impact on the Biosensor community. The price of the instrument was more than 100 times higher than any other electrochemical or optical apparatus. The instrument based on surface plasmon resonance (SPR) technology was a fully automated instrument which monitored the

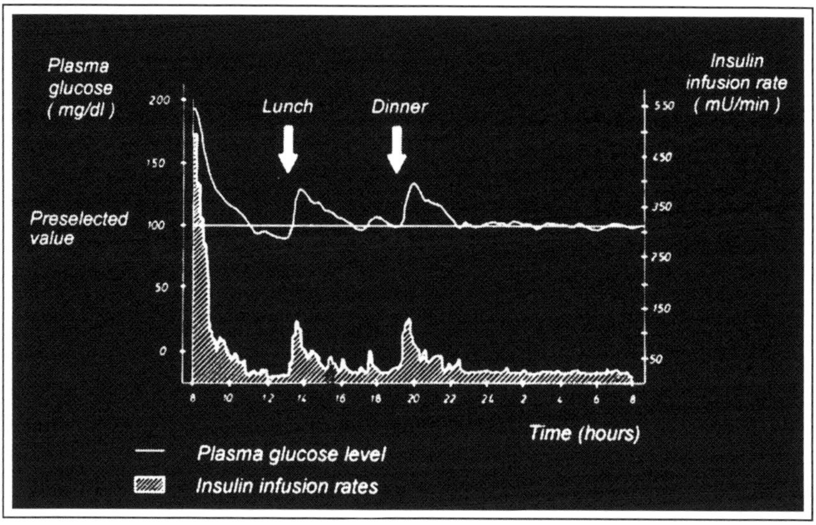

Figure 3. A typical experiment performed by clamping a "normal" value of glucose.

biomolecular interactions and included a sample handling equipment. The instrument performed the immobilization of the biomolecules, the SPR analysis and the regeneration of the sensor surface automatically by a microprocessor and this was a great advantage over the more traditional sensor technologies.

The autosampler was able to handle up to 192 samples without operator assistance. It increased the reproducibility of the analysis and provide a large sample capacity. The instrument stabilized the temperature at 0.1°C and allowed analysis of biomolecular kinetics.

The BIAcore instrument also eliminates the moving components that are generally associated with prism-based instrumentation. However the real advancement of the instrument was the unique sensor chip technology to simplify the immobilization of biomolecules to the sensor surface.

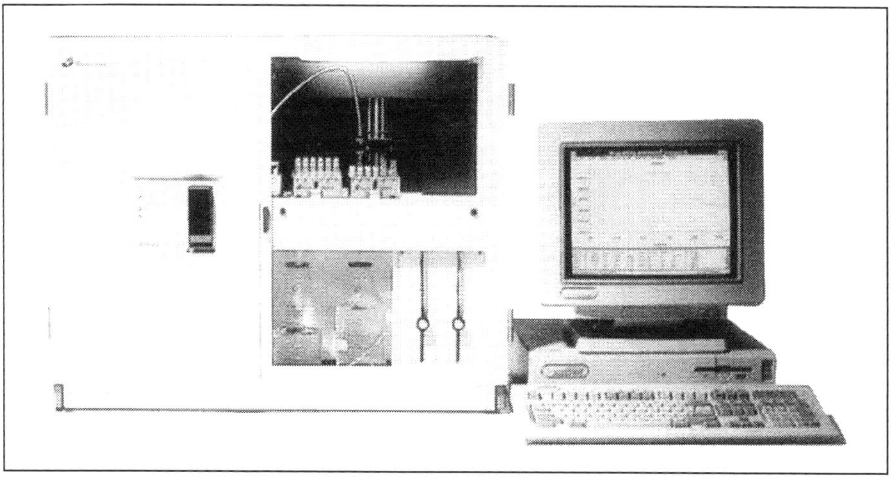

Figure 4. A general view of BIAcore.

Figure 5. Biacore sensor chip.

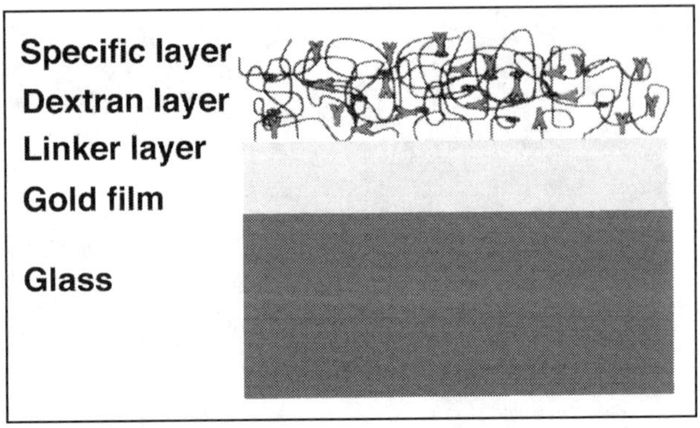

Figure 5, is a scheme of the BIAcore sensor chip. The chip consisted of a glass slide embedded in a plastic support; the glass surface is 1 cm^2 and has approximately 50 nm of gold coated on one side of the glass. The gold layer was then covered with dextran acting as a linking layer in order to facilitate the binding of biomolecules. Dextran, an old Pharmacia product, acted as a support for biomolecules but also protected the gold layer from nonspecific binding which is the main problem of this kind of apparatus. Typical protein concentrations required for immobilization are in the range 10 to 100 μg/ml. Usually, the chips can be used for more than 50 measurements without a notable loss in sensitivity and reproducibility. Moreover the company provided the sensor with alternative linking and binding layers (for protocols see http://www.biacore.com). The flow injection system has been designed with miniaturized sample loops, valves and conduits reducing drastically sample and reagent volumes with help of silicone layers. This is, of course, very important when dealing with valuable biological reagents.

The instrument was sold initially mainly to pharmaceutical companies looking for monoclonal antibodies (antibodies were ranked and selected for specific conditions). The instrument was very powerful in its automatic performance which significantly cut the time for the evaluation of the binding constant between antibodies and antigens. Consequently it was also applied to the study several other affinity reactions and became an important instrument in several research laboratories.

References

1. Clark LC. Monitor and control of blood and tissue oxygenation. Trans Am Soc Artif Intern Organs 1956; 2:41-48.
2. Clark LC, Lyons C. Electrode systems for continuous monitoring cardiovascular surgery. Ann NY Acad Sci 1962; 102:29–45.
3. Updike SJ, Hicks GP. The enzyme electrode. Nature 1967; 214:986-988.
4. Guilbault GG, Montalvo J. A Urea specific enzyme electrode. JACS 1969; 91:2164-2169.
5. Rechnitz GA, Llenado R. Improved enzyme electrode for amygdalin. Anal Chem 1971; 43:1457-61.
6. Cass AEG, Francis DG, Hill HAO et al. Ferrocene-mediated enzyme electrode for amperometric determination of glucose. Anal Chem 1984; 56:667-671.
7. Henning TP, Cunningham DD. Biosensors for personal diabetes management. In: Ramsay G, ed. Commercial Biosensors: Applications to Clinical, Bioprocess and Environmental Samples. Wiley Publisher, 1998.
8. Clemens AH, Chang PH, Myers RW. De Diabetologie de L'Hotel-Dieu, Proc Journees Ann, Paris, 1976.
9. Poscia A, Mascini M, Moscone D et al. A microdialysis technique for continuous subcutaneous glucose monitoring in diabetic patients (part 1). Biosens Bioelectron 2003; 18:891-898.
10. Varalli M, Marelli G, Maran A et al. A microdialysis technique for continuous subcutaneous glucose monitoring in diabetic patients (part 2). Biosens and Bioelectron 2003; 18:899-905.

CHAPTER 3

Photosystem II:
Composition and Structure

Aspasia Spyridaki,* Emmanuel Psylinakis and Demetrios F. Ghanotakis

Introduction

Photosystem II (PSII) is a light driven, water-plastoquinone oxidoreductase which catalyses the most thermodynamically demanding reaction in biology.[1] This highly endergonic reaction splits water into molecular oxygen, protons and electrons, thereby sustaining an aerobic atmosphere on earth and providing the reducing equivalents necessary to fix carbon dioxide to organic molecules, creating biomass, food and fuel.

PSII is a multisubunit pigment-protein complex embedded in the thylakoid membranes of higher plants, algae and cyanobacteria. Its unique properties require an elaborate arrangement of integral membrane proteins, specifically bound pigment moieties, extrinsic proteins and inorganic cofactors. Light energy is absorbed by light harvesting complexes that contain most of the pigments associated with PSII. Excitation energy is transferred from this antenna to the "core" of the PSII complex, where the primary photochemistry takes place. This photochemical part of PSII contains the ultra-fast and very efficient light-induced charge separation and stabilization steps that occur vectorially across the membrane. Finally, the photochemical reactions result in the accumulation of oxidizing equivalents in the oxygen-evolving complex (OEC); four oxidizing equivalents are used to convert two molecules of water into oxygen.

The photochemical and enzymatic reactions catalyzed by PSII are strictly conserved among all oxygenic photosynthetic organisms including cyanobacteria, eukaryotic algae and higher plants, while quite diverse pigment-protein complexes have developed for light-harvesting antenna systems associated with PSII.[2,3] These antenna systems, though similar in function, differ in their structures, with those of higher plants and green algae (LHCP) being located in the thylakoid membrane while those of most classes of cyanobacteria (phycobilisomes) are bound extrinsically to the stromal surface of PSII. With regard to protein structure, the PSII complex differs mostly in peripheral subunits between cyanobacteria and higher plants but shares the core parts in common.[4]

The PSII core is the minimal unit which is capable of catalysing full PSII function.[5,6] It is composed of a reaction center, which consists of the D1 and D2 polypeptides, cytochrome b559 (Cyt b559), the psbI protein and six chlorophyll (Chl) and two pheophytin (Pheo) molecules, an inner antenna of chlorophyll-binding proteins termed CP47 and CP43, and the extrinsic lumenally bound proteins of the OEC, 33 kDa protein (psbO), 23 kDa (psbP) and 17 kDa (psbQ), in higher plants and green algae, whereas in cyanobacteria psbP and psbQ are replaced by the 15 kDa psbV (cytochrome c550) and the 12 kDa psbU.[7] These intrinsic and extrinsic proteins, together with a number of low-molecular weight subunits[8-10] make up the core complex.

*Corresponding Author: Aspasia Spyridaki—Department of Human Nutrition and Dietetics, School of Food Technology and Dietetics, Technological Educational Institute of Crete, 723 00 Sitia, Crete, Greece. Email: saspa@dd.teiher.gr

Biotechnological Applications of Photosynthetic Proteins: Biochips, Biosensors and Biodevices, edited by Maria Teresa Giardi and Elena V. Piletska. ©2006 Landes Bioscience.

Light is funneled from the light harvesting complexes to the photochemically active reaction center (RC). The RC contains a special form of Chl α, P680 (the primary electron donor of PSII), which acts as an exciton trap and is converted to a strong reducing agent after excitation (P680*),[11] P680* is then photooxidized and donates an electron to the primary acceptor of PSII, a protein-bound pheophytin.[12] This charge separation is stabilized by the transfer of an electron to a plastoquinone molecule (Q_A), and subsequently to a second plastoquinone (Q_B). On the lumenal side, the oxidized form of P680 is reduced by a redox active tyrosine residue of the D1 protein, TyrZ.[13,14] The oxidized tyrosine radical TyrZ extracts an electron and a proton from a cluster of four manganese atoms that binds substrate water. The Q_A plastoquinone is a one-electron acceptor, firmly bound to the D2 protein. In contrast, the Q_B plastoquinone is a two-electron acceptor, bound to the D1 protein. A second photochemical turnover leads to the accumulation of two reducing equivalents on Q_B, which is then released from PSII, as plastoquinol, to be subsequently oxidized by PSI via the cytochrome b_6f complex. The empty Q_B site is then occupied by another plastoquinone molecule. Two further photochemical turnovers provide the oxygen-evolving complex with four oxidizing equivalents. The accumulation of four oxidizing equivalents, in a manner consistent with the observed S-state transitions,[15,16] leads to the release of molecular oxygen from the complex. During the accumulation of oxidizing equivalents, protons are released to the lumenal space of the thylakoid membrane.[17]

Electron crystallography of higher plant monomeric PSII[18,19] or dimeric PSII[20-25] and X-ray crystallography of cyanobacterial PSII[26,27] have revealed the locations of the main subunits and the relative positioning of their transmembrane helices, as well as the organization of the redox active cofactors and chlorophyll α molecules. Twenty-two α-helices have been assigned to D1, D2, CP43 and CP47, which are arranged with a local pseudo-two-fold rotation symmetry. The corresponding symmetry axis, named pseudo-C2 axis, is parallel to the local-C2 axis and passes through the non-heme iron located at the stromal side of the membrane.

The Hydrophobic Core

The Reaction Center

D1/D2

At the heart of PSII is the photochemically active reaction center which, when isolated in its most stable form, is composed of the D1 and D2 proteins, cytochrome b559 and the psbI protein.[28-30] The D1 and D2 intrinsic membrane protein components of PSII are encoded by the *psbA* and *psbD* genes, respectively. The mature D1 protein contains 343 amino acid residues and is post-translationally cleaved after Ala-344. The mature D2 protein is 352 amino acid residues long and is not C-terminally processed. After N-terminal formyl-methionine removal, both proteins are acetylated and phosphorylated at Thr-2 (it should be noted that Thr-2 is not conserved in *Euglena gracilis*).[31] Due to the strong local amino acid sequence homologies between these proteins and the L and M subunits of the bacterial reaction center[32] it was assumed that D1 and D2 form the reaction center of PSII.[33] The first evidence to support this idea was provided by the isolation of a D1-D2-Cyt b559-psbI core preparation that was capable of carrying out the primary charge separation.[28,30,34] The psbI protein and Cyt b559 could be removed from this complex by additional detergent treatment yielding a D1-D2 complex, which retained photochemical activity.[35] The three-dimensional structure of a D1/D2/CP47 subcomplex of PSII isolated from spinach, was obtained by electron crystallography at a resolution of 8 Å.[22,23] This electron crystallographic approach was extended to analysis of two-dimensional crystals of the complete PSII core complex isolated from spinach at a resolution sufficient to assign the organization of its transmembrane helices.[25] Higher resolution structures have derived from X-ray crystallography of cyanobacterial PSII.[26,27]

According to the X-ray crystallographic data from cyanobacterial PSII, in the center of the PSII monomer there are two clusters of five transmembrane -helices assigned to the D1 and D2 RC subunits (Fig. 1A), arranged in two semicircles interlocked in a handshake motif related by a pseudo-twofold axis. The α-helices C and D are connected in each of the subunits by a long α-helix CD on the lumenal side, while a short α-helix DE located on the stromal side connects

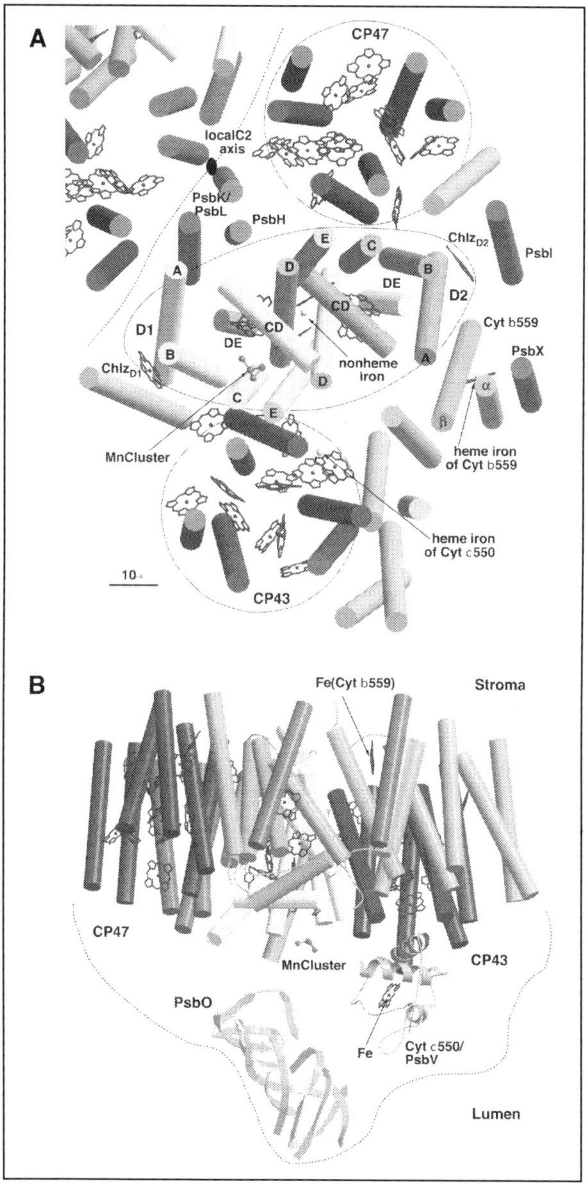

Figure 1. Structure of PSII with assignment of protein subunits and cofactors. A) Arrangement of transmembrane α-helices and cofactors in PSII. One monomer of the dimer is shown completely, with part of the second monomer related by the local-C2 axis (filled ellipse on the dotted interface). Chl α head groups and hemes are indicated by black wire drawings. The view direction is from the lumenal side, perpendicular to the membrane plane. The β-helices of D1, D2 and Cyt b559 are labelled. D1/D2 are highlighted by an ellipse and antennae, and CP43 and CP47 by circles. Seven unassigned β-helices are shown in grey. The four prominent landmarks (three irons and the manganese (Mn) cluster) are indicated by arrows. B) Side view of the PSII monomer looking down the long axis of the D1/D2 subunits. The 33 kDa protein is shown as a α-sheet structure (green), and Cyt c-550 as a helical model (grey). (Reproduced, with permission, from Zouni et al, 2001.[26]) A color version of this figure is available online at http://www.Eurekah.com.

α-helices D and E. Although structurally very similar, the subunits D1 and D2 have been clearly distinguished from each other by the positions of the cofactors. The arrangement of α-helices of D1 and D2 in the cyanobacterial reaction center resembles that of subunits L and M in the purple bacterial reaction center and the PSI RC proteins,[36-38] as well as those of higher plant PSII and PSI. This supports the hypothesis of the evolution of all photosynthetic reaction centers from a common ancestor.[39]

Redox Active Components

The D1-D2 heterodimer binds all essential redox components of PSII. These include the chlorophylls of the primary donor P680 (P_{D1} and P_{D2}), the primary acceptor pheophytin $Pheo_{D1}$, Q_A, the non-heme iron Fe^{2+}, Q_B, the tyrosines TyrZ and TyrD, four additional chlorophylls (Chl_{D1}, Chl_{D2}, $ChlZ_{D1}$ and $ChlZ_{D2}$), the non-photochemical pheophytin $Pheo_{D2}$ and one or two β-carotenes.[35,40] The first direct structural information on the organization of the PSII redox-active cofactors came from electron crystallography,[23] which was confirmed and extended by X-ray diffraction analyses from cyanobacteria.

According to the most recent structure[27] the two RC chlorophyll α molecules (P_{D1} and P_{D2}) are located towards the lumenal side of the heterodimer, their head groups being parallel to each other and perpendicular to the membrane plane (Fig. 2A,B). Due to the large separation of these two molecules (10 Å), excitonic coupling is weak and they can be regarded as monomeric chlorophylls. Therefore the primary electron donor of PSII, P680, is not a "special pair" as in other types of photosynthetic reaction centers, but a monomeric chlorophyll, P_{D1} or P_{D2}. Similarly to the bacterial RC, D1-His-198 and D2-His 197 in helix D of D1 and D2 are coordinated to the two chlorophylls. Two more chlorophyll molecules (Chl_{D1} and Chl_{D2}) are found with their planes tilted in a way analogous to the accessory bacteriochlorophylls in the purple bacterial RC. The four chlorophylls are approximately equidistant from each other, with a center-to-center distance of about 10 Å (similar values were found in Zouni et al's structure).[26] These distances agree with the "multimer of monomers" model proposed for P680.[41,42] However, both cyanobacterial structures suggest a stronger interaction between the two RC Chls than those between the RC Chls and accessory Chls, indicating that each of the monomer Chls within the tetramer is not the same.

The two pheophytin molecules, $Pheo_{D1}$ and $Pheo_{D2}$, are located towards the stromal surface of PSII. The four chlorophylls and the two pheophytins, which are clustered within the D and E transmembrane helices of the D1 and D2 proteins, are arranged in a pseudo-C2-symmetrical fashion around the non-heme iron and constitute the active and inactive electron transfer branches. The two plastoquinones are located towards the stromal side of PSII. The site of the tightly bound Q_A (Fig. 2A) in D2 is at 12.0 Å from $Pheo_{D1}$ and was occupied by a plastoquinone in the crystal structure,[26] the center of its ring being at a distance of 10.5 Å from the non-heme iron. Candidate residues for Q_A binding include His-215, Ala-261 and Trp-254. The putative binding site for the mobile Q_B in D1, including His-215, Tyr-262, Ala-263, Ser-264, Phe-255 and Phe-265,[43] was unoccupied in both cyanobacterial structures. The non-heme iron is located between the two plastoquinones. The D1-His-215, D1-His-272, D2-His-114 and D2-His-268 are close enough to provide coordination to the non-heme iron.[27] Two extra chlorophyll molecules were assigned to the spectroscopically identified species $ChlZ_{D1}$ and $ChlZ_{D2}$, shown to be coordinated to His-118 in D1 and His-117 in D2.[44,45] The redox-active tyrosine 161 of D1 (TyrZ), which gives rise to the EPR Signal IIvf,[13] is located in the last turn at the lumenal side of helix C in D1, 7 Å away from the Mn cluster. Tyr 160 of D2 (TyrD), which gives rise to the EPR Signal IIs,[14] is found at helix C in D2. This position is related to that of TyrZ by the pseudo-C2 axis.

The arrangement of cofactors is similar in both cyanobacterial structures, although the structure from *Thermosynechococus vulcanus* includes two α-carotenes, in agreement with reports that there are two carotenes in PSII-RC preparations.[46] The two carotenes are located close to each other (the closest distance being 5 Å), which agrees with spectroscopic studies.[47] The carotenes are positioned on the D2 side of the RC near Cyt b559 and the RC Chl_{D2} (Fig. 2A,B), consistent with a role mediating electron transport between them.[48]

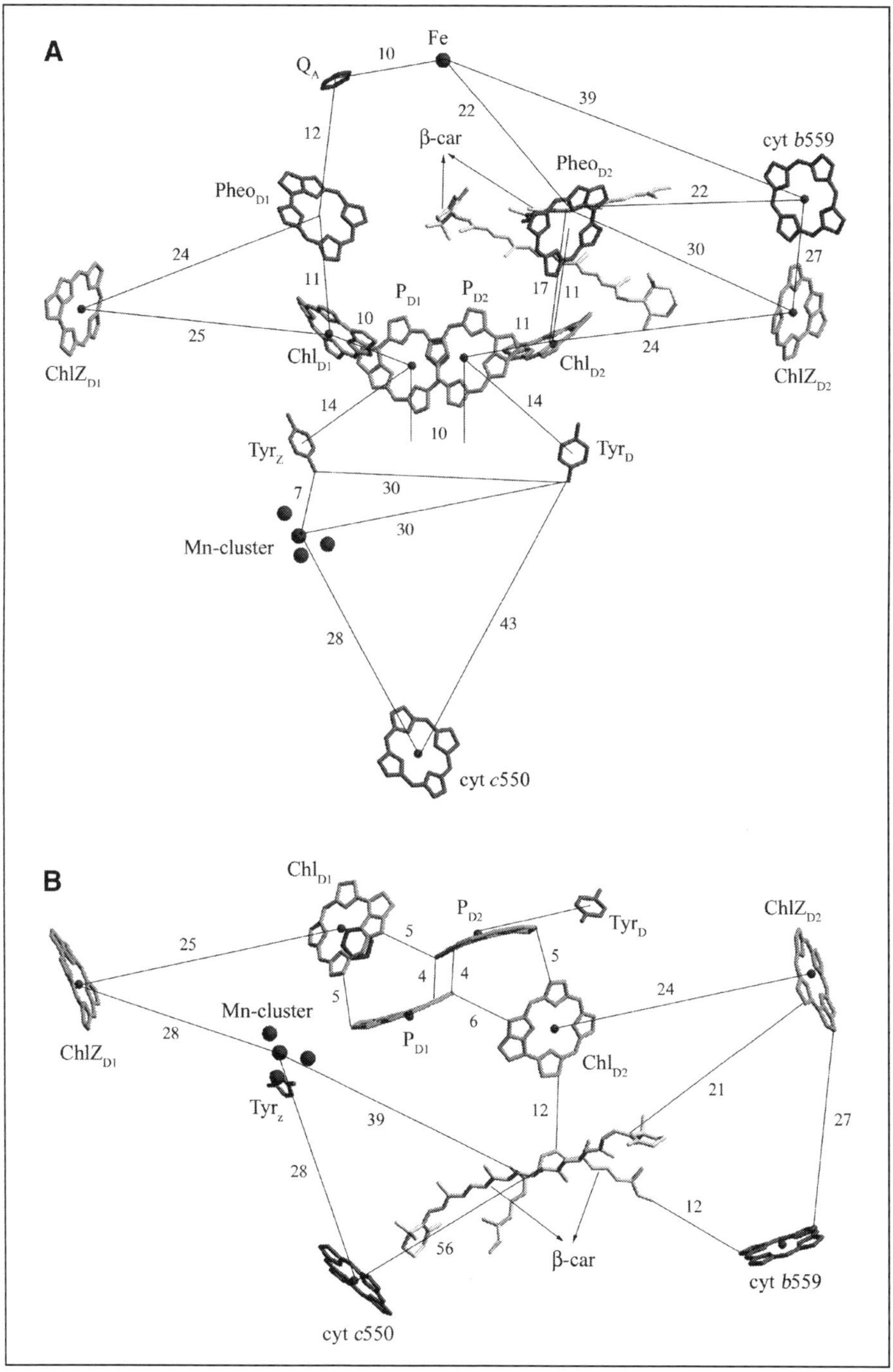

Figure 2. Arrangements of Chls and other PSII cofactors and their relative distances (Å). A) View along the membrane plane. B) View from the lumenal side perpendicular to the membrane plane. The non-heme iron and two pheophytins are omitted for clarity. (Reproduced, with permission, from Kamiya and Shen, 2003.[27])

Cytochrome b559

Cytochrome b559 is an integral component of all PSII complexes.[49] Although this redox center does not play a definite role in the major electron transfer reactions, it has been invoked to a key role in numerous functions, including PSII assembly, oxygen evolution, protection against photoinhibition, cyclic electron transfer. Two redox forms of Cyt b559 are observed. The high potential form has an unusually high redox potential of 370 mV, whereas the low potential form has a redox potential of 60-80 mV. Structurally, Cyt b559 is a novel b-type cytochrome consisting of a heterodimer of two subunits: the 9 kDa, α subunit and the 4.5 kDa, β subunit. These subunits are encoded by the *psbE* and the *psbF* genes, respectively.[50]

Cyt b559 is required for assembly of functional PSII. Deletion of the *psbE* or *psbF* genes in *Synechocystis* sp. PCC 6803 leads to the loss of PSII RC activity. Additionally, these mutants do not accumulate significant amount of the D1 and D2 proteins. Site-directed mutagenesis experiments that alter the putative heme ligands His-22 in either the α or β subunits also lead to dramatic losses of D1, D2 and of both subunits of the cytochrome.[51] Truncation of the carboxyl terminus of the α subunit by 31 amino acid residues leads to an 80-90% loss of PSII centers without the loss of assembled Cyt b559.[52] This result indicates that the C-terminus of the subunit is not required for the formation of functional cytochrome but is required for the assembly of functional and stable PSII centers.

The pseudo-C2 symmetry of the cofactor arrangement is broken by Cyt b559, its heme-iron being 27.0 Å apart from $ChlZ_{D2}$, and about 8 Å apart from the stromal side.[26] As in the case of the purple bacterial reaction center, there remains the preferred directionality of the charge separation along the 'active' branch, which is on the D1 side of the reaction center. Cyt b559 is located near helix A of D2 on the stromal side of PSII (Fig. 1A,B). Each subunit contains a single transmembrane α-helix with its N-terminus exposed to the stromal surface.[53] The α subunit is characterized by having a long C-terminus extending from the lumenal surface of the membrane while the β subunit has essentially no lumenal domain. Each subunit contains a conserved histidyl residue (His-22), located within the transmembrane region towards the stromal surface, which coordinates the heme of the cytochrome.[49]

There had been much controversy over the stoichiometry of this cytochrome within PSII, with values of one and two being suggested.[54] The recent structural studies support the former stoichiometry. The presence of only one Cyt b559 per reaction center has been confirmed by spectroscopic analysis of fresh PSII preparations and of redissolved PSII crystals[26] and by anomalous diffraction data from single crystals.[27] The controversial issue of one or two Cyt b559 per reaction center present in PSII depends on the preparation and/or organism, since most studies are based on determination of the heme content. However, radioactive labeling of PSII RC proteins yielded equimolar amounts of each polypeptide.[55]

PsbI Protein

The psbI protein has a molecular mass of about 4.2 kDa. In most eukaryotes it contains 35 amino acids and is predicted to have a single transmembrane helix with a short N-terminal region at its stromal end. The mature protein retains the initiating N-formyl group at its N-terminal methionine residue.[50] Targeted mutagenesis of the *psbI* gene yields *Synechocystis* sp.[56] and *Chlamydomonas reinhardtii*[57] strains that assemble PSII and evolve oxygen. Thus the function of the psbI protein remains unknown.

The low resolution of the structural models has not allowed the unambiguous assignment of the psbI protein helix. According to the structure from *Synechococcus elongatus* the psbI protein is located in a position close to CP47 and D2 (Fig. 1A), whereas at the *Thermosynechococus vulcanus* structure it is placed close to the helix of Cyt b559. Since it is present in the isolated D1/D2/Cyt b559 complex[58] it must be located close to the RC heterodimer. There are reports that it can be chemically crosslinked with both the D2 protein and the α-subunit of Cyt b559, suggesting it to be on the D2 rather than the D1 side of the RC.[59] The *psbI* gene product appears to be present in a 1:1:1 stoichiometry with the and subunits of Cyt b559.[60,61]

The Inner Antenna Subunits

The chlorophyll binding proteins CP47 (CPa-1) and CP43 (CPa-2) function as an internal antenna system of PSII. They transfer excitation energy from the exterior antenna (LHCP in plants or phycobilisomes in cyanobacteria and red algae) to the chlorophylls of the RC of PSII. Both proteins are integral membrane protein components of PSII. CP47 is encoded by the *psbB* gene while the *psbC* gene encodes CP43. They are highly conserved; 80% and 85% of the residues are conserved or conservatively replaced in CP47 and CP43, respectively.[62] Hydropathy analyses indicate that each protein contains six transmembrane helices with the C- and N-terminal ends exposed at the stromal surface. Both proteins bind about 15 chlorophyll a molecules and 2 to 3 β-carotenes each,[63,64] with the majority or all of the pigments being located in the transmembrane regions where a number of conserved histidyl residues are CP47 and CP43 contain twelve and ten histidines respectively. These histidyl residues are clustered within the predicted membrane spanning regions near both the stromal and lumenal ends of their respective transmembrane helices,[23] to give a distribution reminiscent of the positioning of the conserved histidyl residues in other light harvesting chlorophyll proteins.[65] These histidyl residues, therefore are prime candidates for chlorophyll axial ligands, which is also supported by site-directed mutagenesis studies.[66] In both proteins helices 5 and 6 are joined by a large hydrophilic loop that is located on the lumenal side of PSII. These loops contain about 200 and 150 amino acids for CP47 and CP43, respectively.

Numerous lines of evidence indicate a possible role of the large extrinsic loop of CP47 in water oxidation. A monoclonal antibody (FAC2) was isolated that recognized its antigenic determinant on CP47 only in the absence of the extrinsic 33 kDa protein and the chloride insensitive manganese ions associated with the oxygen-evolving complex of PSII.[67] The epitope for this antibody is in the Pro-360 to Ser-391 domain,[68] which is located in the large extrinsic loop of CP47. A variety of protein crosslinkers are capable of crosslinking CP47 to the extrinsic 33 kDa protein in spinach PSII membranes.[69,70] Studies with the zero-length crosslinking agent EDC, crosslinked the domain Glu-364 to Asp-440 of the large extrinsic loop to the N-terminal domain Glu-2 to Lys-76 of the 33 kDa protein.[71] Additionally, the extrinsic 33 kDa protein shields lysyl residues located on the large extrinsic loop of CP47 from labelling with the amino group-modifying reagent NHS-biotin. Treatments which remove the extrinsic 33 kDa protein from spinach PSII membranes allow the specific labelling of CP47 with this reagent.[69,72] Furthermore, 33 kDa also protects the large extrinsic loop of CP47 from cleavage by trypsin.[67] A study by Hayashi et al[73] showed that cleavage of CP47 at Lys-389 by the endoproteinase Lys-C inhibited oxygen evolution and the ability to rebind the 33 kDa protein.

Site-directed mutagenesis studies within the large extrinsic loop of CP47 have indicated that the residues Arg-384 and Arg-385 affect the stability of the oxygen-evolving complex[74,75] and the oxygen-evolving capability[76] and are crucial for tight binding of the 33 kDa protein.[77] Other site-directed mutagenesis studies within the large extrinsic loop of CP47 yielded mutants with perturbed photoautotrophic growth and oxygen-evolving capability.[75,78-81]

CP43 appears to be more loosely associated with the PSII core complex than is CP47. Chaotropic agents or additional detergent treatments easily remove CP43, which yield a CP47-D1-D2-Cyt b559-psbI complex.[30,82,83] Additionally, partially functional (non-oxygen evolving) PSII RCs can assemble in the absence of CP43[84] whereas this has not been observed in the absence of CP47. It can be concluded, therefore, that CP43 is not absolutely required for the chemistry of water oxidation. CP43 is required for the binding of LHCP to the CP47-D1-D2-Cyt b559-psbI complex and must be present to facilitate excitation energy transfer from the LHCP to the core complex.[85] CP43 also differs from CP47 in that CP43 is post-translationally modified by endoproteolysis and the N-terminal threonine of the mature protein can be reversibly phosphorylated in the case of higher plants and green algae.[31]

Electron crystallography and X-ray crystallography have provided a structural model for the transmembrane helical domains of CP47 and CP43.[22-27,86] The analyses revealed that the six helices of CP47 and CP43 are arranged as a trimer of dimers related by the pseudo-twofold axis of the D1 and D2 heterodimers. According to the structure from *Thermosynechococcus vulcanus* helix 6 of CP47 is located close to helices C and E of D2, and helix 6 of CP43 is located adjacent to helices C and E of D1 (Fig. 1A). The *Thermosynechococus vulcanus* structure contains an extra Chl in CP43 located between $ChlZ_{D1}$ and the antenna Chl of CP43, which could facilitate excitation energy transfer and electron transport between $ChlZ_{D1}$ and the rest of PSII and a novel electron transport pathway has been suggested.[48] The main subunits of the PSII RC core (D1, D2, CP47 and CP43) are present in a 1:1:1:1 stoichiometry and are arranged in a way that the transmembrane helices of the CP47/D2 proteins are related to the transmembrane helices of the CP43/D1 proteins by a pseudo-twofold axis, in both cyanobacterial and higher plant structures.[87,88]

The Hydrophilic Cluster

Mn Cluster

The Mn cluster and its ligands, together with TyrZ and the calcium and chloride ions form a functional unit that is referred to as the oxygen-evolving complex (OEC).[89-92] The oxygen evolving complex (OEC) is the terminal electron donor of PS II. The catalytic site of the OEC consists of a cluster of four manganese ions. Based on EXAFS and ESEEM spectroscopy the first shell Mn ligands are mostly oxygens, one or two nitrogens and possibly one chloride.

During oxygen evolution, the OEC cycles through five intermediate redox S-states, S_0 to S_4,[1,93] each S-state transition representing a one-electron oxidation of the OEC. Oxygen is evolved during the S_4 to S_0-state transition. This process is driven by the energy of four successive photons absorbed by the pigment P680 of the photosystem II (PS II) reaction center.[15] The Mn complex in the OEC couples the four-electron oxidation of water with the one-electron photochemistry occurring at the PS II reaction center by acting as the locus of charge accumulation.

Selective extraction experiments and a series of spectroscopic studies revealed heterogeneity in the manganese population. Seventy-five percent of the total manganese is released by treatment of PSII with 0.8 M Tris at pH 8.5 or hydroxylamine.[94,95] The same amount of manganese is also released by treatment of PSII membranes, first with 2M NaCl and subsequently with a reductant such as hydroquinone.[96] The main difference between the two treatments is that, while Tris releases all three extrinsic proteins, the high salt/reductant treatment results in a PSII system that has been depleted of most of its manganese and the 17 and 23 kDa polypeptides, but retains the extrinsic 33 kDa protein. Exposure of PSII membranes to 1M $CaCl_2$ results in depletion of all three extrinsic proteins but retention of the four manganese ions.[97] Under certain conditions, it is possible to reassemble a functional manganese complex in PSII preparations that have been depleted of manganese through a process, which requires light, known as photoactivation. Calcium and chloride are also required for the assembly of a functional manganese complex.[95]

In PSII, there is only one manganese cluster, located close to the surface helix on the lumenal side joining transmembrane helices C and D of the D1 protein (close to TyrZ) and about 15 Å off the pseudo-C2 axis (Fig. 3).[26] This location was confirmed by measuring X-ray edge anomalous diffraction. Three Mn atoms are located at the corners of an isosceles triangle, with a fourth placed at the center of the triangle, out of the plane of the triangular structure. This arrangement and the interatomic distances are consistent with recent predictions arising from EPR (electron paramagnetic resonance), ENDOR (electron nuclear double resonance) spectroscopy and X-ray spectroscopy.[98-101] The electron transfer between P680 and the manganese cluster is bridged by the redox-active TyrZ, which is 7.0 Å away from the manganese cluster. The center-to-center distance between the P680 chlorophyll molecules P_{D1} and P_{D2} and the manganese cluster is 18.5 Å and 25.1 Å respectively.

According to the X-ray crystallographic data from *Thermosynechococus vulcanus*[27] the shape of the manganese cluster is similar to that reported by Zouni et al.[26] The only difference is that in this structure, all four Mn atoms are located roughly in the same plane, whereas in Zouni et al's structure, the central Mn is protruded toward the lumenal surface of the membrane. The Mn cluster is coordinated by the D1 polypeptide and there are at least four to five connections between the Mn cluster and polypeptide backbones. The C-terminal carboxyl group of Ala-344 provides ligands directly to the Mn cluster. This is in agreement with mutagenesis studies suggesting the possible ligation of the Mn cluster by D1 C terminus.[102] Other residues coordinated to the Mn cluster are Asp-170 and

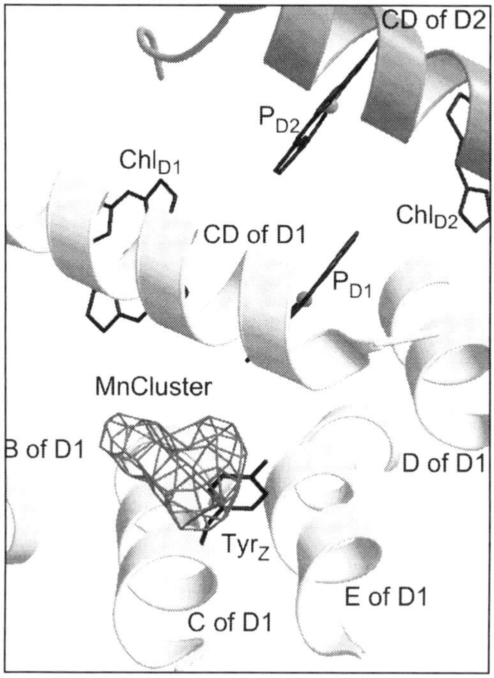

Figure 3. Location and orientation of the manganese cluster. A close-up view of the PSII RC, with the electron density of the $(Mn)_4$ cluster contoured at 5. The view is from the lumenal side onto the membrane plane. (Reproduced, with permission, from Zouni et al, 2001.[26])

Glu-333 (or His-332). His-337 and Asp-189 (or His-190), which have been suggested as possible ligands from mutagenesis[102] or chemical modification studies[103] are also possibly coordinated to the Mn cluster.

Calcium Cofactor

Ca^{2+} is an essential cofactor in oxygen evolution.[1,104] Depleting this cofactor suppresses OEC activity, which can be restored (up to 90%) by replenishing with Ca^{2+}. Various cations compete with calcium for its binding site(s) in PSII. Sodium, potassium and cesium are weakly competitive with calcium, but they do not support oxygen evolution activity. Partial reactivation (up to 40%) results from addition of strontium to Ca-depleted PS II membranes[105,106] and no other metal ions (except VO^{2+}, vanadyl ion)[107] can restore activity.[108-110]

There is debate about the Ca^{2+} cofactor binding site. One set of experiments using EXAFS on Sr-reactivated PS II membranes was interpreted to indicate a 3.4 to 3.5 Å distance between the Ca (Sr) and the Mn cluster.[111] This close link is also supported by FTIR spectroscopic work[112] that is consistent with a carboxylate bridge between Mn and Ca. Analysis of EXAFS spectra from purified PS II membrane preparations also support the proximity of Ca to Mn.[113] However, EPR-based experiments involving Mn^{2+} substitution in Ca-depleted PSII membranes,[114] indicated that the Mn^{2+}-occupied Ca^{2+} binding site was outside the first coordination region of the catalytic cluster. Further investigation on the Ca^{2+} binding site was carried out by X-ray absorption studies on Sr-reconstituted PS II.[115] The results confirmed the proximity of Ca^{2+} (Sr^{2+}) cofactor to the Mn cluster and suggested that the active site is a Mn-Ca heteronuclear cluster.

Most researchers addressing the stoichiometry of the Ca^{2+} cofactor in PS II now conclude that functional water oxidase activity requires one Ca^{2+}, which can be removed by low-pH/citrate or 1.2 M NaCl wash.[109,116-119]

Chloride Cofactor

Chloride plays an essential role in the oxygen-evolving process.[120-122] Chloride depletion of PSII samples results in the inactivation of the OEC. Addition of certain anions restores the oxygen evolution activity. The effectiveness of the anions follows the order: chloride → bromide → iodide → nitrate.[123] The loss of the two polypeptides, 17 kDa and 23 kDa, induces an increased demand for Cl^- in order to retain optimal function of the water-oxidizing reaction, suggesting a role for Cl^- in maintaining the protein organization needed for O_2-evolution.[124-126]

Steady-state kinetic experiments indicate a halide binding site on the Mn cluster.[127] It has been proposed that there is one Cl^--binding site per PSII unit with a high-affinity (K_d = 20 μM) and a low-affinity (K_d = 0.5 mM) state. The high-affinity state is the normal state of binding, but once Cl^- has been removed, it will first rebind as low-affinity, rapidly exchanging followed by conversion into a high-affinity, slowly exchanging mode of binding.[128] Recent studies indicate that the presence of the Cl^- is necessary only for the S_2 to S_3 and S_3 to S_0 transitions of the OEC, while the earlier steps of the cycle can proceed in its absence.[129,130]

Extrinsic Proteins

In all types of oxygenic photosynthetic organisms the manganese cluster is stabilized by the extrinsic 33 kDa protein, known as the manganese-stabilizing protein (MSP).[131] This protein which is encoded by the *psbO* gene, provides a unique protein structural environment where water oxidation takes place. Its dissociation by washing with 1 M $CaCl_2$ or $MgCl_2$[132] or 2.6 M urea plus 0.2 M NaCl[133] leads to paramagnetic uncoupling and a gradual release of two of four Mn^{2+} ions present in the cluster.[134] Oxygen evolution is strongly suppressed by the release of the 33 kDa protein, but the lost activity can be restored by rebinding of the protein.[135,136] The cyanobacterial protein, however, is not essential for oxygen evolution.[137,138] Reconstitution of spinach PSII membranes with heterologous extrinsic proteins from other species, including cyanobacteria, also leads to partial restoration of oxygen evolution.[139,140]

A number of experiments have addressed the topology of the 33 kDa protein with respect to the intrinsic proteins of PS II. Extraction/reconstitution experiments indicate that the 33 kDa protein is exposed on the surface of PS II[135,140,141] in agreement with structural models derived from electron diffraction studies[142] that suggest a surface-located binding site for the subunit. Removal of the manganese cluster weakens the binding of 33 kDa to PSII.[143,144] Removal of the 33 kDa protein exposes lysyl residues on the large lumenal loop of CP47 to labeling with NHS biotin and to monoclonal antibodies directed against this structural feature of the protein.[67,69,72] EDC crosslinks the 33 kDa protein to the extrinsic loop of CP47.[71,145] Site-directed mutations in the extrinsic loop of CP47 confirm the crosslinking results. These mutations are found predominantly in the amino acid sequence region Ala-373 to Arg-385 in the extrinsic loop.[74,78] Defective assembly of the 33 kDa protein is apparent in these mutants; electron transfer is affected and, in some cases, increased concentrations of exogenously added Cl^- are necessary for optimum growth of mutant cyanobacterial cells.[78,146] A possible interaction between the 33 kDa protein and the extrinsic loop of CP43 has also been detected. Experiments employing trypsin digestion of 33 kDa-depleted PS II preparations revealed an enhanced digestion of CP43.[147] Although no crosslinking between the 33 kDa protein and CP43 has been detected, the proteolysis result suggests a strong relationship between the structural integrity of the 33 kDa protein bound to PS II and exposure of the CP43 lumenal loop to the protease.

Various attempts have been made to identify amino acid residues on the 33 kDa protein that are involved in binding to its functional site on the PSII complex. The N-terminal sequence of spinach 33 kDa protein was suggested to have a binding site to PSII because removal of 16 or 18 amino acid residues from its N terminus by protease digestion resulted in total loss of the protein binding.[148] It was also suggested that Asp-9, the only conserved, charged residue in the N-terminal 18-amino acid sequence, might engage in both intra- and intermolecular interactions.[77]

The secondary structure of the 33 kDa protein in solution has been examined by spectroscopic studies.[149] Fourier transform infrared spectroscopy has indicated that the 33 kDa protein is composed predominantly of β structure[150,151] a conclusion reinforced by circular dichroism measurements in the UV region.[152,153] The 33 kDa protein has been proposed to be either a natively unfolded[154] or a molten globule type protein.[155] Both possible structures provide the necessary conformational flexibility to achieve optimal interaction with PSII. It is probable that the 33 kDa protein is a structurally extended protein enriched with β-sheets and random coils.[154] The C-terminal domain contains two long loops, Gly-152 to Gly-163 and Gly-177 to Gln-190.[156] The former includes a highly conserved region (between Pro-148 and Pro-174) containing several charged residues (Asp-157, Lys-159, Arg-161) that have been related to the specific interaction and binding of the 33 kDa protein to PSII and probably to the stabilization of the Mn cluster.[149,157,158] The conserved Pro-148 to Pro-174 region also includes a motif Glu-X-Asp-Glu-Glu-Asp, which is very similar to the calcium-binding motif identified in PROSITE as PS00330. Although a thermostable protein, the secondary structure of spinach 33 kDa protein is affected by temperature[159] and conformational changes occur upon binding to the PSII reaction center,[160] including an increase in β-sheet, that are essential for oxygen evolution.[161]

The amino acid sequences of the 33 kDa protein have been reported from cyanobacteria, *Euglena gracilis*, green algae and higher plants.[162-173] The sequences of the 33 kDa protein show a relatively high homology around 40 to 50% from cyanobacteria to higher plants. The 33 kDa protein has also been reported to be exchangeable in binding to PSII and supporting oxygen evolution from various different organisms.[139,174] Thus, the structure of the 33 kDa protein has been considered to be largely conserved during evolution from cyanobacteria to higher plants. There are, however, some reports suggesting that the structure of the 33 kDa protein may be different, at least in its free form, among different plant species. The cleavage sites of the 33 kDa protein by protease were reported to be different between higher plant and cyanobacterium.[148,157] Another line of evidence suggesting a possible difference in the structure of the 33 kDa protein is that oxygen evolution of cyanobacterial PSII was restored to a larger extent with its own 33 kDa protein than with the 33 kDa protein from other sources in cross-reconstitution experiments.[174] A comparison of the cleavage sites by chymotrypsin or *Staphylococcus aureus* V8 protease of the 33 kDa protein from various species[175] showed that the structure of the 33 kDa protein is different among different organisms, and can be divided into three major groups of higher plant-type, cyanobacterial-type (red algae and cyanobacteria) and their intermediate-type (green algae and *Euglena gracilis*) based on their protease-cleavage sites.

The cyanobacterial 33 kDa protein is characterized by a large amount of β-strand[26,150,152] and contains a short helix in the region close to the C-terminus.[27] According to Kamiya and Shen the structure of the 33 kDa protein has some similarities to the structure of the porin family[176,177] but is not compatible with the model predicted theoretically by Pazos et al[178] or that analyzed by single particle analysis of cryoelectron microscopy images from higher plants.[156]

Also associated with the oxygen evolving complex (OEC) are two other extrinsic proteins, the 23 kDa (psbP) and 17 kDa (psbQ) proteins in plants and green algae and the 15 kDa (psbV) and 11 kDa (psbU) proteins in cyanobacteria.[179] The similarities in binding and properties of these cyanobacterial proteins and those of higher plants may imply an evolutionary connection among the extrinsic proteins of PSII. However, both cyanobacterial proteins psbV and psbU have been described to be involved in the stabilization of oxygen evolution against heat inactivation,[180-182] a role not shown for the respective higher plant proteins. Removal of the 23 kDa and 17 kDa proteins inhibits the rate of oxygen evolution by as much as 80% at lower cofactor concentrations thereby increasing the amount of calcium and chloride required for optimal oxygen evolution by PSII. The function of the 23 kDa protein appears to be associated with a decrease in calcium requirement[133,183,184] while the presence of the 17 kDa protein may lower the requirement of chloride.[124] Site-directed mutagenesis experiments[185] indicate that the N-terminal domain of the 23 kDa protein is important for PSII activity and calcium retention. The presence of the 23 kDa protein also protects the manganese cluster from the effects of bulky reductants.[96,183] Upon removal of the 23 kDa protein, the binding affinity for atrazine at the Q_B site is decreased[186] and high potential Cyt b559 is converted to its low potential form.[187]

The 23 kDa protein requires the 33 kDa protein for binding to PSII and the 17 kDa protein requires the 23 kDa protein for binding to PSII.[143,188] The binding of both proteins to PSII requires their N-terminal domains.[148,189,190] The association of both proteins with PSII is stabilized by the manganese cluster.[136,144,191]

The secondary structures in solution of the extrinsic 23 kDa[192] and 17 kDa[193] proteins of spinach PSII have been examined by FTIR spectroscopy. The 23 kDa protein contains a large proportion of extended β-sheet structure (37%) and only a small amount (5%) of α-helical structure in solution, as shown by FTIR spectroscopy, whereas the prominent feature of the secondary structure of 17 kDa is the relatively large proportion of α-helical structure. The recent resolution of the three-dimensional structure of cytochrome c550 from two cyanobacteria[194,195] has confirmed a bis-histidine heme coordination. The protein shows the typical hydrophobic core of monoheme cytochromes c, with three helices forming a nest for the prosthetic group and a fourth helical segment in the N-terminal domain, protecting the heme from solvent. The 12 kDa protein has an all-α architecture composed of five or more short α-helices, with no homologous structure in the database.[27]

Information on the organization of the extrinsic proteins of higher plants has been derived from single-particle analyses of the PSII supercomplex isolated from spinach.[156,196,197] These studies indicated that the 33 kDa protein is located above the lumenal ends of the helices of the D1 and D2 reaction center proteins and their corresponding CD surface helices. Moreover, due to its elongated shape, the 33 kDa protein also lies above helices 5 and 6 of CP47 and to one side of the transmembrane helices of CP43. The 23 and 17 kDa proteins are located over the lumenal surface of the N-terminal region of the D1 protein and above the lumenal ends of the CP43 transmembrane helices. The 23 kDa extrinsic protein is immediately adjacent to the 33 kDa protein and the 17 kDa protein is attached to the surface of the 23 kDa protein, in line with crosslinking studies[198,199] which is consistent with studies[133] that have shown that the binding of the 23 kDa to PSII requires the presence of the 33 kDa protein and that 23 kDa is required for the binding of 17 kDa.

In the structure from *Thermosynechococus vulcanus,*[27] the 12 kDa protein is located between the 33 kDa protein and cyt c550 but apart from the lumenal surface of the membrane by about 30 Å, in agreement with previous results that this protein has no direct contact with the membrane and cannot bind to PSII in the absence of the 33 kDa protein and cyt c550.[179] The arrangement of the three extrinsic proteins suggests that the 12 kDa protein helps to link the 33 kDa protein and cyt c550. Consequently, the 33 kDa protein and cyt c550 interact not directly but through the 12 kDa protein, which suggests a different organization of the extrinsic proteins between cyanobacteria and higher plants, where the 33 and 23 kDa proteins interact directly.[131]

The three extrinsic proteins together with the lumenal regions of CP47, CP43, D1, and D2 form a large barrier to shield the Mn cluster from the bulk solution (Kamiya and Shen, 2003).[27] When the extrinsic proteins are present, the Mn cluster is buried within a protein matrix by about 30 to 40 Å from the lumenal surface.[26,156] On the basis of the *Thermosynechococus vulcanus* model,[27] the lumenal part of CP47 is close to those of D2, the 33 kDa and 12 kDa proteins, suggesting possible interactions among these subunits. In particular a close location between the large E loop of CP47 and the extrinsic 33 kDa has been found, consistent with many reports suggesting their possible association and interaction.[131,200] Likewise, D2 is close to CP47 and 33 kDa, D1 is close to CP43, D2, 33 kDa and cyt c550, and CP43 is close to D1 and all three extrinsic proteins.

It is generally agreed that the stoichiometry ratio of the three extrinsic proteins (33, 23 and 17 kDa) is 1:1:1.[133,149] However, the question of the stoichiometry of the 33 kDa protein in relation to the PSII reaction center has been an unsettled subject. Early investigations suggested that anywhere from 1 to 3 copies of the protein might be present in PSII.[201-203] Later investigations using biochemical methods to quantify the level of the protein showed that two copies were present in spinach PSII preparations.[141,204,205] However, this ratio was challenged by Nield,[156] who reported on the structure of the lumenal surface of spinach preparations using cryoelectron microscopy. In this structure, at about 17 Å resolution, the authors asserted that a single copy of the 33 kDa protein was present in PSII. This finding was gratifying because it agreed with the stoichiometry presented in the crystal structure of Zouni et al[26] for the cyanobacterium *Synechococcus elongatus.* However, it did not provide an answer to the question of why binding of 2 copies of the protein were necessary for full reconstitution of PSII activity in spinach preparations.[206] The resolution of this question was provided by site-directed mutagenesis experiments aimed at detected the domains of the N-terminus of the 33 kDa protein that are required for binding to PSII. Popelkova et al[207] found that removal of the first 10 amino acids from the spinach protein reduced rebinding to PSII, from 2 copies to 1 copy, with a concomitant loss of some oxygen evolving activity. Removal of 6 additional amino acids caused a near total loss of rebinding of the mutant 33 kDa protein,[207] and of restoration of any activity. When the sequences of the deletions from the spinach protein were analyzed in detail, it was found that there are two binding related tetrapeptide domains in the eukaryotic protein (T(Y/F)XX), but that one of these domains is absent from the N-terminal sequence of cyanobacterial proteins.[208] This fundamental difference in amino acid sequences in the N-termini of eukaryotic and prokaryotic proteins is the likely explanation for the absence of a second copy of the protein in both cyanobacterial crystal structures.[26,27]

Conclusions

In view of its importance, PSII has received extensive studies regarding its protein composition, function, and dynamic regulation. As a result, our knowledge of the function and reaction mechanisms of PSII has advanced significantly. The recent progress towards elucidating the structure of PSII has revealed the structure of most of the subunits and the arrangement of pigments and cofactors within PSII, whereas the assignment of individual aminoacid residues is necessarily speculative at the present resolution. Electron densities on the manganese complex have been identified, although details of the metal site remain invisible. New insights have been provided towards the location and arrangement of the three extrinsic proteins, which is different between cyanobacteria and higher plants. The most interesting outcomes of the recent structural analyses seem to be the location of the manganese complex, the relatively isolated locations of the redox centers, $ChlZ_{D1}$, $ChlZ_{D2}$, Cyt b559 and the fact that the primary oxidant, P680, is not a special pair of chlorophylls, as in other types of photosynthetic RCs.

A higher resolution structure is needed particularly for PSII from higher plants, that will reveal amino acid side-chains and thus allow a detailed understanding of how the various cofactors and chlorophyll molecules within PSII relate with their protein environment. This information is necessary in order to understand the unique features of PSII and the molecular basis of the water oxidation mechanism.

References

1. Debus RJ. The manganese and calcium ions of photosynthetic oxygen evolution. Biochim Biophys Acta 1992; 1102:269-352.
2. Gant E. Pigment protein complexes and the concept of the photosynthetic unit: Chlorophyll complexes and phycobilisomes. Photosyn Res 1996; 48:47-53.
3. Vermaas W. Molecular-biological approaches to analyze photosystem II structure and function. Annu Rev Plant Physiol Plant Mol Biol 1993; 44:457-481.
4. Ikeuchi M. Subunit proteins of Photosystem II. Bot Mag 1992; 105:327-373.
5. Ikeuchi M, Yuasa M, Inoue Y. Simple and discrete isolation of an O2-evolving PSII reaction center complex retaining Mn and the extrinsic 33 kDa protein. FEBS Lett 1985; 185:316-322.
6. Ghanotakis DF, Demetriou DM, Yocum CF. Isolation and characterization of an O2-evolving Photosystem II reaction center core preparation and a 28 kDa chl -binding protein Biochim Biophys Acta 1987; 891:15-21.
7. Shen J-R, Ikeuchi M, Inoue Y. Stoichiometric association of extrinsic cytochrome c_{550} and 12 kDa protein with a highly purified oxygen-evolving photosystem II core complex from Synechococcus vulcanus. FEBS Lett 1992; 301:145-149.
8. Ikeuchi M, Koike H, Inoue Y. N-terminal sequencing of Photosystem II low-molecular-mass proteins 5 and 3.1 kDa components of the O_2-evolving core complex from higher plants. FEBS Lett 1989; 242:263-269.
9. Bricker TM, Ghanotakis DF. Introduction to oxygen evolution and the oxygen-evolving complex. In: Ort DR, Yocum CF, eds. Advances in Photosynthesis: The Light Reactions. Vol. 4. Dordrecht: Kluwer Academic Publishers, 1996:113-136.
10. Hankamer B, Morris EP, Nield J et al. Subunit positioning and transmembrane helix organisation in the core dimer of photosystem II. FEBS Lett 2001; 504:142-151.
11. Diner BA, Babcock GT. Structure, dynamics and energy conversion efficiency in photosystem II. In: Ort DR, Yocum CF, eds. Advances in Photosynthesis: The Light Reactions. Vol. 4. Dordrecht: Kluwer Academic Publishers, 1996:213-247.
12. Klimov VV, Dolan E, Ke B. EPR properties of an intermediary electron acceptor (pheophytin) in Photosystem II reaction centers at cryogenic temperatures. FEBS Lett 1980; 112:97-100.

13. Debus RJ, Berry BA, Sithole I et al. Directed mutagenesis indicates that the donor to P680+ in Photosystem II is Tyr-160 of the D1 polypeptide. Biochemistry 1988; 27:9071-9074.
14. Debus RJ, Berry BA, Babcock GT et al. Site specific mutagenesis identifies a tyrosine radical involved in the photosynthetic oxygen-evolving complex. Proc Natl Acad Sci USA 1988; 85:427-430.
15. Kok B, Forbush B, McGloin M. Cooperation of charges in photosynthetic oxygen evolution. A linear four step mechanism. Photochem Photobiol 1970; 11:457-475.
16. Tommos C, Babcock GT. Proton and hydrogen currents in photosynthetic water oxidation. Biochim Biophys Acta 2000; 1458:199-219.
17. Saphon S, Crofts T. Protolytic reactions in Photosystem II: A new model for the release of protons accompanying the photooxidation of water. Z Naturforch 1977; 32c:617-626.
18. Ford RC, Rosenberg MF, Shepherd FH et al. Photosystem II 3-D structure and role of the extrinsic subunits in photosynthetic oxygen evolution. Micron 1995; 26:133-140.
19. Tsiotis G, Walz T, Spyridaki A et al. Tubular crystals of a Photosystem II core complex. J Mol Biol 1996; 259:241-248.
20. Boekema EJ, Hankamer B, Bald D et al. Supramolecular structure of the photosystem II complex from green plants and cyanobacteria. Proc Natl Acad Sci USA 1995; 92:175-179.
21. Rögner M, Boekema EJ, Barber J. How does photosystem 2 split water? The structural basis of efficient energy conversion. Trends Biochem Sci 1996; 21:44-49.
22. Rhee K-H, Morris EP, Zheleva D et al. Two dimensional structure of plant photosystem II at 8 Å resolution. Nature 1997; 389:522-526
23. Rhee K-H, Morris EP, Barber J et al. Three dimensional structure of the photosystem II reaction centre at 8 Å. Nature 1998; 396:283-286.
24. Hankamer B, Morris EP, Barber J. Cryoelectron microscopy of photosystem two shows that CP43 and CP47 are located on opposite sides of the D1/D2 reaction centre proteins. Nat Struct Biol 1999; 6:560-564.
25. Hankamer B, Morris EP, Nield J et al. Three-dimensional structure of photosystem II core dimer of higher plants determined by electron microscopy. J Struct Biol 2001; 13:262-269.
26. Zouni A, Witt H-T, Kern J et al. Crystal structure of photosystem II from Synechococcus elongatus at 3.8 Å resolution. Nature 2001; 409:739-742.
27. Kamiya N, Shen JR. Crystal structure of oxygen-evolving photosystem II from Thermosynechococcus vulcanus at 3.7-Å resolution. Proc Natl Acad Sci 2003; 100:98-103.
28. Nanba O, Satoh K. Isolation of a Photosystem II reaction center consisting of D1 and D2 polypeptides and cytochrome b559. Proc Natl Acad Sci USA 1987; 84:109-112.
29. Barber J, Chapman DJ, Tefler A. Characterization of a photosystem II reaction center isolated from chloroplasts of Pisum sativum. FEBS Lett 1987; 220:67-73.
30. Ghanotakis DF, de Paula JC, Demetriou DM et al. Isolation and characterization of the CP47 kDa protein and the D1-D2-cytochrome b559 complex. Biochim Biophys Acta 1989; 974:44-53.
31. Michel HP, Hunt DF, Shabanowitz J et al. Tandem mass spectrometry reveals that three photosystem II proteins of spinach chloroplasts contain N-acetyl-O-phosphothreonine at their NH2 termini. J Biol Chem 1988; 263:1123-1130.
32. Michel H, Deisenhofer J. Relevance of the photosynthetic reaction center from purple bacteria to the structure of photosystem II. Biochemistry 1988; 27:1-7.
33. Trebst A, Depka B. The architecture of photosystem II in plant photosynthesis. Which polypeptide carries the reaction center of photosystem II? In: Michel-Beyerle ME, ed. Antennas and Dynamics. Berlin: Springer-Verlag, 1985:216-223.
34. Fotinou C, Ghanotakis DF. A preparative method for the isolation of the 43 kDa, 47 kDa and the D1-D2-Cytb-559 species directly from the thylakoid membranes. Photosyn Res 1990; 37:41-48.
35. Tang X-S, Fushimi K, Satoh K. D1-D2 complex of the Photosystem II reaction center from spinach. Isolation and partial characterization. FEBS Lett 1990; 273:257-260.
36. Krauss N, Schubert W-D, Klukas O et al. Photosystem I at 4 Å resolution represents the first structural model of a joint photosynthetic reaction centre and core antenna system. Nat Struct Biol 1996; 3:965-973.
37. Jordan P, Fromme P, Witt H-T et al. Three-dimensional structure of cyanobacterial photosystem I at 2.5 Å resolution. Nature 2001; 411:909-916.
38. Saenger W, Jordan P, Krauss N. The assembly of protein subunits and cofactors in photosystem I. Curr Opin Struct Biol 2002; 12:244-254.
39. Schubert W-D, Klukas O, Saenger W et al. A common ancestor for oxygenic and anoxygenic photosynthetic systems - a comparison based on the structural model of photosystem I. J Mol Biol 1998; 280:297-314.
40. Kobayashi M, Maeda H, Watanabe T et al. Chlorophyll α and β-carotene content in the D1/D2/ cytochrome b559 reaction center complex from spinach FEBS Lett 1990; 260:138-140.

41. Barber J, Archer MD. P680, the primary electron donor of photosystem II. J Photochem Photobiol 2001; 142:97-106.
42. Diner BA, Rappaport F. Structure, dynamics, and energetics of the primary photochemistry of photosystem II of oxygenic photosynthesis. Annu Rev Plant Biol 2002; 53:551-580.
43. Trebst A. The three-dimensional structure of the herbicide binding niche on the reaction center polypeptides of Photosystem II. Z Naturforsch 1986; 42c:742-750.
44. Schelvis JPM, van Noort PI, Aartsma TJ et al. Energy transfer, charge separation and pigment arrangement in the reaction center of Photosystem II. Biochim Biophys Acta 1994; 1184:242-250.
45. Ruffle S, Hutchison R, Sayre RT. In: Garab G, ed. Photosynthesis: Mechanisms and Effects Dordecht: Kluwer Academic, 1998:1013-1016.
46. Tracewell CA, Vrettos JS, Bautista JA et al. Carotenoid photooxidation in photosystem II. Arch. Biochem Biophys 2001; 385:61-69.
47. Newell WR, van Amerongen H, Barber J et al. Spectroscopic characterization of the reaction center of photosystem-li using polarized-ligh—Evidence for beta-carotene excitons in psi-Ii reaction Centers. Biochim Biophys Acta 1991; 1057:232-238.
48. Vasil'ev S, Brudwig GW, Bruce D. The X-ray structure of photosystem II reveals a novel electron transport pathway between P680, cytochrome b559 and the energy-quenching cation, $Chlz^+$. FEBS Lett 2003; 543:159-163.
49. Whitmarsh J, Pakrasi HB. Form and function of cytochrome b559 In: Ort DR, Yocum CF, eds. Advances in Photosynthesis: The Light Reactions. Vol. 4. Dordrecht: Kluwer Academic Publishers, 1996:249-264.
50. Sharma J, Panico M, Barber J et al. Characterization of the low molecular weight photosystem II reaction center subunits and their light-induced modifications by mass spectrometry J Biol Chem 1997; 272:3935-3943.
51. Pakrasi HB, Ciechi PD, Whitmarsh J. Site-directed mutagenesis of the heme axial ligands of cytochrome b559 affects the stability of the Photosystem II complex. EMBO J 1991; 10:1619-1627.
52. Tae GS, Cramer WA. Truncation of the COOH-terminal domain of the psbE gene product in Synecocystis sp. 6803: Requirements for Photosystem II assembly and function. Biochemistry 1992; 31:4066-4073.
53. Tae GS, Black MT, Cramer WA et al. Thylakoid membrane protein topography: transmembrane orientation of the chloroplast cytochrome b559 psbE gene product. Biochemistry 1988; 27:9075-9080.
54. Buser CA, Diner BA, Brudwig GW. Reevaluation of the stoichiometry of cytochrome b559 in Photosystem II and thylakoid membranes. Biochemistry 1992; 31:11441-11448.
55. Alizadeh S, Morais F, Barber J et al. Isotopic labelling of the polypeptide subunits of the isolated photosystem II reaction-center complex of Chlamydomonas reihnhardtii suggests an αβ heterodimeric structure for cytochrome b-550. J Photochem Photobiol 1999; 48:148-153.
56. Kunstner P, Guardiola A, Takahashi Y et al. A mutant strain of Chlamydomonas reinhardtii lacking the chloroplast photosystem II psbI gene grows photoautotrophically. J Biol Chem 1995; 270:9651-9654.
57. Ikeuchi M, Shukla VK, Pakrasi HB et al. Directed inactivation of the psbI gene does not affect photosystem II in the cyanobacterium Synechocystis sp. PCC 6803. Mol Gen Genet 1995; 249:622-628.
58. Webber AN, Packman LC, Chapman DJ et al. A fifth chloroplast-encoded polypeptide is present in the photosystem II reaction centre complex FEBS Lett 1989; 242:259-262.
59. Tomo T, Enami I, Satoh K. Orientation and nearest neighbor analysis of psbI gene product in the photosystem II reaction center complex using bifunctional cross-linkers. FEBS Lett 1993; 323:15-18.
60. Ikeuchi M, Inoue Y. A new Photosystem II reaction center component (38 kDa) encoded by the chloroplast genome. FEBS Lett 1988; 241:99-103.
61. Ikeuchi M, Inoue Y. A new 38 kDa polypeptide intrinsic to Photosystem II reaction center as revealed by modified SDS-PAGE with improved resolution of the low-molecular-weight proteins. Plant Cell Physiol 1988; 29:1233-1239.
62. Bricker TM. The structure and function of CPa-1 and CPa-2 in Photosystem II. Photosyn Res 1990; 24:1-13.
63. Barbato R, Race H, Friso G et al. Chlorophyll levels in the pigment-binding proteins of photosystem II A study based on the chlorophyll to cytochrome ratio in different photosystem II preparations FEBS Lett 1991; 286:86-90.
64. Zheleva D, Sharma J, Panico M et al. Isolation and characterization of monomeric and dimeric CP47-reaction center photosystem II complexes. J Biol Chem 1998; 273:16122-16127.
65. Zuber H, Brunisholz R, Sidler W. Structure and function of light harvesting pigment protein complexes. In: Amesz J, eds. Photosynthesis. Amsterdam: Elsevier Press, 1987:233-271.

66. Shen G, Eaton-Rye JJ, Vermaas WFJ. Mutation of histidine residues in CP47 leads to destabilization of the photosystem II complex and to impairment of light energy transfer Biochemistry 1993; 32:5109-5115.
67. Bricker TM, Frankel LK. Use of a monoclonal antibody in structural investigations of the 49-kDa polypeptide of photosystem II. Arch Biochem Biophys 1987; 256:295-301.
68. Frankel LK, Bricker TM. Epitope mapping of the monoclonal antibody FAC2 on the apoprotein of Cpa-1 in photosystem II. FEBS Lett 1989; 257:279-282.
69. Bricker TM, Odom WR, Queirolo CB. Close association of the 33 kDa extrinsic protein with the apoprotein of CPal in photosystem II. FEBS Lett 1988; 231:111-117.
70. Enami I, Kaneko M, Kitamura N et al. Total immobilization of the extrinsic 33 kDa protein in spinach Photosystem II preparations. Protein stoichiometry and stabilization of oxygen evolution. Biochim Biophys Acta 1991; 1060:224-232.
71. Odom WR, Bricker TM. Interaction of CPa-1 with the manganese-stabilizing protein of photosystem II: Identification of domains cross-linked by 1-ethyl-3-[3-(dimethylamino)propyl] carbodiimide. Biochemistry 1992; 31:5616-5620.
72. Frankel LK, Bricker TM. Interaction of CPa-1 with the manganese-stabilizing protein of photosystem II Identification of domains on CPa-1 that are shielded from N-hydroxysuccinimide biotinylation by the manganese-stabilizing protein. Biochemistry 1992; 31:11059-11064.
73. Hayashi H, Fujimura Y, Mohanty PS et al. The role of CP 47 in the evolution of oxygen and the binding of the extrinsic 33-kDa protein to the core complex of Photosystem II as determined by limited proteolysis. Photosynth Res 1993; 3:35-42.
74. Putnam-Evans C, Bricker TM. Site-directed mutagenesis of the CPa-1 protein of photosystem II: alteration of the basic residue pair 384,385R to 384,385G leads to a defect associated with the oxygen-evolving complex. Biochemistry 1992; 31:11482-11488.
75. Putnam-Evans C, Bricker TM. Site-directed mutagenesis of the CP47 protein of photosystem II: Alteration of the basic residue ^{448}R to ^{448}G prevents the assembly of functional photosystem II centers under chloride-limiting conditions. Biochemistry 1994; 33:10770-10776.
76. Putnam-Evans C, Burnap RL, Wu J et al. Site-directed mutagenesis of the CP 47 protein of photosystem II: alteration of conserved charged residues in the domain 364E-444R. Biochemistry 1996; 35:4046-4053.
77. Qian M, Al-Khaldi SF, Putnam-Evans C et al. Photoassembly of the photosystem II $(Mn)^4$ cluster in site-directed mutants impaired in the binding of the manganese-stabilizing protein. Biochemistry 1997; 36:15244-15252.
78. Putnam-Evans C, Bricker TM. Site-directed mutagenesis of the basic residues ^{321}K to ^{321}G in the CP 47 protein of photosystem II alters the chloride requirement for growth and oxygen-evolving activity in Synechocystis 6803. Plant Mol Biol 1997; 34:455-463.
79. Morgan TR, Shand JA, Clarke SM et al. Specific requirements for cytochrome c-550 and the manganese- stabilizing protein in photoautotrophic strains of Synechocystis sp. PCC 6803 with mutations in the domain Gly-351 to Thr-436 of the chlorophyll-binding protein CP47 Biochemistry 1998; 37:14437-14449.
80. Clarke SM, Eaton-Rye JJ. Mutation of Phe-363 in the photosystem II protein CP47 impairs photoautotrophic growth, alters the chloride requirement, and prevents photosynthesis in the absence of either PSII-O or PSII-V in Synechocystis sp. PCC 6803 Biochemistry 1999; 38:2707-2715.
81. Wu JT, Masri N, Lee W et al. Random mutagenesis in the large extrinsic loop E and transmembrane -helix VI of the CP 47 protein of Photosystem II. Plant Mol Biol 1999; 39:381-386.
82. Akabori K, Tsukamoto H, Tsukihara J et al. Disintegration and reconstitution of Photosystem II reaction center core complex. Preparation and characterization of three different types of subcomplexes. Biochim. Biophys. Acta 1988; 932:345-357.
83. Yamaguchi N, Takahashi Y, Satoh K. Isolation and characterization of a Photosystem II core complex depleted in the 43 kDa chlorophyll binding subunit. Plant Cell Physiol 1988; 29:123-129.
84. Rögner M, Chisholm DA, Diner B. Site-directed mutagenesis of the psbC gene of Photosystem II: Isolation and functional characterization of CP43-less Photosystem II core complexes. Biochemistry 1991; 30:5387-5395.
85. Bassi R, Hoyer-Hansen G, Barbato R et al. Chlorophyll-proteins of the Photosystem II antenna complex. J Biol Chem 1987; 262:13333-13341.
86. Barber J, Morris EP, Büchel C. Revealing the structure of the photosystem two chlorophyll binding proteins, CP43 and CP47. Biochim Biophys Acta 2000; 145:239-247.
87. de Vitry C, Diner BA, Lemoine Y. Chemical composition of Photosystem II reaction centers: Phosphorylation of PSII polypeptides. In: Biggins J, ed. Progress in Photosynthesis Research, Vol II. Dordrecht: Martinus Nijhoff, 1987:105-108.
88. Gounaris K, Pick U, Barber J. Stoichiometry and turnover of Photosystem II polypeptides. FEBS Lett 1987; 211:94-98.

89. Renger G. Mechanistic and structural aspects of photosynthetic water oxidation. Physiol Plant 1997; 100:828-841.
90. Hansson Ö, Wydrzynski T. Current perceptions of Photosystem II. Photosynth Res 1990; 23:131-162.
91. Nugent JHA. Oxygenic photosynthesis-Electron transfer in photosystem I and photosystem II. Eur J Biochem 1996; 237:519-531.
92. Yachandra VK, Sauer K, Klein MP. Manganese cluster in photosynthesis: where plants oxidize water to dioxygen. Chem Rev 1996; 96:2927-2950.
93. McDermott E, Yachandra VK, Guiles RD et al. Characterization of the manganese O2-evolving complex and the iron-quinone acceptor complex in photosystem II from a thermophilic cyanobacterium by electron paramagnetic resonance and X-ray absorption spectroscopy Biochemistry 1988; 27:4021-4031.
94. Yocum CF, Yerkes CT, Blankenship RE et al. Stoichiometry, inhibitor sensitivity and organization of manganese associated with photosynthetic oxygen evolution. Proc Natl Acad Sci USA 1981; 78:7507-7511.
95. Tamura N, Cheniae G. Photoactivation of the water-oxidizing complex in Photosystem II membranes depleted of Mn and extrinsic proteins. I. Biochemical and kinetic characterization. Biochim Biophys Acta 1987; 890:179-193.
96. Ghanotakis DF, Topper J, Yocum CF. Structural organization of the oxidizing side of Photosystem II. Exogenous reductants reduce and destroy the Mn-complex in Photosystem II membranes depleted of the 17 and 23 kDa polypeptides. Biochim Biophys Acta 1984; 767:524-531.
97. Ono T, Inoue Y. S-state turnover in the O_2-evolving system of $CaCl_2$-washed Photosystem II particles depleted of three peripheral proteins as measured by thermoluminescence. Removal of 33 kDa protein inhibits S3 to S4 transition. Biochim Biophys Acta 1985; 806:331-340.
98. Peloquin JM, Campbell KA, Randall DW et al. ^{55}Mn ENDOR of the S_2-state multiline EPR signal of Photosystem II: Implications on the structure of the tetranuclear Mn cluster. J Am Chem Soc 2000; 122:10926-10942.
99. Peloquin JM, Britt RD. EPR/ENDOR characterisation of the physical and electronic structure of the OEC Mn cluster. Biochim Biophys Acta 2001; 1503:96-111.
100. Robblee JH, Cince RM, Yachandra VK. X-ray spectroscopy-based structure of the Mn cluster and mechanism of photosynthetic oxygen evolution. Biochim Biophys Acta 2001; 1503:7-23.
101. Carrell TG, Tyryshkin AM, Dismukes GC. An evolution of structural models for the photosynthetic water-oxidising complex derived from spectroscopic X-ray diffraction signatures. J Biol Inorg Chem 2001; 7:2-22.
102. Debus RJ. Amino acid residues that modulate the properties of tyrosine Y(Z) and the manganese cluster in the water oxidizing complex of photosystem II. Biochim Biophys Acta 2001; 1503:164-86.
103. Preston C, Seibert M. Protease treatments of Photosystem-II membrane-fragments reveal that there are 4 separate high-affinity mn-binding sites. Biochemistry 1991; 30:9625-9633.
104. Yocum CF. Calcium activation of photosynthetic water oxidation. Biochim Biophys Acta 1991; 1059:1-15.
105. Ghanotakis DF, Babcock GT, Yocum CF. Calcium reconstitutes high rates of oxygen evolution in polypeptide depleted Photosystem II preparations FEBS Lett 1984; 167:127-130.
106. Boussac A, Rutherford AW. Nature of the inhibition of the oxygen-evolving enzyme of photosystem II induced by NaCl washing and reversed by the addition of Ca^{2+} or Sr^{2+} FEBS Lett 1988; 236:432-436.
107. Lockett CJ, Demetriou C, Bowden SJ et al. Studies on calcium depletion of PS II by pH 8.3 treatment. Biochim Biophys Acta 1990; 1016:213-218.
108. Ghanotakis DF, Babcock GT, Yocum CF. Structure of the oxygen-evolving complex of Photosystem II: Calcium and lanthanum compete for sites on the oxidizing side of Photosystem II which control the binding of water-soluble polypeptides and regulate the activity of the manganese complex Biochim Biophys Acta 1985; 809:173-180.
109. Ono T-A, Inoue Y. Discrete extraction of the Ca atom functional for O_2 evolution in higher plant photosystem II by a simple low pH treatment FEBS Lett 1988; 227:147-152.
110. Ono T-A, Inoue Y. Roles of Ca^{2+} in O_2 evolution in higher plant photosystem II: effects of replacement of Ca^{2+} site by other cations Arch Biochem Biophys 1989; 275:440-448.
111. Latimer MJ, DeRose VJ, Mukerji I et al. Evidence for the proximity of calcium to the manganese cluster of photosystem II: determination by X-ray absorption spectroscopy Biochemistry 1995; 34:10898-10909.
112. Noguchi T, Ono T-A, Inoue Y. Direct detection of a carboxylate bridge between Mn and Ca^{2+} in the photosynthetic oxygen-evolving center by means of Fourier transform infrared spectroscopy Biochim Biophys Acta 1995; 1228:189-200.

113. MacLachlan DJ, Hallahan BJ, Ruffle SV et al. An e.x.a.f.s. study of the manganese O_2-evolving complex in purified photosystem II membrane fractions. The S_1 and S_2 states. Biochem J 1992; 285:569-576.
114. Booth PJ, Rutherford AW, Boussac A. Location of the calcium binding site in Photosystem II: a Mn^{2+} substitution study. Biochim Biophys Acta 1996; 1277:127-134.
115. Cinco RM, Robblee JH, Rompel A et al. Strontium EXAFS Reveals the Proximity of Calcium to the Manganese Cluster of Oxygen-Evolving Photosystem II. J Phys Chem B 1998; 102:8248-8256.
116. Cammarata KV, Cheniae GM. Studies on 17,24 kDa depleted Photosystem II membranes. Plant Physiol 1987; 84:587-595.
117. Katoh S, Satoh K, Ohno T et al. In: J Biggins, ed. Progress in Photosynthesis Research. Dordrecht: Martinus Nijhoff, 1987:625-628.
118. Shen J-R, Satoh K, Katoh S. Calcium content of oxygen-evolving Photosystem II preparations from higher plants. Effects of NaCl treatment. Biochim Biophys Acta 1988; 933:358-364.
119. Ädelroth P, Lindberg K, Andréasson L-E. Studies of Ca^{2+} binding in spinach photosystem II using $^{45}Ca^{2+}$. Biochemistry 1995; 34:9021-9027.
120. Sandusky PO, Yocum CF. The chloride requirement for photosynthetic oxygen evolution. Analysis of the effects of chlorie and other anions on amine inhibition of the oxygen-evolving complex. Biochim Biophys Acta 1984; 776:603-611.
121. Wydrzynski T, Baumgart F, MacMillan F et al. Is there a direct chloride cofactor requirement in the oxygen-evolving reactions of photosystem II? Photosynth Res1990; 25:59-72.
122. Rutherford AW, Zimmermann J-L, Boussac A. In: Barber J, eds. Elsevier The Photosystems: Structure, Function and Molecular Biology. Amsterdam, 1992:179-229.
123. Kelley W, Izawa S. The role of chloride ion in Photosystem II. I. Effects of chloride ion in Photosystem II electron transport and on hydroxylamine inhibition. Biochim Biophys Acta 1978; 502:198-210.
124. Miyao M, Murata N. The chloride effect on photosynthetic oxygen evolution: interaction of chloride with 18-kDa, 24-kDa and 33-kDa proteins FEBS Lett 1985; 180:303-308.
125. Homann PH. Chloride relations of photosystem II membrane preparations depleted of, and resupplied with, their 17 and 23 kDa extrinsic polypeptides Photosynth Res 1988; 15:205-220.
126. Homann PH. Structural effects of chloride and other anions on the water oxidizing complex of chloroplast photosystem II. Plant Physiol 1988; 88:194-199.
127. Lindberg K, Wydrzynski T, Vänngard T et al. Slow release of chloride from ^{36}Cl-labeled photosystem II membranes FEBS Lett 1990; 264:153-155.
128. Lindberg K, Andréasson L-E. A One-Site, Two-State Model for the Binding of Anions in Photosystem II. Biochemistry 1996; 35:14259-14267.
129. Wincencjusz H, van Gorkom HJ, Yocum CF. The photosynthetic oxygen evolving complex requires chloride for its redox state $S_2 \rightarrow S_3$ and $S_3 \rightarrow S_0$ transitions but not for $S_0 \rightarrow S_1$ or $S_1 \rightarrow_2$ transitions. Biochemistry 1997; 36:3663-3670.
130. Wincencjusz H, Yocum CF, van Gorkom HJ. Activating anions that replace Cl^- in the O_2-evolving complex of photosystem II slow the kinetics of the terminal step in water oxidation and destabilize the S_2 and S_3 states Biochemistry 1999; 38:3719-3725.
131. Seidler A. The extrinsic polypeptides of Photosystem II. Biochim Biophys Acta 1996; 1277:36-60.
132. Ono T, Inoue Y. Reconstitution of photosynthetic oxygen evolving activity by rebinding of 33-kDa protein to $CaCl_2$-extracted PSII particles. FEBS Lett 1984; 166:381-384.
133. Miyao M, Murata N. Role of the 33 kDa polypeptide in preserving Mn in the photosynthetic oxygen evolution system and its replacement by chloride ions. FEBS Lett 1984; 168:281-286.
134. Mavankal G, McCain DC, Bricker TM. Effects of chloride on paramagnetic coupling of manganese in calcium chloride-washed Photosystem II preparations. FEBS Lett 1986; 202:235-239.
135. Kuwabara T, Miyao M, Murata T et al. The function of 33-kDa protein in the photosynthetic oxygen-evolution system studied by reconstitution experiments. Biochim Biophys Acta 1985; 806:283-289.
136. Ono T, Kajikawa H, Inoue Y. Changes in protein composition and Mn abundance in Photosystem II particles on photoactivation of the latent O_2-evolving system in flash-grown wheat leaves. Plant Physiol 1986; 80:85-90.
137. Burnap LR, Sherman LA. Deletion mutagenesis in Synechocystis sp. PCC6803 indicates that the Mn-stabilizing protein in photosystem II is not essential for O_2 evolution. Biochemistry 1991; 30:440-446.
138. Philbrick JB, Diner BA, Zilinskas B. Construction and characterization of cyanobacterial mutants lacking the manganese-stabilizing polypeptide of photosystem II. J Biol Chem 1991; 266:13370-13376.
139. Koike H, Inoue Y. Properties of a peripheral 34-kDa protein in Synechococcus vulcanus Photosystem II particles. Its exchangeability with spinach 33-kDa protein in reconstitution of oxygen evolution. Biochim Biophys Acta 1985; 807:64-73.

140. Betts S, Hachigian TM, Pichersky RE et al. Reconstitution of the spinach oxygen-evolving complex with recombinant Arabidopsis manganese-stabilizing protein. Plant Mol Biol 1994; 26:117-130.
141. Leuschner C, Bricker TM. Interaction of the 33 kDa extrinsic protein with photosystem II: Rebinding of the 33 kDa extrinsic protein to photosystem II membranes that contain four, two or zero manganese per photosystem II reaction center. Biochemistry 1996; 35:4551-4557.
142. Hankamer B, Barber J, Boekema EJ. Structure and membrane organization of photosystem II in green plants. Annu Rev Plant Physiol Plant Mol Biol 1997; 48:641-671.
143. Miyao M, Murata N. The model of binding of the three extrinsic proteins of 33 kDa, 23 kDa and 18 kDa in the photosystem II complex of spinach. Biochim Biophys Acta 1989; 977:315-321.
144. Kavelaki K, Ghanotakis DF. Effect of the manganese complex on the binding of the extrinsic proteins (17, 23 and 33 kDa) of photosystem II. Photosynth Res 1991; 29:149-155.
145. Enami I, Miyaoka T, Mochizuki Y et al. Nearest neighbor relationships among constituent proteins of oxygen-evolving photosystem II membranes: binding and function of the extrinsic 33 kDa protein. Biochim Biophys Acta 1989; 973:35-40.
146. Tichy M, Vermaas WFJ. Functional analysis of combinatorial mutants altered in a conserved region in loop E of the CP47 protein in Synechocystis sp. PCC 6803. Biochemistry 1998; 37:1523-1531.
147. Enami I, Tohri A, Kamo M et al. Identification of domains on the 43 kDa chlorophyll-carrying protein (CP43) that are shielded from tryptic attack by binding of the extrinsic 33 kDa protein with photosystem II complex. Biochim Biophys Acta 1997; 1320:17-26.
148. Eaton-Rye JJ, Murata N. Evidence that the amino-terminus of the 33 kDa extrinsic protein is required for binding to the photosystem II complex. Biochim Biophys Acta 1989; 977:219-226.
149. Bricker TM, Frankel LK. The structure and function of the 33 kDa extrinsic protein of photosystem II: a critical assessment. Photosynth Res 1998; 56:157-173.
150. Ahmed A, Tajmir-Riahi HA, Carpentier R. A quantitative secondary structure analysis of the 33 kDa extrinsic polypeptide of photosystem II by FTIR spectroscopy. FEBS Lett 1995; 363:65-68.
151. Zhang HM, Fischer G, Wydrznski T. In: Mathis P, ed. Photosynthesis: From Light to Biosphere. Dordrecht: Kluwer Academic Press, 1995:447-450
152. Xu Q, Nelson J, Bricker TM. Secondary structure of the 33 kDa, extrinsic protein of Photosystem II: a far-UV circular dichroism study. Biochim Biophys Acta 1994; 1188:427-431.
153. Shutova T, Irrgang K-D, Shubin V et al. Analysis of pH-Induced Structural Changes of the Isolated Extrinsic 33 Kilodalton Protein of Photosystem II. Biochemistry 1997; 36:6350-6358.
154. Lydakis-Simantiris N, Hutchison RS, Betts SD et al. Manganese stabilizing protein of photosystem II is a thermostable, natively unfolded polypeptide. Biochemistry 1999; 38:404-414.
155. Shutova T, Irrgang K-D, Klimov VV et al. Is the manganese stabilizing 33 kDa protein of photosystem II attaining a 'natively unfolded' or a 'molten globule' structure in solution? FEBS Lett 2000; 467:137-140.
156. Nield J, Balsera M, de las Rivas J et al. Three-dimensional Electron Cryomicroscopy Study of the Extrinsic Domains of the Oxygen-evolving Complex from Spinach. J Biol Chem 2002; 277:15006-15012.
157. Motoki A, Hirano M, Katoh S. Two regions of the Mn-stabilizing protein from Synechococcus elongatus that are involved in binding to Photosystem II complexes. Biochim Biophys Acta 1998; 1365:492-502.
158. Motoki A, Usui M, Shimazu T et al. A Domain of the Manganese-stabilizing Protein from Synechococcus elongatus Involved in Functional Binding to Photosystem II. J Biol Chem 2002; 277:14747-14756.
159. Motoki A, Miura K, Shimazu T et al. Characterization of the manganese-stabilizing protein from Synechococcus elongatus. In: Murata N, ed. Research in Photosynthesis. Netherlands: Kluwer Academic Publishers, 1992:413-416.
160. Hutchison RS, Betts SD, Yocum CF et al. Conformational changes in the extrinsic manganese stabilizing protein can occur upon binding to the photosystem II reaction center: an isotope editing and FTIR study. Biochemistry 1998; 37:5643-5653.
161. Enami I, Kamo M, Ohta H et al. Intramolecular crosslinking of the extrinsic 33-kDa protein leads to loss of oxygen evolution but not its ability of binding to photosystem II and stabilization of the manganese cluster. J Biol Chem 1998; 273:4629-4634.
162. Kuwabara T, Reddy KJ, Sherman LA. Nucleotide sequence of the gene from the cyanobacterium Anacystis nidulans R2 encoding the Mn-stabilizing protein involved in photosystem II water oxidation. Proc Natl Acad Sci USA 1987; 84:8230-8234.
163. Philbrick JB, Zilinskas BA. Cloning, nucleotide sequence and mutational analysis of the gene encoding the Photosystem II manganese-stabilizing polypeptide of Synechocystis 6803. Mol Gen Genet 1988; 212:418-425.

164. Borthakur D, Haselkorn R. Nucleotide sequence of the gene encoding the 33 kDa water oxidizing polypeptide in Anabaena sp. strain PCC 7120 and its expression in Eschericia coli. Plant Mol Biol 1989; 13:427-439.
165. Miura K, Shimazu T, Motoki A et al. Nucleotide sequence of the Mn-stabilizing protein gene of the thermophilic cyanobacterium Synechococcus elongatus. Biochim Biophys Acta 1993; 1172:357-360.
166. Tucker DL, Hirsh K, Li H et al. The manganese stabilizing protein (MSP) and the control of O_2 evolution in the unicellular, diazotrophic cyanobacterium, Cyanothece sp. ATCC 51142. Biochim Biophys Acta 2001; 1504:409-422.
167. Shigemori Y, Inagaki J, Mori H et al. The presequence of the precursor to the nucleus-encoded 30 kDa protein of photosystem II in Euglena gracilis Z induces two hydrophobic domains. Plant Mol Biol 1994; 24:209-215.
168. Mayfield SP, Rahire M, Frank G et al. Analysis of the genes of the OEE1 and OEE3 proteins of the photosystem II complex from Chlamydomonas reinhardtii. Plant Mol Biol 1989; 12:683-693.
169. Tyagi A, Hermans J, Steppuhn J et al. Nucleotide sequence of cDNA clones encoding the complete "33kDa" precursor protein associated with the photosynthetic oxygen evolving complex from spinach. Mol Gen Genet 1987; 207:288-293.
170. Wales R, Newman BJ, Pappin D et al. The extrinsic 33 kDa polypeptide of the oxygen-evolving complex of photosystem II is a putative calcium-binding protein and is encoded by a multigene family in pea. Plant Mol Biol 1989; 12:439-451.
171. Meadows JW, Holford A, Eaines CA et al. Nucleotide sequence of a cDNA clone encoding the precursor of the 33 kDa protein of the oxygen-evolving complex from wheat. Plant Mol Biol 1991; 16:1085-1087.
172. Van Spanje M, Dirkse WG, Nap JP et al. Isolation and analysis of cDNA encoding the 33 kDa precursor protein of the oxygen-evolving complex of potato. Plant Mol Biol 1991; 17:157-160.
173. Görlach J, Schmid J and Amrhein N. The 33 kDa protein of the oxygen-evolving complex: A multi-gene family in tomato. Plant Cell Physiol 1993; 34:497-501.
174. Enami I, Yoshihara S, Tohri A et al. Cross-reconstitution of various extrinsic proteins and photosystem II complexes from cyanobacteria, red algae and higher plants. Plant Cell Physiol 2000; 41:1354-1364.
175. Tohri A, Suzuki T, Okuyama S et al. Comparison of the structure of the extrinsic 33 kDa protein from different organisms. Plant and Cell Physiol 2002; 43:429-439.
176. Vogt J, Schulz GE. The structure of the outer membrane protein OmpX from Escherichia coli reveals possible mechanisms of virulence. Struct Folding Des 1999; 7:1301-1309.
177. Pautsch A, Schulz GE. High-resolution Structure of the OmpA Membrane Domain. J Mol Biol 2000; 298:273-282.
178. Pazos F, Heredia P, Valencia A et al. Threading structural model of the manganese-stabilizing protein 33 kDa protein reveals presence of two possible β-sandwich domains. Proteins Struct Funct Genet 2001; 45:372-381.
179. Shen G, Inoue Y. Binding and function of two new extrinsic components, cytochrome c550 and a 12 kDa protein, in cyanobacterial Photosystem II. Biochemistry 1993; 32:1825-1832.
180. Nishiyama Y, Hayashi H, Watanabe T et al. Photosynthetic oxygen evolution is stabilized by cytochrome c_{550} against heat inactivation in Synechococcus sp. PCC 7002. Plant Physiol 1994; 105:1313-1319.
181. Nishiyama Y, Los DA, Hayashi H et al. Thermal protection of the oxygen-evolving machinery by PsbU, an extrinsic protein of photosystem II, in Synechococcus species PCC 7002. Plant Physiol 1997; 115:1473-1480.
182. Nishiyama Y, Los DA and Murata N. PsbU, a protein associated with photosystem II, is required for the acquisition of cellular thermotolerance in Synechococcus species PCC 7002. Plant Physiol 1999; 120:301-308.
183. Ghanotakis DF, Topper J, Babcock GT et al. Water-soluble 17 and 23 kDa polypeptides restore oxygen evolution activity by creating a high-affinity binding site for Ca^{2+} on the oxidizing side of Photosystem II. FEBS Lett 1984; 170:169-173.
184. Waggoner CM, Yocum CF. Selective depletion of water-soluble polypeptides associated with Photosystem II. In: Biggins J, eds. Progress in Photosynthesis Research. Dordrecht: Martinus Nijhoff, Vol. I. 1987: 685-688.
185. Ifuku K, Sato F. Importance of the N-terminal sequence of the extrinsic 23 kDa polypeptide in Photosystem II in ion retention in oxygen evolution. Biochim Biophys Acta 2001; 1546:196-204.
186. Rashid A, Carpentier R. The 16 and 23 kDa extrinsic polypeptides and the associated Ca^{+2} and Cl^- modify atrazine interaction with the Photosystem II core complex. Photosynth Res 1990; 24:221-227.

187. Briantais J-M, Vernotte C, Miyao M. Relationship between O_2 evolution capacity and cytochrome b559 high potential form in Photosystem II particles. Biochim Biophys Acta 1985; 808:348-351.
188. Miyao M, Murata N. Partial disintegration and reconstitution of the photosynthetic oxygen evolution system. Binding of 24 kDa and 18 kDa polypeptides. Biochim. Biophys Acta 1983; 725:87-93.
189. Kuwabara T, Murata T, Miyao M et al. Partial degradation of the 18 kDa protein of the photosynthetic oxygen-evolving complex: A study of a binding site. Biochim Biophys Acta 1986; 850:146-155.
190. Miyao M, Fujimura Y, Murata N. Partial degradation of the extrinsic 23 kDa protein of the Photosystem II complex of spinach. Biochim Biophys Acta 1988; 936:465-474.
191. Becker B, Callahan F, Cheniae G. Photoactivation of NH_2OH-treated leaves: Reassembly of released extrinsic polypeptides and religation of Mn into the polynuclear Mn catalyst of water oxidation. FEBS Lett 1985; 192:209-214.
192. Zhang H, Ishikawa Y, Yamamoto Y et al. Secondary structure and thermal stability of the extrinsic 23 kDa protein of photosystem II studied by Fourier transform infrared spectroscopy. FEBS Lett 1998; 426:347-351.
193. Zhang H, Yamamoto Y, Ishikawa Y et al. Characterization of the extrinsic 16 kilodalton protein of spinach photosystem II by Fourier transform infrared spectroscopy. J Mol Struct 1999; 513:127-132.
194. Frazao C, Enguita FJ, Coelho R et al. Crystal structure of low-potential cytochrome c549 from Synechocystis sp. PCC 6803 at 1.21 Å resolution. J Biol Inorg Chem 2001; 6:324-332.
195. Sawaya MR, Krogmann DW, Serag A et al. Structures of cytochrome c-549 and cytochrome c6 from the cyanobacterium Arthrospira maxima. Biochemistry 2001; 40:9215-9225.
196. Nield J, Kruse O, Ruprecht J et al. 3D structure of Chlamydomonas reinhardtii and Synechococcus elongatus photosystem II complexes allow for comparison of their OEC organisation. J Biol Chem 2000; 275:27940-27946.
197. Nield JN, Funk C, Barber J. Supermolecular structure of photosystem II and location of the psbS protein. Proc R Soc London 2000; 355:1337-1344.
198. Andersson B, Akurlund H-E. In: Barber J, ed. Topics in Photosynthesis, Vol. 8. Amsterdam: Elsevier Science Publishers, 1987:379-420.
199. Han KC, Shen JR, Ikeuchi M et al. Chemical crosslinking studies of extrinsic proteins in cyanobacterial photosystem II. FEBS Lett 1994; 355:121-124.
200. Bricker TM, Frankel LK. The structure and function of CP47 and CP43 in photosystem II. Photosynth Res 2002; 72:131-146.
201. Andersson B, Larsson C, Jannson C et al. Immunological studies on the organization of proteins in photosynthetic oxygen evolution. Biochim Biophys Acta 1984; 766:21-26.
202. Murata N, Miyao M, Omata T et al. Stoichiometry of components in the photosynthetic oxygen evolution system of Photosystem II particles prepared with TritonX-100 spinach chloroplast. Biochim Biophys Acta 1984; 765:363-369.
203. Milner PA, Gogel G, Barber J. Investigation of the spatial relationships between Photosystem 2 polypeptides by reversible crosslinking and diagonal electrophoresis. Photosynth Res 1987; 13:185-198.
204. Yamamoto Y, Nakayama S, Cohn CL et al. Highly efficient purification of the 33-, 24-, and 18-kDa proteins in spinach photosystem II by butanol/water phase partitioning and high-performance liquid chromatography. Arch Biochem Biophys 1987; 255:156-161.
205. Xu QA, Bricker TM. Structural organization of proteins on the oxidizing side of photosystem II: Two molecules of the 33 kDa manganese-stabilizing protein per reaction center. J Biol Chem 1992; 267:25816-25821.
206. Betts SD, Ross JR, Pichersky E et al. Mutation val35ala weakens binding of the 33-kDa manganese stabilizing protein of photosystem II to one of two sites. Biochemistry 1997; 36:4047-4053.
207. Popelkova H, Im M, Lydakis-Simantiris N et al. N-terminus of the photosystem II manganese stabilizing protein: Effects of sequence elongation and truncation. Biochemistry 2002; 41:2702-2711.
208. Popelkova H, Im M, Yocum CF. N-terminal truncations of manganese stabilizing protein identify two amino acid sequences required for binding of the eukaryotic protein to photosystem II, and reveal the absence of one binding-related sequence in cyanobacteria. Biochemistry 2002; 41:10038-10045.

Chapter 4

Biogenesis and Structural Dynamics of the Photosystem II Complex

Josef Komenda,* Stanislava Kuviková, Lenka Lupínková and Jiri Masojídek

Abstract

Photosystem II (PSII) represents a multicomponent protein complex located in the thylakoid membrane of cyanobacteria, green algae and higher plants. Due to the ability to oxidize water, its development was responsible for the rise of oxygen atmosphere on Earth, which started about 3 billion years ago. The complex consists of more than 20 protein subunits; during its biogenesis, all these subunits—together with pigments, lipids and other prosthetic groups—are brought together in a highly coordinated process resulting in the functional complex. In addition, Photosystem II is intrinsically vulnerable to light-induced damage: this is mediated by reactive oxygen species and other strong oxidants. Both the latter are generated within the complex and may significantly influence its structure and function. To prevent oxidative damage, the complex frequently undergoes a repair cycle consisting in a selective replacement of its central protein subunit, which is accompanied by the partial disassembly and reassembly of the complex. All these features document the high structural variability and dynamics of the Photosystem II, which is the subject of this chapter.

Assembly of the Photosystem II Complex

To provide an optimal environment for energy and electron transfer processes, the Photosystem II (PSII) subunits have to be precisely arranged during their strictly-regulated de novo assembly. This process has been partially studied in the green alga *Chlamydomonas reinhardtii*, in the cyanobacterium *Synechocystis*, and also in isolated chloroplasts. Using *Chlamydomonas* mutants lacking some PSII subunits, it was concluded that CP43 is associated with PSII in the later stage of the assembly process: when CP47, D1 and D2 have already formed an assembly intermediate.[1] The authors also reported a requirement of the D2 protein for the synthesis of the other large subunits CP47 and D1, whereas D1 synthesis was apparently uncoupled from the synthesis of D2 and CP47. The key role of the cytochrome b-559 in the PSII assembly was evidenced in a *Chlamydomonas* mutant lacking the gene for its alpha subunit.[2] Analysis of *Synechocystis* mutants lacking PSII subunits confirmed the results from *Chlamydomonas*.[3-5] In the absence of CP43, the subunits CP47, D1, and D2 accumulated; while in the absence of CP47, only CP43 accumulated and the D1 and D2 proteins became detectable by radiolabeling only. In the absence of D2, only small amounts of CP47, but no D1, were reported and in the mutant lacking the D1 protein,[6] D2 and CP47 were missing while a nearly wild-type level of CP43 was found. In the absence of cytochrome b-559 subunits, the D1 and D2 proteins disappeared from the membrane while CP47 and CP43 remained detectable.[4] Although these data provided information with respect to the synthesis and accumulation of PSII subunits, their state of assembly remained enigmatic. In this regard, more information

*Corresponding Author: Josef Komenda—Institute of Microbiology, Academy of Sciences, Opatovicky mlyn, CZ-37981, Trebon, Czech Republic. Email: komenda@alga.cz

Biotechnological Applications of Photosynthetic Proteins: Biochips, Biosensors and Biodevices, edited by Maria Teresa Giardi and Elena V. Piletska.

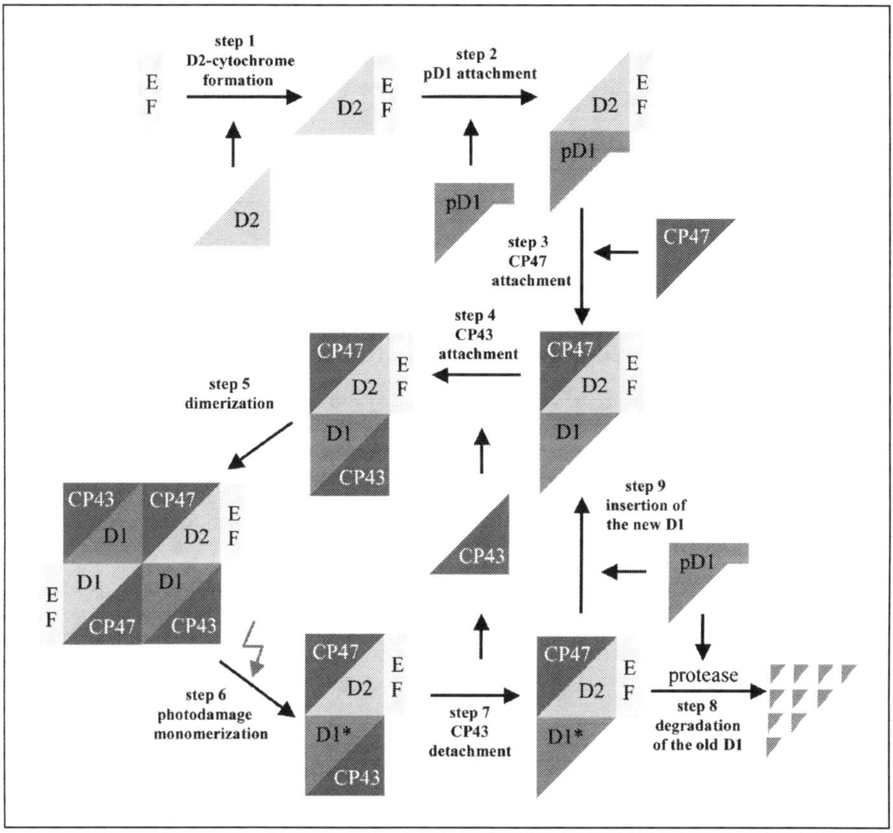

Figure 1. Photosystem II assembly and repair cycle. De novo assembly of PSII starts with association of cytochrome b-559 protein subunits α and β (E/F; step1) with the D2 protein and continues by binding of the D1 precursor pD1 (step 2). Then CP47 is attached and concomitantly pD1 is processed into D1 (step 3). Finally, the CP43 protein is attached (step 4) and resulting PSII monomers form dimer (step 5) that most probably represents the native PSII structure. PSII repair cycle starts with inactivation of the D1 protein (D1*) and parallel monomerization (step 6). Then it continues by detachment of CP43 (step 7), degradation of the old D1 (step 8), insertion of a new pD1 molecule and its prompt processing (step 9). The final two steps, the CP43 attachment and PSII dimerization seem to be common with the de novo assembly pathway. Arrow from pD1 to protease at step 8 indicates synchronization between synthesis of the new D1 and degradation of the old one.

was obtained by radioactive labelling of chloroplast proteins in organelles, followed by analysis of labeled complexes on a sucrose gradient, or by nondenaturing electrophoresis in combination with analysis of the protein complexes by SDS-PAGE and immunodetection.[7,8] This approach led to the postulation of an approximate PSII assembly pathway consisting of the sequential attachment of D1, CP47 and CP43 onto an initially-formed precomplex D2-cytochrome b-559 (Fig. 1). However, details of the process—including attachment of cofactors to apoproteins and the role of a number of small polypeptides and external assembly factors—are still missing. For the PSII chlorophyll-proteins, a stabilization of apoproteins by chlorophylls was detected during greening.[9,10] As a representative of the rarely-identified assembly factors, the hypothetical protein HCF136 was found to be a prerequisite for the formation of the D1-D2 reaction-centre complex.[11]

During PSII biogenesis, a post-translational modification of certain PSII proteins is needed for the correct functioning of the complex. After its synthesis on the membrane-bound ribosomes, the crucial PSII D1 protein contains 352-360 amino acid residues—but during incorporation into PSII it is cleaved at residue 344 (Ala) on the carboxyl terminus, resulting in the removal of a C-terminal extension.[12,13] The length of the extension corresponds roughly with the class of an organism. The only species without the extension is *Euglena*, while in most cyanobacteria the extension consists of 16 amino acid residues: this is seven residues longer than its counterpart in higher plants.[14] The removal of the extension is an essential step for PSII function—and mutants of both the green alga *Scenedesmus* and the cyanobacterium *Synechocystis* unable to cleave the extension, are not able to grow autotrophically.[15,16] The reason is that this extension prevents a proper assembly of the tetranuclear Mn-cluster in PSII, which functions as the site of photosynthetic water oxidation. On the other hand, the absence of the extension in D1 truncation mutants of *Chlamydomonas* or *Synechocystis* did not affect PSII function.[16,17] Nevertheless, recent data of Ivleva et al[18] demonstrated that the extension is required for optimum photosynthetic performance and its removal or elongation results in the decreased viability of these particular *Synechocystis* mutants. The extension may be important for a protection against proteolysis of PSII complexes that have not yet assembled the manganese cluster, or perhaps it stabilizes the D1 precursor during its integration into the thylakoid membrane.[16,18] The processing is accomplished by a specific carboxyl-terminal protease encoded by the ctpA gene.[19,20] The protease is homologous to the periplasmic protease Tsp that is involved in the degradation of proteins of *Escherichia coli* via a peptide-tagging system.[21]

The D1 protein is also processed by the removal of the N-terminal-initiating methionine, followed by N-acetylation of the exposed N-terminal threonine.[22] Moreover, the D1 protein in higher plants undergoes reversible covalent palmitoylation[23] and, together with CP43, D2 and PsbH, phosphorylation of its N-terminal threonine also take place.[24] These PSII proteins can be phosphorylated to various degrees, giving rise to heterogeneous PSII populations that differ in certain functional characteristics.[25] PSII protein phosphorylation may increase the susceptibility of PSII to photodamage[26] and, more specifically, the D1 degradation is affected by its phosphorylation status (see below).[27]

PSII Photoinhibition and Repair Cycle

The unique photochemical properties of PSII required for catalyzing water oxidation probably increase its susceptibility to photoinhibition. After the absorption of a certain number of quanta, unavoidable toxic reactions occur within PSII leading to interruption of PSII electron transfer and irreversible photodamage to the reaction-centre proteins, mainly the D1 protein.[28-31] When the PSII repair is slower than PSII photodamage, a net loss of functional PSII complexes occurs, the efficiency of the photosynthetic apparatus declines and growth is slowed down. This phenomenon has usually been described as photoinhibition of photosynthesis.[32] The susceptibility of photosynthetic organisms to photoinhibition at a given light intensity varies widely with genetic adaptation, physiological status and irradiance history. Besides light, photoinhibition is enhanced by other environmental constraints like extreme temperature, drought, high salinity or presence of heavy metals. Initially, the additional stress frequently results in an acceleration of the D1 turnover rate, however, if the stress conditions persist, the repair becomes insufficient and photoinhibition is enhanced.[33] This modulation of the D1 turnover rate may be considered as a general symptom of stress.[34,35] The reason for the repair inefficiency under conditions of strong light, or in combination with other constraints, seems to be due to both the increased level of PSII photodamage[36,37] and decreased D1 synthesis via an inhibition of psbA transcription and translation.[38,39]

On the basis of in vitro studies, at least three mechanisms of PSII photoinhibition have been suggested: the acceptor-side-induced and donor-side-induced photoinhibition, and the so-called "low light syndrome".[29,31,40] The light requirement, the primary site of PS II electron transport damage, the reactive oxygen chemistry and the degradation pattern of the D1 protein distinguish the pathways. The acceptor-side-induced photoinhibition occurs when thylakoids or PSII preparations with an intact oxygen-evolving complex are exposed to strong illumination.[41] As judged from time-resolved Chl fluorescence kinetics measured after anaerobic photoinactivation, strong illumination leads to a stabilization of the singly-reduced primary quinone acceptor Q_A, which subsequently becomes protonated, double-reduced, and finally leaves the Q_A binding site.[42] All these semi-stable Q_A states are accompanied by an increased probability of light-induced formation of the triplet form of P680 (^{T}P680) that, in the presence of oxygen, readily reacts to produce singlet oxygen.[43,44] Singlet oxygen formation was experimentally confirmed in leaves exposed to high irradiance,[45] in illuminated thylakoid membranes[46] and in isolated reaction centres of PSII.[44,47]

Photoinhibition follows the donor-side-induced route in the case of OEC complex malfunction—when the donor side of PSII is unable to keep pace with the rate of electron withdrawal from P680 and the reduction of P680^{+} and TyrZ^{+} is hampered.[48,49] While TyrZ^{+} is quickly inactivated and does not cause any further injury to the PSII complex,[50,51] P680^{+} becomes long-lived with sufficient oxidizing potential to extract electrons from its surroundings and can cause damage to its molecular neighborhood (accessory chlorophylls, carotenoids or amino acids). During the donor-side mechanism of photoinhibition hydroxyl radicals dominate over other reactive oxygen species (ROS).[46]

An alternative pathway of photoinhibition might be found under low irradiance provided by consecutive flashes of light. According to the model established by Ohad and coworkers, this process is a result of the triplet P680 generation in the dark intervals between flashes via charge recombination of the primary radical pair when singlet oxygen is probably formed.[52]

Despite a good knowledge of photoinhibitory processes in vitro, the question still remains open: which of these mechanisms dominates in vivo? Based on immunodetection of the D1 breakdown products in leaves exposed to high irradiance, evidence has been obtained for both the acceptor-side[53-55] and the donor-side-induced mechanisms.[56] Some reports have questioned the acceptor-side-induced photoinhibition because of the low probability for the accumulation of doubly-reduced Q_A in vivo.[57,58]

In order to restore the PSII activity after photoinhibition, the D1 protein has to be replaced by a newly synthesized copy. This process—called the PSII repair cycle—seems to be the main reason for the observed fast turnover of D1 protein.[28,59] The sequence of events during the PSII repair includes partial disassembly of the PSII core complex, degradation of the photodamaged D1, de novo D1 synthesis and its insertion, reassembly of the PSII core complex and, finally, activation of the electron-transport processes.[40] In higher plants possessing the granal structure of thylakoids, the cycle also involves the movement of inactive complexes from grana (in which PSII normally functions) to stromal lamellae, where the replacement of the D1 protein occurs.[60] The D1 turnover occurs at all light intensities and its rate (similarly as is the rate of PSII photoinactivation) is proportional to light intensity.[57,61-64] Under balanced growth and light conditions, D1 degradation is selective—as the rate of turnover of D1 greatly exceeds that of other PSII proteins. This fact suggests that the PSII photoinhibitory damage is aimed towards D1, which is selectively being turned-over both in eukaryotes[61] as well as in prokaryotes.[65] Data concerning PSII repair in the thermophilic cyanobacterium *Synechococcus elongatus* showed that PSII inactive in oxygen evolution, but still fully assembled, is already marked for D1 replacement.[66] The nature of the triggering signal has been guessed. Many reports have indicated that the degradation of D1 is controlled by the redox and occupancy state of the Q_B pocket.[63,67-69] However, more recent analyses of *Synechocystis* PCC 6803 mutants with a modified Q_B pocket showed that no specific sequences in this D1 region are essential for high rates of D1 degradation.[70,71] The contradictory effect of urea/triazine and phenolic herbicides on D1 degradation has also been explained by a different conformation of the Q_B site induced by herbicide interaction with various amino-acid residues of the binding niche.[61,62,72-75] Recently it has been pointed out that, the binding of herbicides to PSII modulates the recombination pathway within PSII, and thus the degree of $^{1}O_2$ production.[76] So the protective effect of urea herbicides on D1 degradation may be related to their ability to affect the recombination route between P680^{+} and Q_A^{-}, which does not lead to the formation of triplet chlorophyll and singlet oxygen.[77] In contrast, phenolic herbicides increase the probability of recombination between P680^{+} and Pheo^{-}.

Despite the uncertainty about the nature of the signal, its recognition by an external protease results in the prompt and selective degradation of the 'old' D1 protein and insertion of a new molecule.[66] The maximum rate of D1 degradation is reached only under the conditions of its ongoing synthesis[51,78]—indicating synchronization between these two processes.[63,79] Whether selective replacement of D1 requires the same steps as de novo assembly of the complex remains still an open question. Nevertheless, one difference between a repair-related assembly process and a de novo assembly process is the requirement of protease in the former. Several lines of evidence suggest that, initially, in the presence of the newly-synthesized D1 protein, CP43 is detached from the rest of the complex[66,80] and the remaining complex—consisting of at least D2, CP47 and possibly the old D1 copy—acts as the 'base' for the newly-synthesized D1 protein (step 9 in Fig. 1).[81] The association of new D1 with repaired PSII might therefore occur cotranslationally:[82] eliminating the need for complete disassembly and promoting the removal of the old D1 molecule.[51] Following successful initiation of psbA mRNA translation, the nascent D1 chain complexes are targeted—possibly by the chloroplast signal-recognition particle (cpSRP54)—to the thylakoid cpSecY translocation channel.[83,84] A subsequent elongation of *psbA* mRNA depends on photosynthetic electron transport, particularly on the transmembrane pH gradient[85] and the reducing compounds generated by Photosystem I.[82,86]

In higher plants, the removal of D1 from the complex seems to be regulated by its phosphorylation status. Phosphorylation prevents proteolytic degradation of the D1 protein in inactive PSII complexes in the appressed region of grana stacks, where active PSII complexes normally reside.[27] After PSII complexes with nonfunctional D1 protein have migrated to the stroma lamellae, the D1 protein is dephosphorylated and only after this event it is degraded and replaced.[87] Taking into account the need for close coordination between the degradation of the 'old' D1 protein and synthesis of the new one, such a system seems to allow D1 degradation only in stroma membranes in which D1 protein's synthesis and insertion also occur.

In cyanobacteria, thylakoid membranes are not organized into appressed and nonappressed regions as in chloroplasts.[88,89] This feature allows constitutive degradation and resynthesis of the D1 protein, thus its phosphorylation might not be required.[90] In accord with this notion, no phosphorylation of cyanobacterial D1 protein has been observed[91,92]—despite the fact that in all cyanobacteria the position of Thr-2 in D1 is conserved[14] and in many species this Thr-2 is followed by additional threonine residues at position 3 and 4.[93] However, these microorganisms use a different strategy in how to cope with increased rates of photodamage. They usually contain several psbA genes that encode either identical or different D1 forms. In *Synechocystis* PCC 6803, the well-known representative of the first group, there are three psbA genes encoding the identical D1 protein. The cells constitutively use the psbA2 transcript for the D1 synthesis (the psbA1 gene is not transcribed[94]). Under increased visible or UV-B radiation requiring a higher rate of repair, the cyanobacteria rely solely on the increased availability of the same protein via activation of the additional psbA3 gene.[94,95]

The example from the second group is *Synechococcus* PCC 7942, in which the psbAI copy encodes a D1:1 form, while the remaining two gene copies psbAII and psbAIII encode a different D1:2 variant of the protein. The two forms differ in 25 amino-acid residues and there are 12 differences localized among the first 16 residues.[96] As a response to increased irradiance, the D1:1 form synthesized under low irradiance can be quickly exchanged by the D1:2 form and the reverse process occurs when the cells are transferred from high to low irradiance.[97,98] Recent studies from our laboratory showed that, under increased irradiance, the turnover of D1:1 cannot match the rate of PSII photoinactivation and this fact seems to be the primary reason for its replacement by D1:2—the form exhibiting the more efficient PSII repair under high irradiance.[99] Furthermore, in agreement with this, a strain over-expressing D1:2 is less susceptible to photoinhibition under high light conditions,[100] as well as under UV-B radiation.[101] Thus, *Synechococcus* PCC 7942 normally synthesizes D1 protein suitable for low light conditions, whereas high light activates synthesis of the D1 form with the higher repair ability to cope with the increased rate of photodamage.

The putative protease responsible for the D1 degradation was initially localized inside, or in the vicinity of the PSII complex, as judged from the effect of serine-type protease inhibitors that suppressed the light-induced degradation of the D1 protein in isolated PSII core and reaction-centre complexes.[102,103] So far, many proteases homologous to those found in *Escherichia coli* have been identified in plants.[104] Among these, a clp-type protease,[105,106] and especially DegP-type proteases[107] and FtsH-related homologues,[108,109] have been suggested to cleave the D1 protein.

DegP protease belongs to a large family of related Deg/Htr serine proteases which are found in most organisms: including bacteria,[110] humans[111] and plants.[112] The chloroplast DegP1 homologue, that is tightly bound to the lumenal side of the thylakoid membrane,[112] was found to participate in two ATP-independent cleavage events at the lumenal loops of the D1 protein; this occurred during donor-side photoinhibition of PSII. Another DegP homologue, DegP2, was identified in chloroplasts as a peripheral protein attached to the stromal side of the thylakoid membrane. Its biochemical characteristics are in agreement with those reported for an unknown D1 protease (for a review see ref. 29), which include a serine-type ATP-independent proteolytic activity and the stimulatory effect of GTP.[75] The DegP2 has been suggested as recognizing the region close to the Q_B binding site of damaged D1 protein and setting an initial endoproteolytic cut that generates 23 and 10 kDa proteolytic fragments.[113] It was proposed that the distribution of closely-related DegP1 and DegP2 proteases on both the lumenal and stromal sides of the thylakoid membrane, respectively, might be required for an efficient degradation of the polytopic D1 protein.[107]

Even better candidates for the D1 protease are FtsH homologues; they belong to a larger family of proteins called the AAA proteins (ATPases associated with diverse cellular activities).[114] The FtsH proteases have two transmembrane segments with a short cytoplasmic N-terminus and a long C-terminal cytoplasmic region,[115] which includes a zinc-metalloproteinase active site and the ATPase domain participating in the unfolding of substrate polypeptide and its translocation to the protease active sites inside the protease cavity.[116] The *Arabidopsis* genome contains a number of FtsH-related genes. Two of them, FtsH1 and Yellow variegated (*VAR2*) were characterized and appeared to function in the metabolism of D1. Lindahl et al[108] suggested that the DegP2-generated 23 kDa D1 fragment was subject to proteolysis by FtsH1 protease depending on the presence of ATP and Zn ions. However, the mutual "collaboration" of DegP and FtsH during D1 degradation has been questioned: based on results with mutants of *Arabidopsis*[109] and *Synechocystis*[117] lacking genes encoding specific FtsH proteases. The data showed that the *Synechocystis* FtsH homologue encoded by the slr0228 gene and the similar VAR2 *Arabidopsis* homologue are essential for D1 degradation and the PSII repair cycle. However, in both mutants no accumulation of D1 breakdown products was detected, as would be predicted if the primary cleavage was needed for the degradation of D1 protein in vivo. Therefore, the involvement of DegP in D1 protein degradation needs further examination and direct confirmation in vivo.

Role of Reactive Oxygen Species in PSII Dynamics

When proteolytic removal of D1 from inactive PSII cannot proceed at a sufficient rate (in vivo under high stress, or in vitro), D1 degradation becomes much slower and a severe inactivation of PSII charge separation and disassembly of the complex occurs.[66] This process is most probably induced by ROS and radicals produced by aberrant PSII photochemistry, since loss of charge separation activity and PSII disassembly has not been observed under anaerobic conditions.[36] So, besides its function in restoring PSII activity, selective D1 replacement may also represent a means of preventing the formation of ROS and their attack on PSII components.[36]

Proof of ROS-mediated D1 protein damage is a decrease in electrophoretic mobility of the original D1 protein band.[118,119] Most probably it is a consequence of conformational changes in protein secondary structure, mainly in the content of the α-helices and β-sheets, as was detected by FTIR spectroscopy in parallel with degradation of the D1 protein.[120,121] Oxidation of proteins could be confirmed by detection of carbonyl groups that were used as a marker of ROS-mediated protein oxidation.[92]

The formation of small protein cross-linking products is also caused by ROS-induced oxidation of amino acids in the D1 protein that may lead to subsequent covalent association with other PSII polypeptides.[122] The three most-frequently-observed cross-linking products could be detected in illuminated or ROS-treated PSII preparations: the D1-D2 heterodimer; the D1-cytochrome b-559 adduct; and the D1-CP43 cross-linking product. The first two are formed in an oxygen-dependent manner, both in vivo[55,92,123,124] and in vitro, in illuminated PSII preparations.[125-127]

A recent study of D1-D2 heterodimer formation in illuminated spinach leaves has indicated the cross-linking site with the D2 protein: between residues 226-244 of the D1 protein.[124]

The adduct of D1 with the α-subunit of cytochrome b-559 was described in the illuminated reaction-centre complex of higher plants for the first time by Barbato et al.[125] Its light-induced generation was postulated as an initial step in the sequence of events leading to D1 degradation.[128] Recently we have identified the adduct as a product of the reaction between the oxidized residue

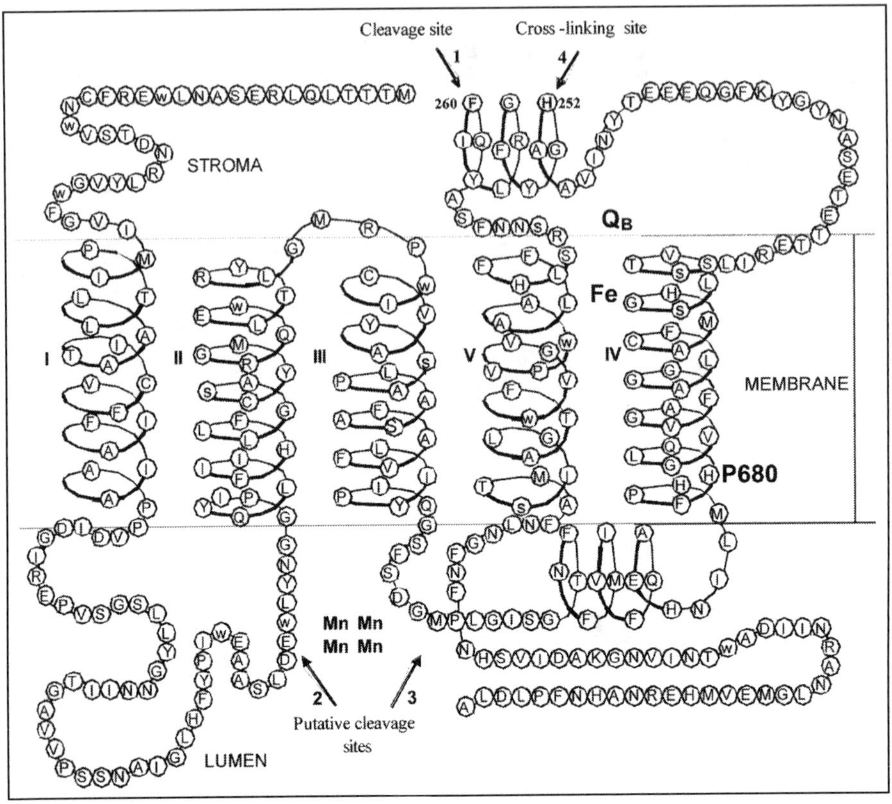

Figure 2. The folding model of the Photosystem II D1 protein from the cyanobacterium *Synechocystis* PCC 6803. The transmembrane helices are designated by Roman numbers, approximate localization of P680, Q_B, nonheme iron (Fe) and manganese cluster (Mn) is shown. Site of ROS-induced D1 cleavages and cross-linking are indicated by arrows and numbers: (1) cleavage between helices IV an V at F260 giving rise to 23 kDa N-terminal and 10 kDa C-terminal fragments; (2) putative cleavage in the loop between helix I and II giving rise to 10 kDa N-terminal and 24 kDa C-terminal fragments; (3) putative cleavage in the loop between helix III and IV giving rise to 16 kDa N-terminal and 16 kDa C-terminal fragments; and (4) cross-linking site between the N-terminal aminogroup of the serine residue of the cytochrome b-559 α subunit and H252 residue of the D1 protein.

His252 of the D1 polypeptide and the N-terminal amino group of the cytochrome α subunit (see Fig. 2).[129] Data also showed that formation of the adduct is not required as an intermediary step in the D1 degradation pathway.

The D1-CP43 aggregate was originally detected in PSII membranes treated with Tris-HCl buffer (pH 9.0) and illuminated with weak light.[130,131] In contrast to the other cross-linking products, it can also be generated in anaerobiosis. Experimental results showed the dependence of cross-linking on the presence of the 33 kDa oxygen-evolving enhancer, which most probably shields the surface of the lumen-exposed loops from ROS attack.[132] As another example of the possibilities for D1 cross-linking, a 160 kDa cross-linked product was observed in illuminated cells of *Dunaliella salina* composed of D1, D2 and other unknown proteins.[64]

The relationship between degradation and cross-linking (aggregation) of the D1 protein is not clear, but it is probable that the protein cross-linking does not allow fast D1 replacement but only slow processing by a protease(s) specific for such protein species. If this is the case, the formation of the D1 adducts competes with the regular degradation of the D1 protein. Therefore, the formation of aggregates might affect the efficiency of the D1 degradation and consequently slow down the repair cycle.[131] In agreement with this hypothesis, a slower degradation of D1 and an inefficient repair cycle were observed in cyanobacterial mutants in which the 41 kDa adduct was detected.[123,133]

Another feature of D1-protein oxidative modifications is the formation of discrete protein fragments. The role of ROS in this process is controversial, and two modes of action have been proposed. One possibility involves enzymatic cleavage by a protease(s) specific for the D1 protein after protein attack by ROS.[29] In this model, D1 modification induced by ROS is recognized by the protease that cleaves the protein. The second possibility is direct chemical cleavage, without any enzymatic activity, by ROS.[124,127,133] The possibility that oxygen-free radicals are involved in nonenzymatic degradation of D1 is supported by the following facts: (i) exposure of purified PSII complexes,[134,135] or the D1 protein alone,[133] to exogenous ROS results in its cleavage; (ii) this cleavage is not inhibited by protease inhibitors;[132,133] and (iii) it can be prevented by the addition of active-oxygen scavengers.[127,133,136] There exists a consensus in the literature that the two mechanisms of photoinhibition yield different D1 breakdown products in vitro.[126] Under conditions of the acceptor-side-induced photoinhibition, D1 is cleaved in the loop-connecting helices IV and V on the stromal side of the thylakoid membrane (see Fig. 2), giving rise to a 23 kDa N-terminal and a 10 kDa C-terminal fragment. These fragments are also generated in the dark—upon treatment of isolated PSII either with hydrogen peroxide,[133,134] or with a phenolic PSII inhibitor *N*-octyl-3-nitro-2,4,6-trihydroxy-benzamide (PNO8).[74] It was shown that this form of cleavage is directly related to the presence of nonheme iron, most probably as a result of the action of hydroxyl radicals formed by the Fenton reaction between the nonheme iron and hydrogen peroxide.[133] Sequencing of the 10 kDa C-terminal fragment[124] resulted in the identification of the residue Phe260 as the cleavage site. In the donor-side-induced photoinhibition, the damage occurs at the lumen-exposed loop between the first two helices of the D1 protein (see Fig. 2): producing a 10 kDa N-terminal and a 24 kDa C-terminal fragment.[126,137] The D1 cleavage also occurs between transmembrane segments III and IV near to P680 or the manganese cluster (see Fig. 2): resulting in two complementary fragments of about 16 kDa.[126,137] The exact localization of these two cleavage sites remains unclear.

When the D1 protein is not replaced, other PSII proteins also undergo oxidative damage during severe photoinhibition.[31,122] The photon-flux saturation curve for D2 protein degradation resembles that of D1, except that the half-life of the D2 protein is about three times longer than that of D1.[138] Mass spectra studies revealed an extensive oxidation—not only of D1, but also the D2 protein isolated from light-treated reaction-centre complexes.[139] D2 specific fragmentation was shown both in vitro[140] and in vivo.[55] Compared with D1 and D2, other PSII proteins have a longer half-life.[141] A partial degradation of cytochrome b-559 has been observed after photoinhibitory treatment of PSII membranes—as the β-subunit of cytochrome b-559 was singly oxidized in the region of Phe27-Phe36 (possibly at the methionine residue in position 34), but no modification of the α-subunit was detected in the illuminated PSII reaction-centre complexes.[142] The α-subunit of cytochrome *b*-559 may also be cross-linked to the D1 protein (see above). Degradation of the inner antenna proteins CP47 and especially CP43 has also been observed.[66,143] The loss of CP43 is frequently related to its cross-linking with the D1 protein, especially after treatment of PSII with Tris-HCl buffer, pH 9.0 (see above).[130,131] Degradation of CP47 is usually slower compared to that of CP43,[66] most probably due to its location at a longer distance from the long-lived strong oxidants created around the D1 protein in light.[144] The turnover of proteins belonging to the oxygen-evolving complex (OEC) is even slower (half-life of more than 8 h in isolated chloroplasts), and may have no direct relation to light conditions.[145] The three OEC subunits are released from PSII concomitantly with the degradation of the D1 protein when PSII-enriched membranes are exposed to high irradiance;[60] but after PSII repair they can be reattached to the PSII core.[146] Based on the current data, the degree of instability of PSII proteins is as follow:

D1 >> D2 > Cyt *b*-559 > CP43 > CP47 > OEC

Conclusion

Photosystem II exhibits high structural dynamics reflecting its complex, multicomponent nature as well as its unique photochemical properties. Such dynamics allows the complex to function as an important regulatory point of photosynthetic light reactions. Moreover, its dynamics also gives a reasonable basis for explaining the functional heterogeneity of PSII, a phenomenon which has been well known for a long time. This functional heterogeneity may simply reflect the presence of specific populations of complexes in particular stages of assembly, photodamage or repair.

References

1. De Vitry C, Olive J, Drapier D et al. Posttranslational events leading to the assembly of photosystem II protein complex - a study using photosynthesis mutants from Chlamydomonas reinhardtii. J Cell Biol 1989; 109:991-16.
2. Morais F, Barber J, Nixon PJ. The chloroplast-encoded alpha subunit of cytochrome b-559 is required for assembly of the photosystem two complex in both the light and the dark in Chlamydomonas reinhardtii. J Biol Chem 1998; 273:29315-20.
3. Vermaas WFJ, Ikeuchi M, Inoue Y. Protein composition of the photosystem II core complex in genetically engineered mutants of the cyanobacterium Synechocystis sp. PCC 6803. Photosynth Res 1988; 17:97-113.
4. Pakrasi HB, Diner BA, Williams JGK et al. Deletion mutagenesis of the cytochrome b559 protein inactivates the reaction center of photosystem II. Plant Cell 1989; 1:591-597.
5. Yu J, Vermaas W. Transcript levels and synthesis of photosystem 2 components in cyanobacterial mutants with inactivated photosystem II genes. Plant Cell 1990; 2:315-322.
6. Nilsson FJ, Andersson B, Jansson C. Photosystem-II characteristics of a constructed Synechocystis 6803 mutant lacking synthesis of the D1 polypeptide. Plant Mol Biol 1992; 14:1051-1054.
7. van Wijk KJ, Roobol-Boza M, Kettunen R et al. Synthesis and assembly of the D1 protein into photosystem II: Processing of the C-terminus and identification of the initial assembly partners and complexes during photosystem II repair. Biochemistry 1997; 36:6178-86.
8. Müller B, Eichacker LA. Assembly of the D1 precursor in monomeric photosystem II reaction center precomplexes precedes chlorophyll a-triggered accumulation of reaction center II in barley etioplasts. Plant Cell 1999; 11:2365-2378.
9. Eichacker LA, Soll J, Lauterbach P et al. In vitro synthesis of chlorophyll-A in the dark triggers accumulation of chlorophyll-a apoproteins in barley etioplasts. J Biol Chem 1990; 265:13566-13571.
10. Kim J, Eichacker LA, Rudiger W et al. Chlorophyll regulates accumulation of the plastid-encoded chlorophyll proteins P700 and D1 by increasing apoprotein stability. Plant Physiol 1994; 104:907-916.
11. Plücken HB, Muller B, Grohmann D et al. The HCF136 protein is essential for assembly of the photosystem II reaction center in Arabidopsis thaliana. FEBS Lett 2002; 532:85-90.
12. Marder JB, Goloubinoff P, Edelman M. Molecular architecture of the rapidly metabolized 32-kilodalton protein of photosystem II – indications for COOH-terminal processing of a chloroplast membrane polypeptide. J Biol Chem 1984; 259:3900-3908.
13. Takahashi M, Shiraishi T, Asada K. COOH-terminal residues of D1 and the 44-kDa CPa-2 at spinach photosystem II core complex. FEBS Lett 1988; 240:6-8.
14. Svensson B, Vass I, Styring S. Sequence analysis of the D1 and D2 reaction center proteins of photosystem II. Z Naturforsch 1991; 46c:765-776.
15. Taylor MA, Packer JCL, Bowyer JR. Processing of the D1 polypeptide of the photosystem-II reaction center and photoactivation of a low fluorescence mutant (LF-1) of Scenedesmus-obliquus. FEBS Lett 1998; 237:229-233.
16. Nixon JP, Trost JT, Diner BA. Role of the carboxy terminus of polypeptide D1 in the assembly of a functional water-oxidizing manganese cluster in photosystem I1 of the cyanobacterium Synechocystis sp. PCC 6803: Assembly requires a free carboxyl group at C-terminal position 3447. Biochemistry 1992; 31:10859-10871.
17. Lers A, Heifetz PB, Boynton JE et al. The carboxyl-terminal extension of the D1 protein of photosystem II is not required for optimal photosynthetic performance under CO2- and light-saturated growth conditions. J Biol Chem 1992; 267:17494-17497.
18. Ivleva NB, Shestakov SV, Pakrasi HB. The carboxyl-terminal extension of the precursor D1 protein of photosystem II is required for optimal photosynthetic performance of the cyanobacterium Synechocystis sp. PCC 6803. Plant Physiol 2000; 124:1403–1411.
19. Anbudurai PR, Mor TS, Ohad I et al. The CtpA gene encodes the C-terminal processing protease for the D1 protein of the photosystem-II reaction-center complex. Proc Natl Acad Sci USA 1994; 91:8082-8086.

20. Inagaki N, Yamamoto Y, Mori H et al. Carboxyl-terminal processing protease for the D1 precursor protein: Cloning and sequencing of the spinach cDNA. Plant Mol Biol 1996; 30:39-50.
21. Keiler KC, Waller PRH, Sauer RT. Role of a peptide tagging system in degradation of proteins synthesized from damaged messenger RNA. Science 1996; 271:990-993.
22. Michel H, Hunt DF, Shabanowitz J et al. Tandem mass spectrometry reveals that three photosystem II proteins of spinach chloroplasts contain N-acetyl-O-phosphothreonine at their N-termini. J Biol Chem 1988; 263:1123-1130.
23. Mattoo AK, Edelman M. Intramembrane translocation and posttranslational palmitoylation of the chloroplast 32-kDa herbicide-binding protein. Proc Natl Acad Sci USA 1987; 78:1572-1576.
24. Elich TE, Edelman M, Mattoo AK. Identification, characterization, and resolution of the in vivo phosphorylated form of the D1 photosystem II reaction center protein. J Biol Chem 1992; 267:3523-3529.
25. Giardi MT, Cona A, Geiken B. Photosystem II core phosphorylation heterogeneity and the regulation of electron transfer in higher plants: A review. Biolectrochem Bioenergetics 1995; 38:67-75.
26. Giardi MT, Komenda J, Masojídek J. Involvement of protein phosphorylation in the sensitivity of photosystem II to strong illumination. Physiol Plant 1994; 92:181-187.
27. Rintamäki E, Kettunen R, Tyystjärvi E et al. Light-dependent phosphorylation of D1 reaction center protein of Photosystem II: Hypothesis for the functional role in vivo. Physiol Plant 1995; 93:191-195.
28. Prasil O, Adir N, Ohad I. The Photosystems: Structure, function and molecular biology. In: Barber J, ed. Topics in Photosynthesis. Elsevier 1992:295-348.
29. Aro E-M, Virgin I, Andersson B. Photoinhibition of photosystem II. Inactivation, protein damage and turnover. Biochim Biophys Acta 1993; 1143:113-134.
30. Barber J. Molecular-basis of the vulnerability of photosystem-II to damage by light. Aust J Plant Physiol 1994; 22:201-208.
31. Mattoo AK, Giardi MT, Raskind A et al. Dynamic metabolism of photosystem II reaction center proteins and pigments. Physiol Plant 1999; 107:454-461.
32. Osmond CB, Grace SC. Perspectives on photoinhibition and photorespiration in the field: Quintessential inefficiencies of the light and dark reactions of photosynthesis? J Experiment Botany 1995; 46:351-1362.
33. Geiken B, Masojidek J, Rizzuto M et al. Incorporation of [S-35] methionine in higher plants reveals that stimulation of the D1 reaction centre II protein turnover accompanies tolerance to heavy metal stress. Plant Cell Environment 1998; 21:1265-1273.
34. Giardi MT, Masojidek J, Godde D. Effects of abiotic stresses on the turnover of the D1 reaction center II protein. Physiol Plant 1997; 101:635-642.
35. Franco E, Alessandrelli S, Masojidek J et al. Modulation of D1 protein turnover under cadmium and heat stresses monitored by [S-35] methionine incorporation. Plant Sci 1999; 144:53-61.
36. Komenda J, Masojídek J. Oxygen retards recovery from photoinhibition due to a secondary damage to the PSII complex. In: Mathis P, ed. Photosynthesis: From Light to Biosphere. Kluwer: Academic Publishers, 1995:203-206.
37. Komenda J, Koblízek M, Prásil O. Characterization of processes responsible for the distinct effects of herbicides DCMU and BNT on Photosystem II photoinactivation in cells of the cyanobacterium Synechococcus PCC 7942. Photosynth Res 2000b; 63:135-144.
38. Nishiyama Y, Yamamoto H, Allakhverdiev SI et al. Oxidative stress inhibits the repair of photodamage to the photosynthetic machinery. EMBO J 2001; 20:5587-5594.
39. Allakhverdiev SI, Nishiyama Y, Miyairi S et al. Salt stress inhibits the repair of photodamaged photosystem II by suppressing the transcription and translation of psbA genes in Synechocystis. Plant Physiol 2002; 130:1443-1453.
40. Melis A. Photosystem II damage and repair cycle in chloroplasts: What modulates the rate of photodamage in vivo? Trends Plant Sci 1999; 4:130-135.
41. Setlík I, Allakhverdiev SI, Nedbal L et al. Three types of photosystem II photoinactivation. 1.Damaging processes on the acceptor side. Photosynth Res 1990; 23:39-48.
42. Vass I, Styring S, Hundal T et al. Reversible and irreversible intermediates during photoinhibition of photosystem II: Stable reduced QA species promote chlorophyll triplet formation. Proc Natl Acad Sci USA 1992; 89:1408-1412.
43. Durrant JR, Giorgi LB, Barber J et al. Characterization of triplets states in isolated photosystem II reaction centers: Oxygen quenching as a mechanism for photodamage. Biochim Biophys Acta 1990; 1017:167-175.
44. Telfer A, Bishop SM, Phillips O et al. The isolated photosynthetic reaction center of photosystem II as a sensitizer for the formation of singlet oxygen: Detection and quantum yield determination using a chemical trapping technique. J Biol Chem 1994; 269:13244-13253.

45. Hideg É, Kálai T, Hideg K et al. Photoinhibition of photosynthesis in vivo results in singlet oxygen production detection via nitroxide-induced fluorescence quenching in broad bean leaves. Biochemistry 1998; 37:11405-11411.
46. Hideg É, Spetea C, Vass I. Singlet oxygen and free radical production during acceptor- and donor-induced photoinhibition. Studies with spin trapping EPR spectroscopy. Biochim Biophys Acta 1994; 1186:143-152.
47. Macpherson AN, Telfer A, Barber J et al. Direct detection of singlet oxygen from isolated photosystem II reaction centers. Biochim Biophys Acta 1993; 143:301-309.
48. Theg SM, Filar LJ, Dilley RA. Photoinactivation of chloroplasts already inhibited on the oxidizing side of photosystem II. Biochim Biophys Acta 1986; 49:104-111.
49. Jegerschöld C, Virgin I, Styring S. Light dependent degradation of the D1 protein in photosystem II is accelerated after inhibition of the water splitting reaction. Biochemistry 1990; 29:6179-6186.
50. Chen GX, Blubaugh DJ, Homann PH et al. Superoxide contributes to the rapid inactivation of specific secondary donors of the photosystem II reaction center during photodamage of manganese-depleted photosystem II membranes. Biochemistry 1995; 34:2317-2332.
51. Komenda J, Hassan HAG, Diner BA et al. Degradation of the Photosystem II D1 and D2 proteins in different strains of the cyanobacterium Synechocystis PCC 6803 varying with respect to the type and level of psbA transcript. Plant Mol Biol 2000; 2:635-645.
52. Keren N, Berg A, van Kann PJM et al. Mechanism of photosystem II photoinactivation and D1 protein degradation at low light: The role of back electron flow. Proc Natl Acad Sci USA 1997; 94:1579-1584.
53. Greenberg BM, Gaba V, Mattoo AK et al. Identification of a primary in vivo degradation product of the rapidly turning-over 32 kDa protein of photosystem II. EMBO J 1987; 6:2865-2869.
54. Cánovas PM, Barber J. Detection of a 10 kDa breakdown product containing the C-terminus of the D1 protein in photoinhibited wheat leaves suggests an acceptor side mechanism. FEBS Lett 1993; 324:341-344.
55. Shipton CA, Barber J. In vivo and in vitro photoinhibition gives rise to similar degradation fragments of D1 and D2 photosystem-II reaction-center proteins. Eur J Biochem 1994; 20:801-808.
56. Kettunen R, Tyystjärvi E, Aro E-M. Degradation pattern of photosystem II reaction center protein D1 in intact leaves. Plant Physiol 1996; 111:1183-1190.
57. Tyystjärvi E, Aro E-M. The rate constant of photoinhibition, measured in lincomycin-treated leaves, is directly proportional to light intensity. Proc Natl Acad Sci USA 1996; 93:2213-2218.
58. Anderson JM, Park Y-I, Chow WS. Unifying model for the photoinactivation of photosystem II in vivo under steady-state photosynthesis. Photosynth Res 1998; 56:1-13.
59. Ohad I, Kyle DJ, Arntzen CJ. Membrane protein damage and repair: Removal and replacement of inactivated 32-kD polypeptides in chloroplast membranes. J Cell Biol 1984; 99:481-485.
60. Hundal T, Virgin I, Styring S et al. Changes in the organization of photosystem-II following light-induced D1-protein degradation. Biochim Biophys Acta 1990; 1017:235-241.
61. Mattoo AK, Hoffman-Falk H, Marder JB et al. Regulation of protein metabolism; coupling of photosynthetic electron transport to in vivo degradation of the rapidly metabolized 32-kDa protein of chloroplast membranes. Proc Natl Acad Sci USA 1984; 81:1380-1384.
62. Kyle DJ, Ohad I, Arntzen CJ. Membrane-protein damage and repair – selective loss of a quinone-protein function in chloroplast membranes. Proc Natl Acad Sci USA 1984; 81:4070-4074.
63. Komenda J, Barber J. Comparison of psbO and psbH deletion mutants of Synechocystis PCC 6803 indicates that degradation of the D1 protein is regulated by the Q(B) site and dependent on protein synthesis. Biochemistry 1995; 32:1454-1465.
64. Baroli I, Melis A. Photoinhibition and repair in Dunaliella salina acclimated to different growth irradiances. Planta 1996; 98:640-646.
65. Goloubinoff P, Brusslan J, Golden SS et al. Characterization of the photosystem II 32 kDa protein in Synechococcus PCC 7942. Plant Mol Biol 1988; 11:441-447.
66. Komenda J, Masojídek J. Functional and structural changes of the photosystem II complex induced by high irradiance in cyanobacterial cells. Eur J Biochem 1995a; 233:677-682.
67. Kirilovsky D, Vernotte C, Astier C et al. Reversible and irreversible photoinhibition in herbicide-resistant mutants of Synechocystis 6714. Biochim Biophys Acta 1988; 933:124-131.
68. Ohad I, Koike H, Shochat S et al. Changes in the properties of reaction center-II during the initial-stages of photoinhibition as revealed by thermo-luminescence measurements. Biochim Biophys Acta 1988; 933:288-298.
69. Gong H, Ohad I. The PQ/PQHB_{2B} ratio and occupancy of photosystem II QB_{BB} site by plastoquinone control the degradation of the D1 protein during photoinhibition in vivo. J Biol Chem 1991; 266:21293-21252.

70. Nixon PJ, Komenda J, Barber J et al. Deletion of the PEST-like region of photosystem two modifies the QB_{BB}-binding pocket but does not prevent rapid turnover of D1. J Biol Chem 1995; 270:14919-14927.
71. Mulo P, Laakso S, Mäenpää P et al. Stepwise photoinhibition of photosystem II. Studies with Synechocystis species PCC 6803 mutants with a modified D-E loop of the reaction center polypeptide D1. Plant Physiol 1998; 17:483-490.
72. Kirilovsky D, Rutherford AW, Etienne A-L. Influence of DCMU and ferricyanide on photodamage in photosystem-II. Biochemistry 1994; 33:3087-3095.
73. Jansen MAK, Depka B, Trebst A et al. Engagement of specific sites in the plastoquinone niche regulates degradation of the D1 protein in photosystem II. J Biol Chem 1993; 268:21246-21252.
74. Nakajima Y, Yoshida S, Inoue Y et al. Occupation of the Q_B-binding pocket by a photosystem II inhibitor triggers dark cleavage of the D1 protein subjected to brief preillumination. J Biol Chem 1996; 71:17383-17389.
75. Spetea C, Keren N, Hundal T et al. GTP enhances the degradation of the photosystem II D1 protein irrespective of its conformational heterogeneity at the Q_B site. J Biol Chem 2000; 275:7205-7211.
76. Fufezan C, Rutherford AW, Krieger-Liszkay A. Singlet oxygen production in herbicide-treated photosystem II. FEBS Lett 2002; 532:407-410.
77. Rutherford AW, Krieger-Liszkay A. Herbicide-induced oxidative stress in photosystem II. Trends Biochem Sci 2001; 26:648-653.
78. Komenda J, Koblízek M, Masojídek J. The regulatory role of photosystem II photoinactivation and de novo protein synthesis in the degradation and exchange of two forms of the D1 protein in the cyanobacterium Synechococcus PCC 7942. J Photochem Photobiol B: Biology 1999; 48:114-119.
79. Schnettger B, Leitsch J, Krause GH. Photoinhibition of photosystem 2 in vivo occurring without net D1 protein degradation. Photosynthetica 1992; 27:261-265.
80. Barbato R, Friso G, Rigoni F et al. Structural changes and lateral redistribution of photosystem II during donor side photoinhibition of thylakoids. J Cell Biol 1992c; 119:325-335.
81. Adir N, Shochat S, Ohad I. Light-dependent D1 protein synthesis and translocation is regulated by reaction center II. Reaction center II serves as an acceptor for the D1 precursor. J Biol Chem 1990; 265:12563-12568.
82. Zhang L, Paakkarinen V, van Wijk KJ et al. Biogenesis of the chloroplast-encoded D1 protein: Regulation of translation elongation, insertion, and assembly into photosystem II. Plant Cell 2000; 12:1769-1781.
83. Tyystjärvi T, Herranen M, Aro EM. Regulation of translation elongation in cyanobacteria: Membrane targeting of the ribosome nascent-chain complexes controls the synthesis of D1 protein. Mol Microbiol 2001; 40:76-484.
84. Tyystjärvi T, Maanpaa P, Mulo P et al. D1 Polypeptide degradation may regulate psbA gene-expression at transcriptional and translational levels in Synechocystis sp PCC-6803. Photosynth Res 1996; 47:111-120.
85. Mühlbauer SK, Eichacker LA. Light-dependent formation of the photosynthetic proton gradient regulates translation elongation in chloroplasts. J Biol Chem 1998; 273:20935-20940.
86. Kuroda H, Kobashi K, Kaseyama H et al. Possible involvement of a low redox potential component(s) downstream of photosystem I in the translational regulation of the D1 subunit of the photosystem II reaction center in isolated pea chloroplasts. Plant Cell Physiol 1996; 37:754-761.
87. Rintamäki E, Kettunen R, Aro EM. Differential D1 dephosphorylation in functional and photodamaged Photosystem II centers. J Biol Chem 1996; 71:14870-14875.
88. Andersson B, Anderson JM. Lateral heterogeneity in the distribution of chlorophyll-protein complexes of the thylakoid membranes of spinach chloroplasts. Biochim Biophys Acta 1980; 593:472-440.
89. Stanier G, Cohen-Bazire. Fine structure of cyanobacteria. Methods Enzymol 1988; 167:157-172.
90. Kanervo E, Mäenpää P, Aro EM. D1 protein degradation and psbA transcript levels in Synechocystis PCC 6803 during photoinhibition in vivo. J Plant Physiol 1993; 42:669-675.
91. Allen JF. Protein phosphorylation in regulation of photosynthesis. Biochim Biophys Acta 1995; 1098:275-335.
92. Komenda J, Lupínková L, Kopecky J. Absence of the psbH gene product destabilizes photosystem II complex and bicarbonate binding on its acceptor side in Synechocystis PCC 6803. Eur J Biochem 2002; 269:610-619.
93. Jansson C, Mäenpää P. Site-directed mutagenesis is for structure-function analysis of the Photosystem II reaction center protein D1. In: Esser K, ed. Progress in Botany. 58. Heidelberg: Springer-Verlag, 1997:352-367.
94. Mohamed A, Eriksson J, Osiewacz MD et al. Differential expression of the psbA genes in cyanobacterium Synechocystis 6803. Mol Gen Genet 1993; 238:161-168.

95. Mate Z, Sass L, Szekeres M et al. UV-B-induced differential transcription of psbA genes encoding the D1 protein of photosystem II in the cyanobacterium Synechocystis 6803. J Biol Chem 1998; 273:17439-17444.
96. Golden SS, Brusslan J, Haselkorn R. Expression of a family of psbA genes encoding a photosystem II polypeptide in the cyanobacterium Anacystis nidulans R2. EMBO J 1986; 5:2789-2798.
97. Schaefer MR, Golden SS. Light availability influences the ratio of two forms of D1 in cyanobacterial thylakoids. J Biol Chem 1998; 264:7412-7417.
98. Clarke AK, Soitamo A, Gustafsson P et al. Rapid interchange between two distinct forms of cyanobacterial photosystem II reaction center protein D1 in response to photoinhibition. Proc Natl Acad Sci USA 1993; 90:9973-9977.
99. Komenda J. Role of two forms of the D1 protein in the recovery from photoinhibition of photosystem II in the cyanobacterium Synechococcus PCC 7942. Biochim Biophys Acta 2000; 1457:243-252.
100. Soitamo A, Zhou G, Clarke AK et al. Over-production of the D1:2 protein makes Synechococcus cells more tolerant to photoinhibition of Photosystem II. Plant Mol Biol 1996; 30:467-478.
101. Campbell D, Eriksson MT, Öquist G et al. The cyanobacterium Synechococcus resists UV-B by exchanging photosystem II reaction-center D1 proteins. Proc Natl Acad Sci USA 1998; 95:364-369.
102. Shipton CA, Barber J. Photoinduced degradation of the D1-polypeptide in isolated reaction centers of photosystem-II evidence for an autoproteolytic process triggered by the oxidizing side of the Photosystem. Proc Natl Acad Sci USA 1991; 88:6691-6695.
103. Salter AH, Virgin I, Hagman A et al. On the molecular mechanism of light-induced D1 protein degradation in PSII core particles. Biochemistry 1992; 31:3990-3998.
104. Adam Z. Protein stability and degradation in chloroplasts. Plant Mol Biol 1996; 32(5):773-783.
105. Gong H. Light-dependent degradation of the photosystem II D1 protein is retarded by inhibitors of chloroplast transcription and translation: A possible involvement of a chloroplast-encoded proteinase. Biochim Biophys Acta 1994; 1188:422-426.
106. Trebst A, Soll-Bracht E. Cycloheximide retards high light driven D1 protein degradation in Chlamydomonas reinhardtii. Plant Sci 1996; 155:191-197.
107. Haussühl K, Andersson B, Adamska I. A chloroplast DegP2 protease performs the primary cleavage of the photodamaged D1 protein in plant photosystem II. EMBO J 2001; 20:713-722.
108. Lindahl M, Spetea C, Hundal T et al. The thylakoid FtsH protease plays a role in the light-induced turnover of the photosystem II D1 protein. Plant Cell 2000; 12:419-432.
109. Bailey S, Thompson E, Nixon PJ et al. A critical role for the Var2 FtsH homologue of Arabidopsis thaliana in the photosystem II repair cycle in vivo. J Biol Chem 2002; 277:2006-2011.
110. Gottesman S. Proteases and their targets in Escherichia coli. Annu Rev Genet 1996; 30:465–506.
111. Spiess C, Beil A, Ehrmann N. A temperaturedependent switch from chaperone to protease in a widely conserved heat shock protein. Cell 1999; 97:339–347.
112. Itzhaki H, Naveh L, Lindahl M et al. Identification and characterization of DegP, a serine protease associated with the lumenal side of the thylakoid membrane. J Biol Chem 1998; 273:7094–7098.
113. Preiss S, Schrader S. Johanningmeier U. Rapid, ATP-dependent degradation of a truncated D1 protein in the chloroplast. Eur J Biochem 2000; 268:4562-4569.
114. Estelle M. Proteases and cellular regulation in plants. Curr Opin Plant Biol 2001; 4:254-260.
115. Tomoyasu T, Yamanaka K, Murata K et al. Topology and subcellular localization of FtsH protein in Escherichia coli. J Bacteriol 1993; 175:1352-1357.
116. Langer T. AAA proteases: Cellular machines for degrading membrane proteins. Trends Biochem Sci 2000; 25:247-251.
117. Silva P, Thompson E, Bailey S et al. FtsH is involved in the early stages of repair of Photosystem II in Synechocystis sp PCC 6803. Plant Cell 2003; 15:2152–2164.
118. Shipton CA, Barber J. Characterization of photoinduced breakdown of the D1-polypeptide in isolated reaction centers of photosystem II. Biochim Biophys Acta 1992; 1099:85-90.
119. Allakhverdiev SI, Komenda J, Feyzijev YM et al. Photoinactivation of Isolated D1/D2/cytochrome b-559 complex under aerobic and anaerobic conditions. Photosynthetica 1993; 28:281-288.
120. He WZ, Newell WR, Haris PI et al. Protein secondary structure of the isolated photosystem II reaction center and conformational changes studies by Fourier transform infrared spectroscopy. Biochemistry 1991; 30:4552-4559.
121. Xiang R, Xu Q, Mao HB et al. Strong-light photoinhibition treatment accelerates the changes of protein secondary structures in triton-treated photosystem I and photosystem II complexes. J Prot Chem 2001; 20:247-254.
122. Yamamoto Y. Quality control of photosystem II. Plant Cell Physiol 2001; 42:121-128.
123. Dalla Chiesa M, Friso G, Deak Z et al. Reduced turnover of the D1 polypeptide and photoactivation of electron transfer in novel herbicide mutants of Synechocystis sp. PCC 6803. Eur J Biochem 1997; 248:731-740.

124. Mizusawa N, Tomo T, Satoh K et al. Degradation of the D1 protein of Photosystem II under illumination in vivo. Two different pathways involving cleavage or intermolecular cross-linking. Biochemistry 2003; 33:9722-9730.
125. Barbato R, Friso G, Rigoni F et al. Characterization of a 41 kDa photoinhibition adduct in isolated photosystem II reaction centers. FEBS Lett 1992b; 309:165-169.
126. De Las Rivas J, Andersson B, Barber J. Two sites of primary degradation of the D1-protein induced by acceptor or donor side photoinhibition in PSII core complexes. FEBS Lett 1992; 301:246-252.
127. Miyao M. Involvement of active oxygen species in degradation of the D1 protein under strong illumination in isolated subcomplexes of Photosystem-II. Biochemistry 1984; 3:9722-9730.
128. Barbato R, Friso G, Ponticos M et al. Characterization of the light-induced cross-linking of the α-subunit of cytochrome b-559 and the D1 protein in isolated photosystem II reaction centers. J Biol Chem 1995; 270:24032-24037.
129. Lupínková L, Metz JG, Diner BA et al. Histidine residue 252 of the Photosystem II D1 polypeptide is involved in a light-induced cross-linking of the polypeptide with the α-subunit of cytochrome b-559: Study of a site-directed mutant of Synechocystis PCC 6803. Biochim Biophys Acta 2002; 1554:192-201.
130. Mori H, Yamamoto Y. Deletion of antenna chlorophyll-a-binding proteins CP43 and CP47 by Tris-treatment of PSII membranes in weak light: Evidence for a photodegradative effect on the PSII components other than the reaction center-binding proteins. Biochim Biophys Acta 1992; 100:293-298.
131. Mori H, Yamashita Y, Akasaka T et al. Further characterization of the loss of antenna chlorophyll-binding protein CP43 from photosystem II during donor-side photoinhibition. Biochim Biophys Acta 1995; 1228:37-42.
132. Henmi T, Yamasaki H, Sakuma S et al. Dynamic interaction between the D1 protein, CP43 and OEC33 at the lumenal side of photosystem II in spinach chloroplasts: Evidence from light-induced cross-linking of the proteins in the donor-side photoinhibition. Plant Cell Physiol 2003; 44(4):451-456.
133. Lupínková L, Komenda J. Oxidative modifications of the Photosystem II D1 protein by reactive oxygen species: From isolated protein to cyanobacterial cells. Photochem Photobiol 2004; 79:152-162.
134. Miyao M, Ikeuchi M, Yamamoto N et al. Specific degradation of the D1 protein of photosystem II by treatment with hydrogen peroxide in darkness: Implications for the mechanism of degradation of the D1 protein under illumination. Biochemistry 1995; 34:10019-10026.
135. Mishra NP, Ghanotakis DF. Exposure of a photosystem-II complex to chemically generated singlet oxygen results in D1 fragments similar to the ones observed during aerobic photoinhibition. Biochim Biophys Acta 1994; 87:296-300.
136. Sopory SK, Greenberg BM, Mehta RA et al. Free radical scavengers inhibit light-dependent degradation of the 32 kDa Photosystem II reaction center protein. Z Naturfors C-A Journal of Biosciences 1990; 45:412-417.
137. Barbato R, Frizzo A, Rigoni F et al. Photoinduced degradation of the D1 protein in isolated thylakoids and various photosystem II particles after donor-side inactivations. Detection of a C-terminal 16 kDa fragment. FEBS Lett 1992d; 304:136-140.
138. Jansen MAK, Mattoo AK, Edelman M. D1-D2 protein degradation in the chloroplast. Complex light saturation kinetics. Eur J Biochem 1999; 260:527-532.
139. Sharma J, Panico M, Shipton CA et al. Primary structure characterization of the photosystem II D1 and D2 subunits. J Biol Chem 1997a; 272:33158–33166.
140. Barbato R, Friso G, de Laureto PP et al. Light-induced degradation of D2 protein in isolated photosystem II reaction center complex. FEBS Lett 1992a; 311:33-36.
141. Schuster G, Timberg T, Ohad I. Turnover of thylakoid photosystem II proteins during photoinhibition of Chlamydomonas reinhardtii. Eur J Biochem 1988; 177:403-410.
142. Sharma J, Panico M, Barber J et al. Characterization of the low molecular weight Photosystem II reaction center subunits and their light-induced modifications by mass spectrometry. J Biol Chem 1997; 272:3935–3943.
143. Zer H, Ohad I. Photoinactivation of photosystem II induces changes in the photochemical reaction center II abolishing the regulatory role of the QB site in the D1 protein degradation. Eur J Biochem 1995; 231:448-453.
144. Zouni A, Witt HT, Kern J et al. Crystal structure of photosystem II from Synechococcus elongatus at 3.8 Å resolution. Nature 2001; 409:739-742.
145. Hashimoto A, Yamamoto Y, Theg SM. Unassembled subunits of the photosynthetic oxygen-evolving complex present in the thylakoid lumen are long-lived and assembly-competent. FEBS Lett 1996; 391:29-34.
146. Eisenberg-Domovich Y, Oelmuller R, Herrmann RG et al. Role of the RCII-D1 protein in the reversible association of the oxygen-evolving complex proteins with the lumenal side of Photosystem-II. J Biol Chem 1995; 270:30181-30186.

Chapter 5

Engineering the D1 Subunit of Photosystem II:

Application to Biosensor Technology

Udo Johanningmeier,* Ivo Bertalan, Lydia Hilbig, Jana Schulze, Stefan Wilski, Edda Zeidler and Walter Oettmeier

Dedicated to Prof. Dr. Drs. h.c. mult. Achim Trebst on the occasion of his 75th birthday.

Abstract

Photosystem II (PSII) is a light driven machine, which supplies our atmosphere with oxygen and, if properly engineered, can be developed into a specific sensor for various pollutants. Its reaction center subunit D1 has long been a target for genetic engineering. It is known to bind a variety of herbicides in a pocket which is naturally occupied by the plastoquinone molecule Q_B. Amino acid residues lining this binding niche provide ligands to a diverse set of inhibitors, which can loose or gain affinity upon substitution of certain amino acids. One can exploit this and other properties of the D1 protein by changing the corresponding *psbA* gene which is located on the chloroplast genome in algae and higher plants. We have developed fast site-specific and random mutagenesis techniques specifically adjusted for *psbA*-gene manipulation in the unicellular green alga *Chlamydomonas reinhardtii*. Using these protocols mutant collections were generated which have the innate potential of becoming an array of sensitive and specific biosensors.

Introduction

Photosystem II (PSII) is part of the photosynthetic apparatus in cyanobacteria, algae and higher plants and catalyzes the light-induced transfer of electrons from water to plastoquinone via a set of delicately arranged cofactors. It has a well known binding site for diverse chemical compounds in its so-called D1 subunit and the ability to convert such a binding event into signals which can be easily detected by optical, potentiometric or amperometric systems. Due to these inherent properties PSII can be considered as a natural biosensor and has consequentially been used for the detection of herbicides and other pollutants in pilot studies.[1,2]

Recent progress both in chloroplast engineering and in crystal structure analysis has increased our manipulative possibilities as well as our knowledge of structure-function relationships in PSII considerably. Therefore, the development of more stable, more specific and more sensitive PSII-based biosensors appears feasible now by making use of various mutagenesis techniques and in vivo expression of modified PSII complexes. Even redesigning PSII such that the sensor element gains novel properties appears to be within reach. By applying a molecular "Lego" approach[3,4] small protein modules with the desired properties can be fused with the D1 subunit of PSII without compromising its function.

*Corresponding Author: Udo Johanningmeier—Institut für Pflanzenphysiologie, Martin-Luther Universität Halle-Wittenberg, Weinbergweg 10, D-06120 Halle, Germany.
Email: johanningmeier@pflanzenphys.uni-halle.de

Biotechnological Applications of Photosynthetic Proteins: Biochips, Biosensors and Biodevices, edited by Maria Teresa Giardi and Elena V. Piletska. ©2006 Landes Bioscience.

Structure of the D1 Protein

PSII has a complex architecture consisting of at least 20 protein subunits, most of which are integrated into the thylakoid membrane. Its inner reaction center core is made up of two subunits, the D1 and D2 proteins. Both are intimately associated with each other and provide ligands to cofactors like chlorophylls, phaeophytins, carotenoids, plastoquinones and the metal ions manganese and iron. Initially, sequence homologies, crystallization and X-ray analysis of the bacterial reaction center[5] provided a structural model for PSII reaction centers in plants, algae and cyanobacteria.[6] Only recently, the 3-D structure of a cyanobacterial PSII was resolved at 3.8Å,[7] 3.7Å[8] and 3.5Å,[50] revealing new information and verifying many predictions made earlier. Although at this resolution the exact position of individual atoms is not visible yet, organization, location and gross orientation of polypeptide chains and cofactors can be identified.

The D1 subunit of PSII has, like its sister protein D2, five transmembrane helices (Fig. 1A-E) with the N- and C-terminus facing the stroma and lumenal side, respectively. There are also discrete

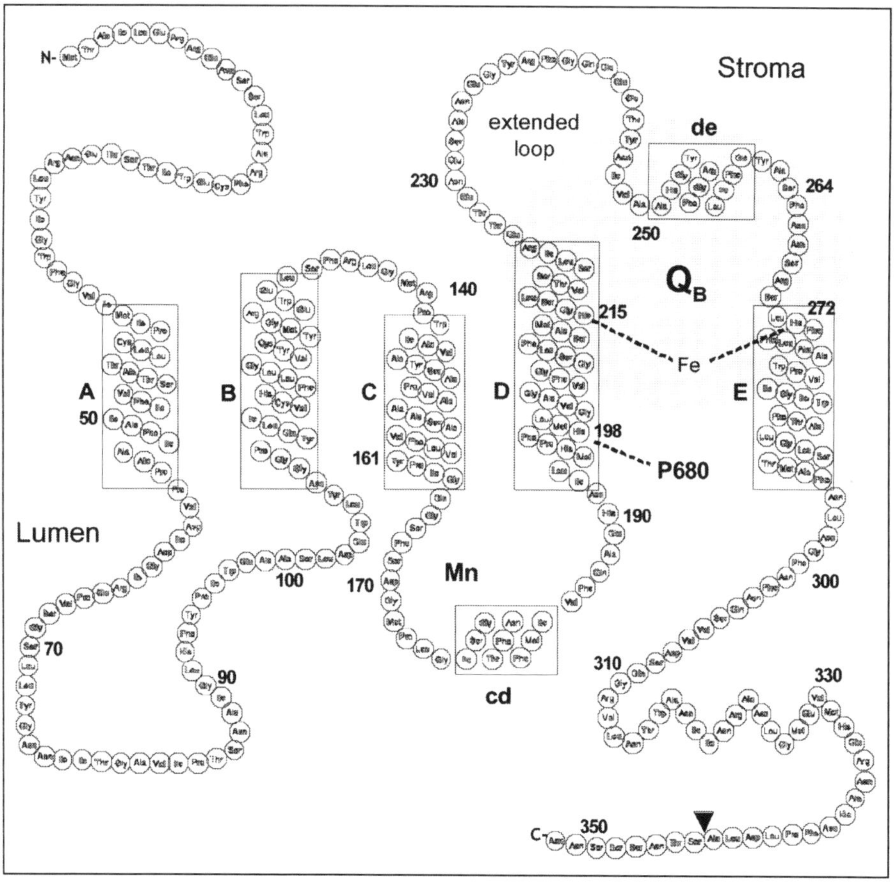

Figure 1. Secondary structure profile of the *C. reinhardtii* D1 protein. Transmembrane helices A-E and parallel helices cd and de are shown together with locations for the primary donor P680, the nonheme iron, Fe, the manganese, Mn, and the secondary plastoquinone Q_B. The shaded area roughly outlines the Q_B and herbicide binding niche. A few prominent amino acids have been highlighted: the redox active Tyr161 is electron donor to P680, His198 is one ligand to P680 chlorophyll, His215 and His272 are ligands to Fe, Ser264 is i.a. participating in quinone and herbicide binding. The arrow near the C-terminus indicates the processing site.

parallel helices in the connecting loops between helices CD and DE and close to the C-terminus.[7,8] The extended loop between helix D and the parallel helix de (Fig. 1) has been implicated to represent a contact site between the D1 and D2 proteins[9] and to play a role in the light-triggered degradation of D1 by providing a primary cleavage site for proteolysis.[10]

The detection of amino acid substitutions specifically in the D1 reaction center subunit was greatly supporting our understanding of PSII structure and function. A prominent discovery was a mutation at position 264 from Ser to Gly in the D1 protein,[11,12] which turned out to be the molecular basis for triazine resistance in *Amaranthus hybridus*. Shortly before, the D1 protein was tagged by photoaffinity labeling and thus shown to be the target for PSII herbicides.[13] These discoveries identified the D1 subunit as the "herbicide binding protein". Since the inhibitors tested were known to compete with the native plastoquinone Q_B for its binding site this protein was likely to be also the Q_B binding protein. Furthermore, a specific region within D1 emerged which we now address as herbicide binding niche and which is located between transmembrane helices D and E (Fig. 1). Abundant biochemical and genetic evidence has accumulated which places the binding niche between amino acids 211 and 275. However, at the present state of resolution X-ray structure analysis does not yield a detailed picture from its inner architecture. Here, modification of amino acids using mutagenesis techniques not only provide valuable information about the role of specific side chains but also offer the possibility to add novel properties which are useful for various applications such as molecular sensing.

D1 Protein Engineering in an Eukaryotic Alga

While engineering the D1 protein in prokaryotes like cyanobacteria has been a comparatively simple task due to the ease of transformation, manipulation of D1 in chloroplasts of green algae or higher plants resisted transformation for quite some time. The first stable transformation of a chloroplast genome only became possible with the newly developed particle gun[14] using the eukaryotic green alga *C. reinhardtii*. Boynton and colleagues used this technique to complement an *atpB* deletion mutant of *C. reinhardtii* with an intact *atpB* gene present on the transforming plasmid.[15] The wild type gene integrated into the chloroplast genome by homologous recombination and restored photoautotrophic growth. Since then many chloroplast genes have been modified, most frequently using *C. reinhardtii* as a model organism.[16,17]

C. reinhardtii is particularly instrumental in engineering the D1 subunit of PSII. Among the first modified genes was the *psbA* gene[18,19] which encodes D1 and is located within the inverted repeat region of the algal plastome.[20] However, interrupted by 4 large introns, its manipulation is a tedious task including the work with large DNA constructs and subtle selection procedures involving herbicide or antibiotic resistance markers.[18-22] A first step towards a more simple manipulation was the construction of mutants with intronless *psbA* genes,[23,24] followed by the generation of the Del1 mutant with a tailor made deletion encoding that part of the D1 coding region which also contains the herbicide binding site.[25] This mutant is unable to grow photoautotrophically, but grows normally on media containing acetate as a carbon source. Its gene product is a truncated D1 protein which does not accumulate due to its rapid, ATP-dependent proteolysis.[26] Upon transformation of the deletion mutant with a plasmid carrying an intact *psbA* gene, photosynthetic growth is restored. This is naturally also true for plasmids with *psbA* genes, which have been modified in vitro and still encode functional D1 subunits. Unlike other procedures, this selection for photosynthetic growth represents a very robust method that can be easily extended to also screen for e.g., herbicide-, temperature or radiation-tolerance. A PCR-based mutagenesis protocol (Fig. 2) had been developed to successfully introduce various point mutations into the D1 protein.[25]

Apart from the generation of suitable recipient cells like those with intron-free *psbA* genes, the deletion mutant Del1 and the application of appropriate mutagenesis protocols, there has recently been a significant technical improvement in molecular engineering of the D1 protein in *C. reinhardtii*.[27] Without any cloning or purification steps the deletion mutant described above can be complemented with PCR fragments directly by precipitating the linear DNA onto carrier particles and delivering them into the chloroplast by biolistic transformation. It was shown that for homologous recombination the size of the flanking regions bordering the deletion breakpoint could be as small as 50 bp upstream and 120 bp downstream of the breakpoint in order to obtain reasonable transformation

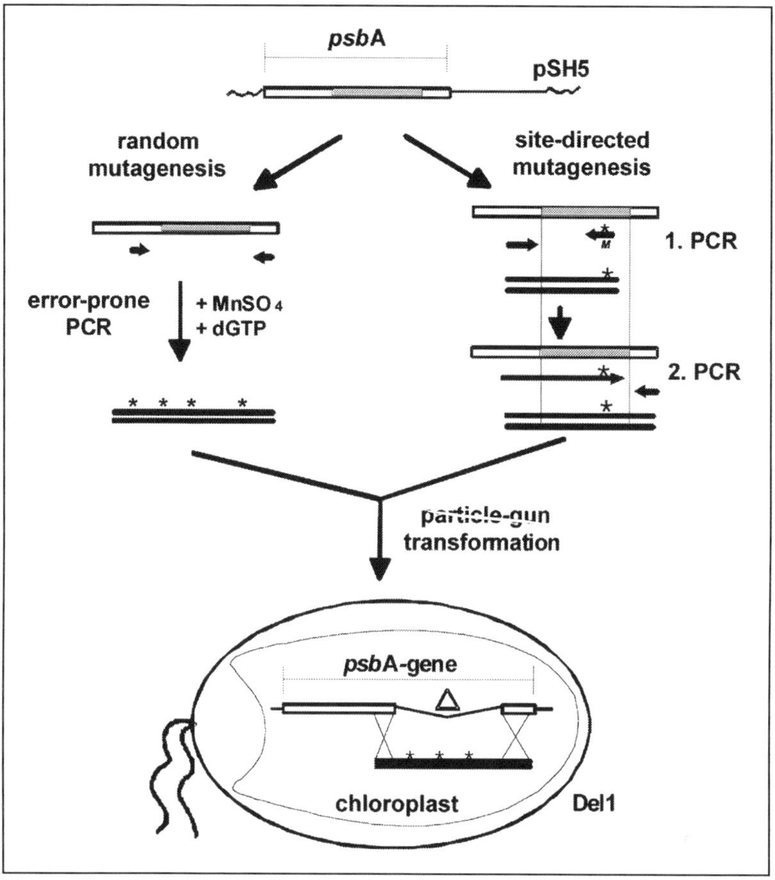

Figure 2. PCR-based mutagenesis procedures for the introduction of random and site-directed mutations into the recipient strain Del1. For random mutagenesis an error-prone PCR in the presence of $MnSO_4$ and dGTP was used. For site-directed mutagenesis a mutagenic primer M was used. Template for PCR was the intronless psbA gene in vector pSH5.[25] PCR fragments were precipitated directly onto tungsten particles and introduced by particle gun transformation without further cloning or purification steps.[27] Homologous recombination in the recipient cell Del1 is indicated by crosses. Stars indicate point mutations. The shaded area and Δ indicate the deleted sequence in the Del1 mutant.

frequencies. Although this method is so far restricted to a specific section of the D1 protein, i.e., that part of D1 not encoded due to the deletion, it considerably speeds up the in vitro site- directed mutagenesis process and, most importantly, random mutagenesis can be accomplished very efficiently (Fig. 2). We used commercially available kits to generate pools of PCR fragments under error-prone reaction conditions but controlled mutation frequencies. These pools, representing a complex library of amino acid changes (and other mutations), were directly delivered into the recipient strain Del1. In order to largely avoid transformants with wild-type *psbA* genes, algal colonies growing under photoautotrophic conditions were initially screened directly on plates for their fluorescence characteristics with the help of an imaging fluorometer. Colonies with significant deviations from the wild type fluorescence characteristics were further analyzed by sequencing that part of the *psbA* gene which could have been modified by the incoming PCR fragment. Using this procedure more than 60 mutations resulting in amino acid substitutions were identified so far (Fig. 3). Roughly one half of the mutants contain single and the other half double, triple or quadruple mutations.

Silent mutations have not been included and only a small fraction of all transformants obtained have been analyzed. From this still growing mutant library we only know that all individual modifications support D1 function. However, the library likely includes mutant cells with novel properties some of which can be useful for biosensor purposes. Given the proper selection conditions, the appropriate mutants within this collection can be easily detected and analyzed in more detail.

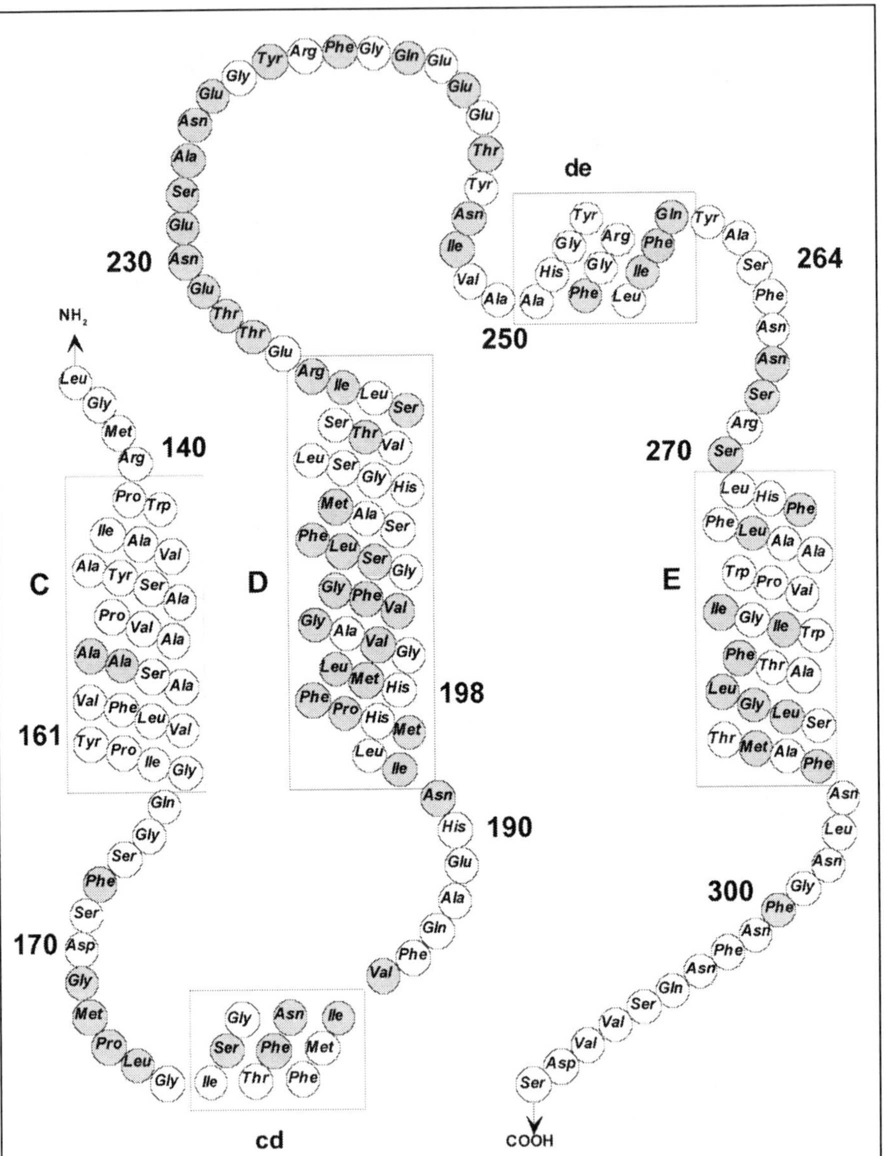

Figure 3. Secondary structure profile of part of the *C. reinhardtii* D1 protein with mutations obtained by random mutagenesis. Amino acid positions which were substituted are indicated by filled circles. Due to the selection for photoautotrophic growth all modifications must basically encode a functional D1 protein.

Herbicide Binding Niche

Many commercial herbicides inhibit photosynthesis by displacing Q_B from its binding site in D1 and thus block electron transport from Q_A to Q_B.[28] They belong to various chemical classes like triazines, ureas or phenols (For a review see ref. 29) and pollute soil and water due to their massive use in agriculture. This in turn can be harmful for human and animal health and necessitates the development of fast and sensitive detection methods. Coincidentally, the herbicidal target itself is part of the PSII complex, which represents a reporter system directly coupled to an analyte binding site. Thus the most obvious application of the D1 protein in association with other central PSII proteins is its use as a biosensor for herbicides.

The herbicide binding niche has been examined to some extent by various mutagenesis procedures.[30,31] The niche is rather spacious, accommodating the plastoquinone molecule and diverse chemical compounds.[32] Single point mutations can have dramatic effects on inhibitor binding, resulting in either resistance or supersensitivity ("negative cross resistance"). In case of resistance, the I_{50}-value of the inhibitor is higher in the mutant as compared to the "susceptible" wild type; the ratio of the R/S-values (I_{50}-value resistant versus I_{50}-value susceptible) is >1. Contrary, in the case of supersensitivity the I_{50}-value of the inhibitor is lower in the mutant as compared to the wild type. Consequently, the R/S-value is <1.

Table 1 lists some selected D1 mutants together with their R/S-values for various herbicides (a comprehensive list of D1 mutants is given in ref. 31).

The Val_{219}→Ile mutant is resistant against triazines and triazinones, benzthiazuron and ioxynil. However, it is supersensitive against the urea derivative lenacil. The Ala_{251}→Val mutant shows high resistance against triazines, triazinones and ureas, but supersensitivity against ketonitrile. In most of the mutants obtained so far, resistance is high against triazines and triazinones, but much less pronounced against ureas.[31] The Phe_{255}→Tyr mutant is an exception to this rule, because both triazinones metamitron and metribuzin exhibit supersensitivity.

As already stressed, the mutation Ser_{264}→Gly is one which has been observed in nature. It is found by now in all countries and in a variety of weeds which are rendered resistant against triazines and triazinones (see Table 1). It should be noted, that atrazine-resistant rape with a modified D1 protein (Ser_{264}→Gly) is used as a crop in Canada. Resistance against ureas is comparably small and against phenolic herbicides like i-dinoseb, ioxynil and DNOC negative cross resistance is observed.

Contrary to the natural occurring mutation Ser_{264}→Gly, the mutations Ser_{264}→Lys and Ile have been generated by site-directed mutagenesis. Like in the Ser_{264}→Gly mutant, in both mutants high resistance against the triazines atrazine, ametryn and prometryn is observed (Table 1). This resistance is much more pronounced in the Lys mutant as compared to the Ile mutant. Obviously, the amino group in the lysine moiety leads to a positive charge repulsion of the triazine nitrogen and, hence, less tight bindung occurs. Remarkably, both mutants exhibit high resistance against the carbamate herbicide phenisopham. For phenolic herbicides, in the Lys mutant in all cases supersesitivity is observed, whereas the Ile mutant only shows supersensitivity aginst bromoxynil (Table 1).[33]

Changes in binding affinities for herbicides in the Leu_{275}→Phe mutant are marginal. It should only be stressed that triazinones are rendered resistant in this mutant and supersensitivity is observed against the phenolic herbicide ioxynil.

It has already been noted that the herbicide binding region of the D1 protein up to now ranges from Phe_{211} to Leu_{275}. We have recently generated a D1 mutant of *Chlamydomonas reinhardtii* where by site-directed mutagenesis Phe_{206} has been exchanged against Ile, Glu, Ser, Ala, His, Asn, Val and Lys.[34] Two of these mutants, Ser and Lys, together with their R/S-values are listed in Table 1. The herbicides atrazine, diuron and phenmedipham all exhibited supersensitivity in these mutants. It has to be concluded, therefore, that Phe_{206} is either part of or its substitution has a long range effect on the herbicide binding niche. This new finding could be important for the development of more sensitive herbicide biosensors.

Table 1. List of selected D1 mutants together with their R/S-values for various herbicides

Mutation	Organism	Herbicide/Inhibitor R/S-Value
Val_{219}→Ile	*Chlamydomonas*	Atrazine 2
		Benzthiazuron 16
		Ioxynil 50
		Metribuzin 200
		Lenacil 0.8
Ala_{251}→Val	*Chlamydomonas*	Atrazine 25
		Lenacil 160
		Diuron 5-8
		Metribuzin 1000
		Ketonitril 0.5
Phe_{255}→Tyr	*Chlamydomonas*	Atrazine 15
		Dinoseb 3
		Cyanoacrylate 39
		Diuron 0.6-0.8
		Metabenzthiazuron 0.3
		Metamitron 0.3
		Metribuzin 0.6
Ser_{264}→Gly	*Amaranthus retroflexus*	Atrazin 250-1100
		Atraton 1000
		Metamitron 40
		Metribuzin 260->1500
		Diuron 1-4
		Lenacil 50 – 590
		i-Dinoseb 0.5
		Ioxynil 0.64 – 1.6
		DNOC 0.14 – 0.5
Ser_{264}→Lys	*Chlamydomonas*	Atrazine 6244
		Ametryn 3934
		Prometryn 7952
		Phenisopham 252
		Dinoterb 0.2
		DNOC 0.1
		Bromoxynil 0.03
Ser_{264}→Ile	*Chlamydomonas*	Atrazine 50
		Ametryn 16
		Prometryn 13
		Pheniospham 400
		Dinoterb 2.1
		DNOC 1
		Bromoxynil 0.6
Leu_{275}→Phe	*Chlamydomonas*	Atrazine 1
		Diuron 5
		Metamitron 63
		Metribuzin 20 – 26
		Ioxynil 0.2
Phe_{206}→Ser	*Chlamydomonas*	Atrazine 0.32
		Diuron 0.6
		Phenmedipham 0.25
Phe_{206}→Lys	*Chlamydomonas*	Atrazine 0.32
		Diuron 0.8
		Phenmedipham 0.32

Peptide Insertions

Insertion of particular peptide motifs into proteins is not only a common method for protein analysis and purification but also for the development of new biotechnologies.[35,36] Engineered proteins with controlled binding to other molecules include e.g., short metal-binding peptides in bacterial surface proteins for bioaccumulation and bioremediation purposes[37,38] or epitope-tagged enzymes as a molecular sensor system, whose activity is modulated by anti-epitope antibodies.[39]

One strategy for the development of biosensors is the introduction of new binding sites into a protein with an intrinsic signal-transduction function.[40] Since D1 is such a protein, it was worthwhile to probe it for its ability to accommodate foreign peptide insertions. There were clues already concerning the region of insertion. Kless and Vermaas[41] reported on drastic changes of the *D-E* region in the D1 protein that have been observed in pseudorevertants of the cyanobacterium *Synechocystis* sp. PCC 6803. Up to 15 amino acids were inserted without significantly affecting photoautotrophic growth. This surprising flexibility of a conserved section prompted us to construct D1 mutants of *C. reinhardtii* containing various short peptide motifs with different properties. Initially, a hemagglutinin (HA) epitope was inserted near the N-terminus and another one within the extended *D-de* region of the stromal loop (Fig. 4). Both mutants, termed Epi1 and Epi2, grow photosynthetically and were neither markedly affected in growth rates nor oxygen evolution capacities. A monoclonal antibody directed against the HA epitope is able to detect the tagged D1 protein with high specificity. Obviously, the D1 protein contains at least two permissive loops which tolerate insertions without a loss of function.

Encouraged by these results small metal-binding domains were inserted into the extended *D-de* loop structure. The rationale behind this was that such domains would undergo conformational changes upon metal binding and that binding-induced changes at one (allosteric) site can be propagated over considerable distances.[43] Although it is not yet possible to reliably predict the transmission of conformational changes from an allosteric site through the protein structure,[36] it appeared possible that the structural effect upon metal binding would extend to the nearby Q_B binding niche

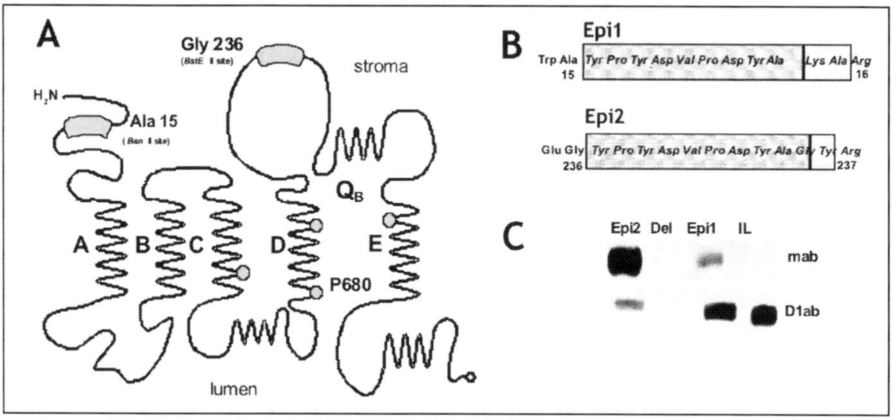

Figure 4. Epitope-tagging of the D1 protein in *C. reinhardtii.* A) The hemagglutinine (HA) epitope was inserted between Ala15 and Arg16 (corresponding to a unique *Ban*II site in the intronless *psb*A gene) and between Gly236 and Tyr237 (corresponding to a unique *BstE* II site in the intronless *psb*A gene) to create mutant strains Epi1 and Epi2, respectively. Both epitopes are stroma exposed. Transmembrane helices are indicated by A-E, the plastoquinone binding niche by Q_B and the reaction center chlorophylls by P680. B) Insertion sequences of mutants Epi1 and Epi2. The grey boxes specify the amino acid sequence of the HA-epitope recognized by the monoclonal antibody 12CA5. White boxes indicate amino acids which were introduced due to the cloning strategy. C) Western blot analysis of total cellular protein obtained from mutants Epi1, Epi2, the deletion mutant Del1 and the mutant IL containing an intronless *psb*A gene. The monoclonal epitope antibody mab strongly reacts with the D1 protein from Epi2 and less strongly with Epi1. As expected, neither IL nor Del react with mab. The antiserum raised against the C-terminal two-thirds of D1[42] detects the D1 protein in all mutants except Del.

Table 2. Metal binding domains introduced into the extended loop of the C.reinhardtii *D1 protein*

Mutant	Motif
CD1	CGCCGCGCCG
CD6	$(CGCCGCGCCG)_2$
CadA	PGCTCACAPI
CP9	GCGCPCGCG
Cd	HSQKVF
His 4.1	SHHHHHH
His 4.3	$(SHHHHHH)_2$
His 10	HHHHHHHHHH

Insertion was between Gly236 and Tyr237 (cf. Fig.4) using appropriate oligonucleotides cloned into a single BstEII site in the intronless *psbA* gene. Insertions were verified by DNA sequencing.

and change PSII fluorescence. Towards this aim a set of mutants has been generated in which the D1 protein contained different metal-binding motifs (Table 2).

Two of them, CD1 and CD6, accumulated Cd 20-fold and 80-fold, respectively, as compared with the wild type strain. This is comparable with Cd accumulation in *E. coli* cells expressing metal-binding peptides incorporated e.g., in maltose-binding protein[44] or in the outer membrane protein LambB.[45] Furthermore, the integrated His-tags will be helpful in the preparation of PSII complexes and for immobilization purposes in nanodevices.[46]

Conclusions and Future Aspects

PSII—either in vivo or in its isolated form—can be envisioned as a natural bioanalytical device which can be hooked up to physical instruments in order to detect environmental pollutants with high specificity and sensitivity. The "sensing" element is the Q_B binding niche within the D1 subunit which accommodates a broad spectrum of diverse chemical compounds[29] and is accessible for extensive genetic manipulation. We are able to generate different, unnatural constellations of functional amino acid residues which in turn can lead to mutants with new binding properties. Selection for photoautotrophic growth assures that only mutants emerge which, despite more or less drastic modifications, still contain a functional PSII. Sustained functionality, even if reduced compared to wild type activities, is a precondition for its use a biosensor.

Two basic approaches for the generation of PSII mutants for biosensor purposes have been followed. The first one is site-directed mutagenesis and relies mostly on the presumed and less frequently on the computed function of a particular amino acid residue or a group of amino acids. Although useful in many instances, such work also produces a plethora of mixed results indicating that our current ability for prediction of structure-function relationships in D1 is still insufficient. The second approach is random mutagenesis which circumvents the need for detailed structural and mechanistic information. Ideally, it identifies all possible changes that satisfy a particular functional selection and can be envisioned as a first step in a directed protein evolution strategy.[47,48] This step has been taken here by creating a library of variants (Fig. 3) which can now be further selected for specific properties (like e.g., increased sensitivity or resistance towards herbicides, temperature, radiation) or recycled to a next round of directed evolution.

The D1 subunit of PSII is highly conserved in organisms ranging from cyanobacteria to higher plants,[49] and it came as a surprise that there appears to be room for ample manipulations without compromising its basic functions. As a new feature presented here for the first time, the sequence space around the binding niche can be reconfigured with peptide modules conferring entirely new properties like metal or antibody binding. Although a coupling between antibody or metal binding and activity change has not been demonstrated yet, peptide insertion appears to be an auspicious first step to greatly expand the recognition capability of a PSII biosensor.

References

1. Koblizek M, Masojidek J, Komenda J et al. A sensitive photosystem II-based biosensor for detection of a class of herbicides. Biotechnol Bioeng 1998; 60(6):664-9.
2. Giardi MT, Koblizek M, Masojidek J. Photosystem II-based biosensors for the detection of pollutants. Biosens Bioelectron 2001; 16(9-12):1027-33.
3. Perham RN. Structural aspects of biomolecular recognition and self-assembly. Biosens Bioelectron 1994; 9(9-10):753-60.
4. Gilardi G, Fantuzzi A. Manipulating redox systems: Application to nanotechnology. Trends Biotechnol 2001; 19(11):468-76.
5. Deisenhofer J, Epp O, Miki K et al. Structure of the protein subunits in the photosynthetic reaction centre of Rhodopseudomonas viridis at 3 Å resolution. Nature 1985; 318:618-624.
6. Trebst A. The topology of the plastoquinone and herbicide binding peptides of photosystem II in the thylakoid membrane. Z Naturforsch 1986; 41c:240-245.
7. Zouni A, Witt HT, Kern J et al. Crystal structure of photosystem II from Synechococcus elongatus at 3.8 A resolution. Nature 2001; 409:739-43.
8. Kamiya N, Shen JR. Crystal structure of oxygen-evolving photosystem II from Thermosynechococcus vulcanus at 3.7-A resolution. Proc Natl Acad Sci USA 2003; 100:98-103.
9. Trebst A. A contact site between the two reaction center polypeptides of photosystem II is involved in photoinhibition. Z Naturforsch 1991; 46c:557–562.
10. Greenberg BM, Gaba V, Mattoo AK et al. Identification of a primary in vivo degradation product of the rapidly-turning-over 32 kd protein of photosystem II. EMBO J 1987; 6(10):2865-9.
11. Hirschberg J, McIntosh L. Molecular basis of herbicide resistance in Amaranthus hybridus. Science 1983; 222:1346-1349.
12. Hirschberg J, Bleecker A, Kyle DJ et al. The molecular basis of triazine-herbicide resistance in higher plant photosynthesis. Z Naturforsch 1984; 39c:412-420.
13. Pfister K, Steinback KE, Gardner G et al. Photoaffinity labelling of an herbicide receptor protein in chloroplast membranes. Proc Natl Acad Sci USA 1981; 78:981-985.
14. Klein TM, ED, Wolf R et al. High-velocity microprojectiles for delivering nucleic acids into living cells. Nature 1987; 327:70-73.
15. Boynton JE, Gillham NW, Harris EH et al. Chloroplast transformationin Chlamydomonas with high velocity microprojectiles. Science 1988; 240:1534–1541.
16. Rochaix JD, Goldschmidt-Clermont M, Merchant S, eds. The Molecular Biology of Chloroplasts and Mitochondria in Chlamydomonas. Advances in Photosynthesis, 1998:7.
17. Rochaix JD. The three genomes of Chlamydomonas. Photosynthesis Research 2002; 73(1):285-293.
18. Przibilla E, Heiss S, Johanningmeier U et al. Site-specific mutagenesis of the D1 subunit of Photosystem II in wildtype Chlamydomonas. Plant Cell 1991; 3:169–174.
19. Roffey RA, Golbeck JH, Hille CR et al. Photosynthetic electron transport in genetically altered Photosystem II reaction centers of chloroplasts. Proc Natl Acad Sci USA 1991; 88:9122–9126.
20. Erickson JM, Rahire M, Rochaix J-D. Chlamydomonas reinhardtii gene for the Mr 32000 protein of photosystem II contains four large introns and is located entirely within the chloroplast inverted repeat. EMBO J 1984; 3:2753-2762.
21. Schrader S, Johanningmeier U. The carboxy-terminal extension of the D1-precursor protein is dispensable for a functional photosystem II complex in Chlamydomonas reinhardtii. Plant Mol Biol 1992; 19(2):251-6.
22. Heiss S, Johanningmeier U. Analysis of a herbicide resistant mutant obtained by transformation of the Chlamydomonas chloroplast. Photosynth Res 1992; 34:311-317.
23. Johanningmeier U, Heiss S. Construction of a Chlamydomonas reinhardtii mutant with an intronless psbA gene. Plant Mol Biol 1993; 22(1):91-9.
24. Minagawa J, Crofts AR. A robust protocol for site-directed mutagenesis of the D1 protein of Chlamydomonas reinhardtii: A PCR-spliced psbA gene in a plasmid conferring spectinomycin resistance was introduced into a psbA deletion strain. Photosynth Research 1994; 42:121-132.
25. Johanningmeier U, Sopp G, Brauner M et al. Herbicide resistance and supersensitivity in Ala_{250} or Ala_{251} mutants of the D1 protein in Chlamydomonas reinhardtii. Pesticide Biochem Physiol 2000; 66:9–19.
26. Preiss S, Schrader S, Johanningmeier U. Rapid, ATP-dependent degradation of a truncated D1 protein in the chloroplast. Eur J Biochem 2001; 268(16):4562-9.

27. Dauvillee D, Hilbig L, Preiss S et al. Minimal extent of sequence homology required for homologous recombination at the psbA locus in Chlamydomonas reinhardtii chloroplasts using PCR-generated DNA fragments. Photosynth Res 2004; 79:219-224.
28. Velthuys BR. Electron dependent competition between plastoquinone and inhibitors for binding to photosystem II. FEBS Lett 1981; 126:277–281.
29. Oettmeier W. Herbicides and photosystem II. In: Barber J ed. Topics in Photosynthesis, Vol. 11. London New York Tokyo: Elsevier Amsterdam, 1992:349-408.
30. Narusaka Y, Narusaka M, Kobayashi H et al. The herbicide-resistant species of the cyanobacterial D1 protein obtained by thorough and random in vitro mutagenesis. Plant Cell Physiol 1998; 39(6):620-6.
31. Oettmeier W. Herbicide resistance and supersensitivity in photosystem II. Cell Mol Life Sci 1999; 55(10):1255-77.
32. Trebst A. The molecular basis of resistance of photosystem II herbicides. In: Caseley JC, Cussans GW, Atkin RK, eds. Herbicide Resistance in Weeds and Crops. Oxford: Butterworth-Heinemann, 1991:145-164.
33. Wilski S. Ph. D. thesis, Ruhr-Universität, Bochum, Germany:2004.
34. Orawski G. Ph. D. thesis, Ruhr-Universität, Bochum, Germany: 2001.
35. Benson DE, Wisz MS, Hellinga HW. The development of new biotechnologies using metalloprotein design. Curr Opin Biotechnol 1998; 9:370-376.
36. Mizoue LS, Chazin WJ. Engineering and design of ligand-induced conformational change in proteins. Curr Opin Struct Biol 2002; 12(4):459-63.
37. Mejare M, Bulow L. Metal-binding proteins and peptides in bioremediation and phytoremediation of heavy metals. Trends Biotechnol 2001; 19(2):67-73.
38. Valls M, de Lorenzo V. Exploiting the genetic and biochemical capacities of bacteria for the remediation of heavy metal pollution. FEMS Microbiol Rev 2002; 26(4):327-38.
39. Brennan CA, Christianson K, La Fleur MA et al. A molecular sensor system based on genetically engineered alkaline phosphatase. Proc Natl Acad Sci USA 1995; 92(13):5783-7.
40. Hellinga HW, Marvin JS. Protein engineering and the development of generic biosensors. Trends Biotechnol 1998; 16(4):183-9.
41. Kless H, Vermaas W. Many combinations of amino acid sequences in a conserved region of the D1 protein satisfy photosystem II function. J Mol Biol 1995; 246(1):120-31.
42. Johanningmeier U. Expression of the psbA gene in E.coli. Z Naturforsch 1987; 42c:755-757.
43. Yu EW, Koshland Jr DE. Propagating conformational changes over long (and short) distances in proteins. Proc Natl Acad Sci USA 2001; 98(17):9517-20.
44. Mauro JM, Pazirandeh M. Construction and expression of functional multi-domain polypeptides in Escherichia coli: Expression of the Neurospora crassa metallothionein gene. Lett Appl Microbiol 2000; 30(2):161-6.
45. Kotrba P, Doleckova L, de Lorenzo V et al. Enhanced bioaccumulation of heavy metal ions by bacterial cells due to surface display of short metal binding peptides. Appl Environ Microbiol 1999; 65(3):1092-8.
46. Gilardi G, Fantuzzi A, Sadeghi SJ. Engineering and design in the bioelectrochemistry of metalloproteins. Curr Opin Struct Biol 2001; 11(4):491-9.
47. Giver L, Gershenson A, Freskgard PO et al. Directed evolution of a thermostable esterase. Proc Natl Acad Sci USA 1998; 95(22):12809-13.
48. Jaeger KE, Eggert T, Eipper A et al. Directed evolution and the creation of enantioselective biocatalysts. Appl Microbiol Biotechnol 2001; 55(5):519-30.
49. Svensson B, Vass I, Styring S. Sequence analysis of the D1 and D2 reaction center proteins of photosystem II. Z Naturforsch 1991; 46c(9-10):765-76.
50. Ferreira KN, Iverson TM, Maghiaoui K et al. Architecture of the photosynthetic oxygen-evolving center. Science 2004; 303:1831-8.

CHAPTER 6

Chloroplast Genomics of Land Plants and Algae

Margarita S. Odintsova and Nadezhda P. Yurina*

Abstract

This review summarizes recent data from chloroplast genomics research, namely the structure and gene content in completely sequenced chloroplast genomes of land plants and algae. It aims to highlight the structural similarity of chloroplast DNAs (cpDNA) gene content and arrangement in various lineages of land plants. It is noteworthy that chloroplast genomes of algae show significantly less structural similarity than those of land plants. Algae contain several unique genes, which are not found in cpDNAs of land plants. The organization of genes on the plastid chromosome differs drastically in land plants and algae. The problems of origin and evolution of plastids are briefly discussed.

Introduction

Chloroplasts, the sites of photosynthesis within plant cells, comprise a prominent and well-known class of plastids, subcellular organelles with diverse functions in plant and algal cells. As endosymbiotic remnants of free-living cyanobacterial progenitor chloroplasts have, over evolutionary time, lost the vast majority of their genes. Depending on the organism, contemporary chloroplast genomes (so-called plastomes) contain only 60-200 open reading frames (ORFs).[1,2] The photosynthetically active chloroplasts are characterized by high rates of transcription and translation, allowing the synthesis of large amounts of the enzyme ribulose bisphosphate carboxylase (Rubisco) and a rapid renewal of electron transfer components, features necessary for efficient photosynthetic CO_2 fixation. Technological developments in the genomics, including the sequencing of chloroplast genomes from 33 different plant species and algae have provided much information on the functions of plastome and of the origin and evolution of plastids.

Below it shows the list of organisms whose plastome has been completely sequenced, as obtained from the NCBI database (Table 1).

Chloroplast Genome of Land Plants

Chloroplasts of land plants contain multiple identical circular double-stranded DNA molecules, whose size, according to different data, varies from 120 to 160 or to 220 kb.[3,4] In the population of cpDNA molecules obtained by chloroplast lysis, monomeric circles prevail (-60% of all circular DNA molecules in tobacco). There also are oligomeric forms, which are concatemers or head-to-tail associates that most probably arose during DNA recombination and/or replication, as well as atypical molecules, presumably replicative intermediates. Accordingly, the chloroplast genome is not uniform and exists in the cell as a DNA population heterogeneous in molecule size and conformation.[5] The first complete nucleotide sequences of cpDNA were determined in 1986 for tobacco[6]

*Corresponding Author: Nadezhda P. Yurina—A.N. Bach Institute of Biochemistry, Russian Academy of Sciences, Moscow 119071, Russia. Email: NYurina@inbi.ras.ru

Biotechnological Applications of Photosynthetic Proteins: Biochips, Biosensors and Biodevices, edited by Maria Teresa Giardi and Elena V. Piletska.

Table 1. Organisms whose plastome has been completely sequenced, as obtained from the NCBI database

Alveolata	**Bryophyta**
Eimeria tenella	*Anthoceros formosae*
Toxoplasma gondii	*Marchantia polymorpha*
Cryptophyta	*Physcomitrella patens*
Guillardia theta	**Psilophyta**
Euglenozoa	*Adiantum capillus-veneris*
Astasia longa	*Psilotum nudum*
Euglena gracilis	**Coniferophyta**
Glaucocystophyceae	*Pinus thunbergii*
Cyanophora paradoxa	**Anthophyta**
Rhodophyta	*Amborella trichopoda*
Cyanidium caldarium	*Arabidopsis thaliana*
Porphyra purpurea	*Atropa belladonna*
Stramenopiles	*Epifagus virginiana*
Odontella sinensis	*Calycanthus fertilis* var. *ferax*
Chlorophyta	*Lotus corniculatus* var. *japonicus*
Chaetosphaeridium globosum	*Medicago truncatula*
Chlamydomonas reinhardtii	*Nicotiana tabacum*
Chlorella vulgaris	*Oenothera elata subsp.hookeri*
Mesostigma viride	*Oryza sativa*
	Spinacia oleracea
	Triticum aestivum
	Zea mays

and liverwort *Marchantia* polymorpha,[7] then rice followed.[8] The cpDNA structure, gene composition and arrangement in tobacco are held to be most typical of land plants. This genome is supposed to preserve the traits of the cpDNA present in the ancestors of land plants.

A characteristic feature of cpDNAs in land plants is the presence of an inverted repeat (IR), averaging from 20-30 kb in size and varying from 5 to 76 kb in different species. This repeat results in duplication of the IR-located genes and accounts for most of the size variation in cpDNAs among land plants (Table 2).[9]

Thus the unusually large size of the *Pelargonium hortorum* chloroplast genome (217 kb) is due to its long 76-kb IR.[10] Only cpDNAs of some legumes and conifers lack IRs. It has been hypothesized that the IR was present in the common ancestor of land plants. In some species, one IR segment was lost during evolution, sometimes the IR was only partly lost (for example, a 495-bp IR stretch was preserved in black pine) (Table 2). The IR segments divide cpDNA into the large and small single-copy regions (LSC and SSC, respectively). In some plant species, the IR segments are shifted toward the unique sequences, as compared with the tobacco cpDNA. As a result, the genes located in the unique genome sequences in the marker DNA, are in these species found partly in the IRA and IRB segments. Thus in the chloroplast genomes of *Arabidopsis thaliana*[11] and *Lotus japonicus*,[12] the *rps19* gene is partly in the IRB and partly in LSC; the *ndhF* gene is in the IRB and SSC. In the tobacco chloroplast genome, both genes are located in the single-copy regions.

The comparison of cpDNA structure in *Lobelia thuliniana* and tobacco showed that an 11-kb IR segment in *L. thuliniana* is shifted toward the SSC. The expansion of the IR into the SSC is retained in all other members of the families Lobeliaceae, Cyphiaceae, and Campanulaceae. At the same time, the IR location in the chloroplast genome in the species of the *Sphenoclea* genus is similar to that in tobacco. This finding supports the exclusion of this genus from the family Campanulaceae.[13]

The overall cpDNA structure is conserved. The IR belongs to the most conserved region of the molecule containing all rRNA and two tRNA genes ($tRNA^{Ile}$ (GAU) and $tRNA^{Ala}$ (UGC)). The IR

Table 2. Structure of several completely sequenced chloroplast genomes of land plants

	Plant Species	cpDNA Size, bp	IR, bp	Gene Number	Number of Introns	Number of Editing Sites
Monocots	*Oryza sativa*	134525	20799	110	18/16	21/11
	Zea mays	140387	22748	110	18/16	27/14
Dicots	*Triticum aestivum**	134540	20702	110	18/16	28/21
	Nicotiana tabacum	155939	25341	113	21/18	31/16
	Spinacia oleracea	150725	25073	113	20/17	22/21
	Arabidopsis thaliana	154478	26264	113	21/18	19/21
	Oenothera elata ssp. hookeri	159443	27807	112	19/17	28/21
	Lotus japonicus	150519	25156	111	21/18	
	*Epifagus virginiana***	70028	19799	42	8/6	***
	Medicago truncatula	124033				
Gymnosperms	*Pinus thunbergii*	119707	495	108	16/14	26/12
Ferns	*Psilotum nudum*	138829	18954	118	19/17	***
Mosses	*Marchantia polymorpha*	121024	10058	122	20/18	0/0

*Updated[9] cpDNA size in wheat is 134,545 bp, the length of the IR is 20,703. **Nonphotosynthetic parasitic plant. ***Editing sites were determined experimentally; their number is unknown. Slashes separate the number of introns and intron-containing genes, as well as the number of editing sites and transcripts. Modified from Wakasugi T et al. The genomics of land plant chloroplasts: Gene content and alteration of genomic information by RNA editing. Photosynthesis Res 70:107-118, ©2001 with kind permission of Springer Science and Business Media.

sequences diverge two-three times more slowly than the unique sequences. The repeats account for the stability of the chloroplast genome, preventing recombination. The large single-copy region is the most variable. Although the cpDNA of land plants is a stable genetic system, certain rearrangements in its structure have occurred during evolution. Sequencing of the chloroplast genomes of land plants revealed inversions, translocations, insertions/deletions even in related species, and allowed one to identify the 'hot spots' of mutations. Thus, the comparison of cpDNA structure in *Oenothera elata*[14] and some dicots (tobacco,[6] *A. thaliana*,[11] and spinach[15]) demonstrated that they differ in a large 54-kb inversion in the LSC between the *accD* and *rps16* genes present in *Oenothera*. The spinach chloroplast genome, retaining the quadripartite structure typical of land plants, is 5,214 bp smaller than the tobacco cpDNA. This results from shorter intergenic regions both in the IR segments and in the entire genome. In the LSC of the *Lotus japonicus* cpDNA, a 51-kb inversion between the *rbcL* and *rps16* genes was identified. A similar inversion was detected in the soybean cpDNA, suggesting that it is typical of legumes.[12]

A comparison of the chloroplast genome structure in several monocots with that of tobacco revealed structural rearrangements in cereal cpDNAs (intron loss in the *rpoC1* gene, an insertion in the *rpoC2* gene, a deletion of an ORF 2280 in the IR, a translocation of the *rpl23* gene, and rearrangements in ORF512). In all probability, structural alterations in the cpDNA of the ancestor(s) of cereals occurred simultaneously.[9,16] Thus, the variation in cpDNA structure may serve as a phylogenetic marker.

A highly rearranged chloroplast genome was reported in *Trachelium caeruleum* (Campanulaceae). Its cpDNA had from seven to ten inversions, one or two transpositions, insertions/deletions, and several dispersed repeat families not found in chloroplast genomes of other land plants. Together with other rearrangements, these alterations in the *Trachelium caeruleum* cpDNA led to considerable changes in gene order, orientation, and composition. As a result, the *T. caeruleum* chloroplast genome significantly differs from that of tobacco.[4]

As regards gene order and composition, chloroplast genomes of land plants are highly conserved.[17,18] They contain genes for all rRNA species, most plastid tRNA genes, genes for RNA polymerase subunits, many ribosomal proteins and translation factors. In addition, they encode the components of the photosynthetic apparatus including the large subunit of ribulose-1,5-bisphosphate carboxylase/oxygenase, the components of photosystems I and II (PS I and PS II), cytochrome *b/f* complex, and ATP synthase. The most striking feature of cpDNA is the presence of 11 *ndh* genes coding for the subunits of the respiratory chain NADH dehydrogenase complex. In addition, the chloroplast genome contains a number of ORFs some of which are conserved (*ycfs*) while others are species-specific. Most reading frames are genes; however, their identification is complicated. Thus originally 82 genes were identified in the tobacco chloroplast genome.[6] In the following 12 years, the combined efforts of 10 research groups resulted in identification of 24 more genes (one gene coding for the low-molecular-weight RNA and 23 protein-coding genes). However, about 20 ORFs are still to be identified in the chloroplast genome of tobacco.[19]

Chloroplast genes (including putative genes) identified in 12 completely sequenced cpDNAs of land plants are presented in Table 3.

In chloroplast genomes of 11 species (except *Epifagus*), from 108 to 122 unique genes were detected,[3] 95 of which are present in all species. Other genes exhibit interspecific variation. For example, in contrast to tobacco cpDNA, the spinach cpDNA lacks the *sprA* gene coding for a small 218-bp structural RNA.[15] In the maize chloroplast genome, the *accD* gene for a subunit of the prokaryotic acetyl-CoA-carboxylase is missing.[17] The spinach *infA* is a pseudogene in cpDNA of tobacco, while *rpl23* functional in the tobacco cpDNA is a pseudogene in spinach.[15] The genes involved in chlorophyll biosynthesis are present in *Marchantia* and *Pinus* in contrast to angiosperms, while the genes for NADH dehydrogenase subunits are found in angiosperms and *Marchantia* but are partially missing from the *Pinus* cpDNA.[20] In *O. elata*, in contrast to tobacco, there are no *clpP* and *sprA* genes.[14] The plastid genome of *Epifagus virginiana*, a nonphotosynthetic parasitic flowering plant, which carries only 42 genes, is the only exception. The *E. virginiana* cpDNA lacks all genes for the components of the photosynthetic apparatus and respiration.[21] The genome size is reduced to 70 kb (Table 2). Similar genome structure was described for a related parasitic plant *Conopholis americana*, a member of the Orobanchaceae family. Different stages of plastid genome degeneration were reported for two holoparasitic flowering plants of the *Cuscuta* genus (Convolvulaceae): *C. reflexa* and *C. europaea*. The latter is devoid of any photosynthetic activity and the corresponding genes, while the plastid DNA of *C. reflexa* is less degenerated than in *Epifagus* and *Conopholis*. In its cpDNA, most photosynthetic genes are preserved, except for a 6.5-kb fragment covering *rpl2*, *rpl3*, *trnI*, and *trnK*. This species occupies an intermediate position between the photosynthetic and holoparasitic plants.[22]

According to their functions, chloroplast genes may be divided into three main groups: genes for the transcription/translation system, genes involved in photosynthesis, and genes related to photosynthetic metabolism (participating in biosynthesis of amino acids and fatty acids, pigments, etc). The transcripts and translation products of the latter group are poorly studied. The genes of the first two groups are most conserved, with the rRNA genes being the extremes (98-100% homology among different species).

Most chloroplast genes are located in clusters of similar polarity. Cluster expression results in large polycistronic primary transcripts processed to oligo- and monocistronic mRNAs and subject to splicing and editing. Editing often restores conserved amino acid residues in the corresponding proteins and thus plays a significant role, eliminating errors in genome sequences at the mRNA level. Editing was reported for the chloroplast gene transcripts of all land plants studied so far but not for those of algae and cyanobacteria. In mitochondrial genomes, editing events take place much more frequently than in chloroplasts. Since editing modifies the information stored in cpDNA, the structure of cpDNA-encoded proteins cannot always be predicted from gene sequences. Furthermore, editing may lead to formation of new ORFs.[3,23]

In land plants, editing most frequently causes C→U substitutions. Reverse U→C editing was observed in the *rpcL* and *atpB* transcripts from the hornwort *Anthoceros formosae*. The genomic *rbcL* sequence includes two stop codons in the protein-coding region, while in the transcript, owing to

Table 3. Genes in some completely sequenced chloroplast genomes of land plants

Gene	Product	Note
	Genes for the Genetic Apparatus	
rrn23	23S rRNA	
rrn16	16S rRNA	
rrn5	5S rRNA	
rrn4.5	4.5S PPHK	In land plants
**trnA-UGC*	Ala-tRNA (UGC)	In the *rrn* spacer
trnR-ACG	Arg-tRNA (ACG)	
trnR-UCU	Arg-tRNA (UCU)	
trnR-CCG	Arg-tRNA (CCG)	In pine/*Psilotum*/*Marchantia*
trnN-GUU	Asn-tRNA (GUU)	
trnD-GUC	Asp-tRNA (GUC)	
trnC-GCA	Cys-tRNA (GCA)	
trnQ-UUG	Gln-tRNA (UUG)	
trnE-UUC	Glu-tRNA (UUC)	Heme/chlorophyll biosynthesis
trnG-GCC	Gly-tRNA (GCC)	
**trnG-UCC*	Gly-tRNA (UCC)	Intron in the D stem
trnH-GUG	His-tRNA (GUG)	
**trnI-GAU*	Ile-tRNA (GAU)	In the *rrn* spacer
trnI-CAU	Ile-tRNA (CAU)	
**trnL-UAA*	Leu-tRNA (UAA)	Group I intron
trnL-CAA	Leu-tRNA (CAA)	
trnL-UAG	Leu-tRNA (UAG)	
**trnK-UUU*	Lys-tRNA (UUU)	2.5-kb intron
trnfM-CAU	FMet-tRNA (CAU)	
trnM-CAU	Met-tRNA (CAU)	
trnF-GAA	Phe-tRNA GAA)	
trnP-UGG	Pro-tRNA (UGG)	
trnP-GGG	Pro-tRNA (GGG)	In pine/*Psilotum*, pseudogene in *Marchantia*
trnS-GGA	Ser-tRNA (GGA)	
trnS-CGA	Ser-tRNA (CGA)	In *Psilotum*
trnS-UGA	Ser-tRNA (UGA)	
trnS-GCU	Ser-tRNA (GCU)	
trnT-GGU	Thr-tRNA (GGU)	
trnT-UGU	Thr-tRNA (UGU)	
trnW-CCA	Trp-tRNA (CCA)	
trnY-GUA	Tyr-tRNA (GUA)	
trnV-GAC	Val-tRNA (GAC)	
**trnV-UAC*	Val-tRNA UAC)	
sprA	Small plastid RNA;	In dicots
30S ribosomal proteins		
rps2	CS2	
rps3	CS3	
rps4	CS4	
rps7	CS7	
rps8	CS8	
rps11	CS11	
**rps12*	CS12	Two introns, trans-splicing
rps14	CS14	
rps15	CS15	

continued on next page

Table 3. Continued

Gene	Product	Note
**rps16*	CS16	In angiosperms
rps18	CS18	
rps19	CS19	
50S ribosomal proteins		
**rpl2*	CL2	No intron in spinach
rpl14	CL14	
**rpl16*	CL16	
rpl20	CL20	
rpl21	CL21	In *Psilotum/Marchantia*
rpl22	CL22	Not in legumes
rpl23	CL23	Pseudogene in spinach
rpl32	CL32	
rpl33	CL33	
rpl36	CL36	Initially designated *secX*
RNA polymerase subunits		
rpoA	α	
rpoB	β	
**rpoC1*	β′	No intron in monocots
rpoC2	β″	Extra sequence in monocots
infA	Initiation factor 1	Pseudogene in tobacco/*Oenothera*, absent in *Arabidopsis/Lotus*
matK	Maturase	In *trnK* intron
	Genes for Photosynthesis	
rbcL	Rubisco large subunit	
Photosystem I		
psaA	P700 apoprotein A1	
psaB	P700 apoprotein A2	
psaC	Subunit VII	Fe-S 9-kDa protein
psaI	4-kDa protein	
psaJ	5-kDa protein	
psaM	Protein M	Not in angiosperms
***ycf3*	Assembly/stability	Two introns, one intron in *Marchantia*
***ycf4*	Assembly/stability	
Photosystem II		
psbA	Protein D1	
psbB	47-kDa protein	
psbC	43-kDa protein	
psbD	Protein D2	
psbE	Cytochrome *b559* (8 kDa)	
psbF	Cytochrome *b559* (4 kDa)	
psbH	10-kDa phosphoprotein	
psbI	Protein I	
psbJ	Protein J	
psbK	Protein K	
psbL	Protein L	
psbM	Protein M	
psbN	Protein N	
psbT	Protein T	
psbZ	PS II core subunit	Initially designated *lhbA*

continued on next page

Table 3. Continued

Gene	Product	Note
b/f complex		
petA	Cytochrome *f*	
**petB*	Cytochrome *b6*	
**petD*	Subunit IV	
petG	Subunit V	Initially designated *petE*
petL	3.5-kDa subunit	
petN	3.2-kDa subunit	
ccsA	*c*-type cytochrome synthesis	
ATP synthase		
atpA	CF_1 subunits, α	
atpB	β	
atpE	ε	
**atpF*	CF_O subunits, I	
atpH	III	
atpI	IV	
NADH dehydrogenase subunits		
**ndhA*	ND1	
**ndhB*	ND2	
ndhC	ND3	
ndhD	ND4	
ndhE	ND4L	
ndhF	ND5	
ndhG	ND6	
ndhH	49-kDa protein	
ndhI	18-kDa protein	
ndhJ	30-kDa protein	
ndhK	27-kDa protein	Initially designated *psbG*
	Other Genes	
chlB	(Protochlorophyllid reduction)	Absent from angiosperms/*Psilotum*
chlL	(Protochlorophyllid reduction)	Absent from angiosperms/*Psilotum*
chlN	(Protochlorophyllid reduction)	Absent from angiosperms/*Psilotum*
cysA	(Transport protein)	In *Marchantia* (*mbpX*)
cysT	(Transport protein)	In *Marchantia* (*mbpY*)
**clpP*	ATP-dependent protease, proteolytic subunit	Two introns in tobacco, no intron in monocots/*Oenothera*/*pine*
accD	Acetyl-CoA carboxylase, β-subunit	Not in monocots, initially designated zfpA
***ycf10*	Inorganic carbon uptake	Initially designated *hbp*
***ycf1*	Unknown	Not in monocots
***ycf2*	Unknown	Not in monocots
***ycf12*	Unknown	In pine/*Psilotum*/*Marchantia*, not in angiosperms
***ycf15*	Unknown	In maize/tobacco/spinach/*Arabidopsis*/*Oenothera*
***ycf66*	Unknown	In *Marchantia*, one intron

*Intron-containing genes (split genes). **Reading frames found in several cpDNAs. Gene transcripts not identified conclusively are given in parantheses. Modified from Wakasugi T et al. The genomics of land plant chloroplasts: Gene content and alteration of genomic information by RNA editing. Photosynthesis Res 70:107-118, ©2001 with kind permission of Springer Science and Business Media.

reverse RNA editing, the UGA and UAA codons are converted to CGA (Arg) and CAA (Gln), respectively, thus generating the ORF for the large Rubisco subunit. The number of editing sites identified and predicted so far in the transcripts of chloroplast genes from the sites identified in rice, maize, and tobacco cpDNAs are shown in Table 2. Of 31 editing sites found in tobacco cpDNA, two sites create the initiation codons, 28 sites lead to amino acid substitutions, and one site is located at the third position of a serine codon that causes no amino acid substitution.[3] RNA editing is probably involved in speciation.[24] Thus, the primary structure comparison of the cpDNAs from *Atropa belladonna* and *Nicotiana tabacum* showed that the regulatory regions (promoters, etc.) in both species are very similar, as well as genes including introns (variations concern only functionally insignificant regions), but these species differ in editing sites, indicating that editing could lead to reproductive isolation of populations, which caused divergence resulting in the formation of the *Atropa* and *Nicotiana* species.

The chloroplast genome of land plants contains from 14 to 18 split genes with the total number of introns averaging 20 (Table 2). The intron size varies from 0.3 to 2.5 kb. Introns are conserved and belong predominantly to group II.[25,26] The majority of chloroplast genes contain only one intron.

Among the intron-containing chloroplast genes, *rps12* is of particular interest owing to its peculiar structure. This gene contains two introns and consists (in tobacco) of one copy of exon 1 called 5'-*rps12* (38 codons), two copies of exon 2 (78 codons), intron (536 bp), and one copy of exon 3 called 3'-*rps12* (seven codons). Both parts of the gene-5'-*rps12* and 3'-*rps12*-are located apart on the complementary cpDNA strands, transcribed independently, and subjected to trans-splicing to produce a mature mRNA for the CS12 protein.[27]

In plastids of the parasitic plant *C. europaea*, a cis-spliced intron is missing; the *rps12* gene is shorter than the homologous chloroplast genes of typical land plants; however, its functions and expression, including trans-splicing of the first two exons, are preserved. Intron loss in the cis-spliced *rps12* from the cpDNA of three *Anemone* species was reported. Intron loss is a rare event in chloroplast genes of nonparasitic land plants (examples are *rpoC1* of maize and other cereals, *clpP* of cereals, *trnI* of *Campanula garganica*, *rpl16* and *rpl2* of angiosperms).[28]

Genes for Proteins of Transcription and Translation

This group of chloroplast genes was initially identified by similarity with the corresponding genes of *Escherichia coli*. Later, these data were confirmed by analysis of gene products. The chloroplast genome of land plants contains four genes coding for the polypeptides homologous to *E. coli* RNA polymerase subunits. The *rpoC* homolog in cpDNA is split into two parts, *rpoC1* and *rpoC2*, and only *rpoC* of dycots carries an intron. The *rpoC1* gene has no introns, and the *rpoC2* of monocots carries an additional sequence (-400 bp) in the middle of the coding region.[29] The cpDNA of *Epifagus* carries no functional *rpo* genes; in all likelihood, nuclear-encoded RNA polymerase participates in transcription of plastid genes in this species.

One-third (20-21) of the chloroplast ribosomal proteins are encoded by cpDNA. The *rps16* gene was found only in angiosperms and gingko; the *rpl21* gene was detected only in *Psilotum* and *Marchantia*. The gene for the initiation factor 1 (*infA*) is a pseudogene in cpDNA of tobacco and *Oenothera* and is absent from the plastid DNA of *Arabidopsis* and *Lotus*.[30]

In land plants, cpDNA encodes all four rRNA species (23S, 16S, 5S, and 4.5S). The low-molecular-weight 4.5S rRNA was found only in the chloroplasts of land plants. It is a component of a large (50S) ribosomal subunit and is supposed to be functionally homologous to the 3' terminus of the prokaryotic 23S rRNA.

The tRNA genes (30-33) identified in cpDNA code for functional tRNAs, which form the typical cloverleaf structure. It is of particular interest that $tRNA^{Glu}$ (UUC) not only participates in protein synthesis, but also acts as a cofactor for the formation of 5-aminolevulinic acid, which is a common precursor in heme and chlorophyll biosynthesis. cpDNA-encoded tRNAs suffice for translation of 61 codons if normal wobble base pairing occurs in codon-anticodon recognition. Although this hypothesis is experimentally supported, transport of the nuclear-encoded tRNA into the chloroplasts cannot be excluded. Thus, the *Epifagus* plastid DNA contains only 17 tRNA genes, which is indirect evidence for the transport of nuclear-encoded tRNAs involved in translation.[21]

In the chloroplasts of land plants, besides rRNA and tRNA genes, the *sprA* gene for the plastid 218-bp RNA was reported. Targeted deletion of this gene has no effect on the phenotype of the plant in a growth chamber; however, in the field the plants exhibit stunted growth.[3]

Genes for Proteins of the Photosynthetic Apparatus

It is common knowledge that Rubisco consists of eight identical large and eight identical small subunits. In land plants, the large subunit is encoded by cpDNA (*rbcL*) and the small subunit is nuclear-encoded (*rbcS*).

The thylakoid membranes of plastids contain four different complexes: PS I, PS II, the cytochrome *b*/*f* complex, and ATP synthase. Of about 60 membrane polypeptides, half is encoded by cpDNA. The cpDNA of land plants carries six *psa* genes for PS I components, 15 *psb* genes for the PS II components, six *pet* genes for the cytochrome *b*/*f* complex components, and six *atp* genes for the ATP synthase subunits. In addition, they contain two genes (*ycf3* and *ycf4*) involved in assembly and/or stability of the PS I complex. Angiosperms lack the *psaM* gene. In the chloroplast genome, 11 ORFs were provisionally identified as the *ndh* genes of the NADH dehydrogenase complex. Recently, this complex has been isolated from land plants, characterized,[31] and the role of the *ndh*-encoded products was analyzed by site-specific mutagenesis of tobacco cpDNA. These experiments confirmed that the *ndh* genes are functional, and chloroplasts have an active NADH dehydrogenase complex. The chloroplasts of most algae lack the *ndh* genes (see below). It should be noted that the cpDNA of black pine contains no intact *ndh*, although seven sequences (B, C, D, E, H, I, and K) are present as pseudogenes.[29]

Genes for Biosynthetic Processes

The information on the genes participating in biosynthetic processes and their products is scarce. Three genes (*chlB*, *L*, *N*) involved in chlorophyll biosynthesis were identified by similarity with the corresponding genes in photosynthetic bacteria. Targeted deletion of these genes supported the idea that they are responsible for light-independent protochlorophyllide reduction in *Chlamydomonas*. These genes were found in the chloroplast genomes of *Marchantia* and *Pinus*; however, they were not detected in cpDNA of angiosperms and *Psilotum*. The *accD* gene encodes the β-subunit of the prokaryotic-type acetyl-CoA carboxylase, which is involved in fatty acid biosynthesis and is light-activated via a redox cascade.[3] The chloroplast genomes of cereals have no *accD*. They also lack the prokaryotic-type acetyl-CoA carboxylase. The *clpP* gene encodes the proteolytic subunit of ATP-dependent protease. This enzyme contains a two-component regulatory subunit encoded by the nuclear genome and a proteolytic subunit, which is chloroplast-encoded. The *ccsA* gene of chloroplasts participates in *c*-type cytochrome biosynthesis. The *cysA* and *cysT* genes were found in cpDNA of *Marchantia* but not of other land plants. Although the predicted gene products are similar to sulfate transporters in cyanobacteria, the function of these genes in chloroplasts remains unclear.[32]

Chloroplast Genome of Algae

To date, the cpDNA sequences of all major algal lineages have been determined. Algae represent a group of oxygen-evolving organisms, whose evolutionary relationships and morphological variation have been summarized in several reviews.[33,34] The monophyletic origin of plastids is no longer a matter of debate.[35] It is commonly accepted that in most cases, chloroplasts arose from the primary endosymbiosis between a cyanobacterium and a primitive eukaryotic cell. As a result, three lineages originated: land plants (embryophytes) and green algae, red algae, and glaucocystophytes (cyanelle-containing algae). The remaining groups of algae carry secondarily derived chloroplasts, resulting from the engulfment of a primary endosymbiont by a eukaryotic host. The euglenoids and the chlorarachniophytes bear plastids secondarily derived from green algae. Lineages derived from secondary endosymbiosis of red algae are the cryptophytes, the brown algae or chromophytes (also known as heterokonts), some dinoflagellates, and the haptophytes.[36] Haptophyte cpDNA sequence data are currently limited.

According to size, structure, and gene composition, algal plastid genomes are more heterogeneous than those of land plants. Genome size averages 140 kb and contains about 110 genes. Genome size is consistent within lineages. Thus the size of red algal cpDNAs ranges from 150 to 191 kb.[36] The green algae are the exception. In addition to the classical 100- to 200-kb genomes, an extremely large genome was found in the Ulvophyceae group. In *Acetabularia mediterranea*, which belongs to this group, a 10-cm uninucleate cell possesses many hundreds of chloroplasts, and its cpDNA size amounts to 1500 kb. This is only 2.4 times smaller than the *Synechocystis* sp. PCC 6803 genome.[37] However, much of the *A. mediterranea* DNA is repeated; it is therefore doubtful that it contains substantially more genes than other cpDNAs. *Codium fragile* also belongs to the Ulvophyceae family with the smallest chloroplast genome of 89 kb. Rather small genomes were reported for the chloroplasts of other green algae: the picoplankton *Nanochlorum eukaryotum* (90 kb), *Pedinomonas minor* (98 kb) (Prasinophyceae), and *Coleochaete orbicularis* (100 kb) (Charophyceae). Within the *Chlamydomonas* genus, high variation in cpDNA size mainly results from different length of intergenic sequences. For example, *Chlamydomonas moewusii* (292 kb) has two large insertions as compared with its close relative *Ch. pitchmannii* (187 kb), which also has smaller intergenic spacers. In two closely related species, *Ch. reinhardtii* (203 kb) and *Ch. gelatinosa* (285 kb), cpDNAs differ not only in the length of intergenic spacers, but in gene order as well. The most significant feature of the *Ch. reinhardtii* chloroplast genome is the presence of more than 1000 short dispersed repeat sequences, which can be clustered into 40 families, accounting for nearly 20 % of the genome sequence.[38] The cpDNA of the unicellular alga *Euglena gracilis* (Euglenozoa) is similar in size to the chloroplast genomes of land plants (143 kb) but it lacks the IR. *Euglena gracilis* has four tandem direct rRNA repeats (three complete repeats containing 16S, 23S, and 5S rRNA genes, and one truncated, carrying only the 16S rRNA gene).[39] *Astasia longa*, a colorless heterotrophic alga closely related to the phototrophic *E. gracilis*, has a similar genome structure.[40] Since in *A. longa* all the proteins involved in photosynthesis are missing (except for the *rbcL* gene), its cpDNA is approximately two times shorter (73 kb) than that of *Euglena*. Unique features of cpDNA in these species include: very small group III introns averaging 100 bp in size, rather low GC content (on average, 22%), and the presence of a region with a variable number of tandem repeats, presumably a replication origin. The smallest plastid genomes were also reported for the heterotrophic green algae *Prototheca wickerhamii* and *Polytoma uvella*.

Algal cpDNAs are circular. Monomers predominate, multimers and atypical forms are very rare. In the apicomplexan parasite *Toxoplasma gondii*, the plastid 35-kb genome is represented by a linear DNA molecule. An unusual chloroplast genome structure was described in dinoflagellates (*Heterocapsa* sp., *Amphidinium carterae*, and *Pyrocystis lunula*). Instead of single genomes, they have fragmentary genomes: each gene is contained in its own separately replicated minicircle of 2-3 kb.[41] This organization is reminiscent of kinetoplast (mitochondrial) DNA in trypanosomes, which have one main genome and numerous catenated minicircles, which encode guide RNAs required for RNA editing. Dinoflagellates have no main genome or RNA editing. The fragmentation of the plastid genome of dinoflagellates is accompanied by rapid sequence divergence and gene loss. A close relationship of dinoflagellate plastids to apicomplexan plastids is obvious: in both instances, their genomes exhibit unusual architecture and reduced size. Both features are characteristic of their mitochondria as well.[41]

While the IR is a characteristic feature of cpDNA of land plants, the chloroplast genomes of several algal species often stray from this rule. They contain only one IR segment (one rRNA operon) similar to that shown for legumes. For example, the chloroplast genome of a unicellular green alga *Chlorella vulgaris* possesses one IR segment and correspondingly one copy of the rRNA gene cluster.[25] In the red algae studied thus far, the classical IR-LSC-SSC structure is found only in *Porphyra yezoensis*.[36] Large IRs are absent from the chloroplast genome of the unicellular red alga *Cyanidium caldarium*.[42]

In contrast to the typical rRNA operon of land plants containing 16S and 23S rRNA genes (separated by a spacer encoding two tRNAs) and followed by the genes for the low-molecular-weight rRNAs (4.5S and 5S), operons of the large and small rRNA components in the algal plastid genomes may be located apart from each other, transcribed from the complementary DNA strands, and separated by large intergenic spacers. The IR-located nonribosomal genes found in land plants were identified in green algae, accounting for an increased IR size in this group. In cpDNA of the green alga *Nephroselmis olivacea*, the IR carries additional large noncoding sequences, increasing the size of each IR segment to 46,137 bp.[26] Since the IR was found in the *Synechocystis* sp. PCC 6301 genome,[37] this trait is considered to be one of the ancient characteristics of the chloroplast genome.

In contrast to the chloroplast genome of land plants, characterized by conserved gene clusters and nearly uniform gene transfer to the nucleus among different species (see below), in the chloroplast genomes of green algae, including *Chlorella vulgaris*, these traits were not preserved and remained only in the early diverged green alga *Mesostigma viride*.[43,44] The chloroplast 118,360-bp genome of this flagellate alga is very similar in gene arrangement and structure to cpDNA of land plants. It has about 150 genes, 81% of which are located in clusters found in land plants.

Among green and brown algae, chloroplast genome structure is often different even in closely related species, although in general it is comparable with that of land plants. The absence of one of the IR segments, fragmentation of the rRNA operon, and insertion of foreign DNA fragments are among the factors that cause chloroplast genome instability in algae.[20]

Outside the green lineage (Chlorophyta), gene density is high in all algal cpDNAs. Large noncoding sequences, introns, and pseudogenes are exceptional. The *Guillardia theta* (Cryptophyta) cpDNA is a spectacular example of genome compactness. At the genome size of 122 kb, it has 183 genes, which account for 90% of its DNA.[45] Plastid genomes of other nongreen algae (all algal groups except green algae) are also gene-rich owing to short intergenic spacers, overlapping genes, large gene clusters (for example, ribosomal protein gene cluster) forming single transcription units, and complete or partial lack of introns (no introns were found in the plastid genomes of *Porphyra*, *Guillardia* and *Odontella*; one intron was detected in the *Cyanophora* cyanelle genome). These features contribute to economic genome structure. For example, the plastid genome of *Porphyra purpurea* (Rhodophyta), at the genome size of 191 kb, carries 252 genes; the 136-kb genome of *Cyanophora paradoxa* (Glaucocystophyceae) has 191 genes, and the 120-kb genome of *Odontella sinensis* (Stramenopiles) has 175 genes (the data on gene number in cpDNAs of *Cyanophora paradoxa* and *Odontella sinensis* are taken from the NCBI database). At the same time, the cpDNA of *E. gracilis* (143 kb) and *Ch. reinhardtii* (203 kb) contain 87 and 96 genes, respectively, including the tRNA genes, and both species have an atypical polar gene distribution.[36] In the chloroplast genes of *E. gracilis*, abundant introns (150) were found; in *Ch. reinhardtii*, many hundreds of small dispersed repeats were detected. Gene composition in the *Ch. reinhardtii* cpDNA differs considerably from that of the green alga *Chlorella vulgaris* (C-27). The cpDNA of *Chlamydomonas* lacks *ndh*-containing regions and has a unique rearrangement of the *rpo* genes encoding the plastid RNA polymerase.[25,37]

Algae contain the same groups of genes in cpDNA as land plants (genes for the transcription/translation system, for photosynthesis and photosynthetic metabolism). However, in many algal species, especially in nongreen algae, the plastid DNA often contains unique genes not found in other cpDNAs. Thus 30 to 40 unique genes were identified in the sequenced plastid genomes of the red algae *C. caldarium*[42] and *P. purpurea*.[46]

Among the unique genes of *C. caldarium* chloroplasts, there are five genes involved in bacterial cell envelope biogenesis, and two genes whose products are supposed to stabilize the photosynthetic apparatus under salt stress and chloroplast detoxification.[42] However, even the largest plastid genomes of red algae, for example the cpDNA of *P. purpurea*, lack the *infA* and *ndh* genes, which are always present in cpDNA of land plants. In the plastid genome of *P. purpurea*, the *clpP* gene is also missing, although it has been retained in the markedly reduced cpDNA of the holoparasite *E. virginiana* that lost all chloroplast genes for the RNA polymerase subunits, many tRNA and ribosomal protein genes.[21,36]

Among the unusual functions of the algal plastid genome, mention should be made of the two-component transcriptional regulators encoded by nongreen algal cpDNAs, which are widespread in bacteria including *Synechocystis*. These regulators mediate responses to environmental changes (such as osmolarity and nutrient deficiency). Two-component genes have not been found in cpDNA of land plants, nor have nucleus-encoded chloroplast-localized transcripts of these genes been identified. Therefore, the nongreen algae may have an unknown regulatory mechanism, which has not yet been recognized in land plants.[36] A canonical *E. coli* two-component system is the EnvZ-OmpR, where EnvZ is a cytoplasmic membrane sensor kinase, and OmpR is its substrate and the regulatory transcription factor. EnvZ homologs have been found in red algal cpDNA, whereas OmpR homologs have been discovered in all phycobilin-containing algae, i.e., red algae, cryptomonads, and *C. paradoxa.* Another putative transcriptional regulator is the *cfxQ* product. In most nongreen cpDNAs, *cfxQ* is a component of the *rbcLS* operon. Again, *cfxQ* is similar to the purple bacterial *cbbX* genes, which encode ATP-binding polypeptides necessary for photoautotrophic growth and thought to be involved in expression of the Rubisco genes. In all probability, in the nongreen algal plastids, the CfxQ protein transcriptionally regulates *rbcL* and/or *rbcS.*

The plastids of some algae contain genes involved in RNA maturation and degradation. The cpDNA of *N. olivacea, C. paradoxa, C. caldarium,* and *P. purpurea,* have the *rnpB* gene coding for the RNA subunit of RNase P, a 5'-endonuclease participating in tRNA processing.[47] The cyanelle RNase P holoenzyme in *C. paradoxa* is destroyed by nuclease treatment, indicating an essential role of the RNA component in the enzyme function.[48]

Small noncoding RNAs were found in algal plastids. A small RNA, the transcript of the chloroplast *tscA* gene participating in trans-splicing of *psaA,* namely in trans-splicing of exons 1 and 2 (the mature *psaA* mRNA is formed by trans-splicing of three distantly located transcripts) was found in Chlamydomonas.[36] Trans-splicing was reported for the colorless euglenoid alga *Entosiphon sulcatum* (Euglenophyta)[49] and is likely to be present in the red alga *Rhodella violacea;* however, there is no evidence for the presence of other *tscA*-type bridging molecules.[36]

Polyadenylation of mRNAs preceded by specific endonuclease cleavage has been demonstrated in *Ch. reinhardtii* chloroplasts. The plastid gene named rne for its similarity to the *E. coli* RNase E is involved in this process. The chloroplast *rne* genes were discovered in the same algal genomes that contain the *rnpB* gene. In land plants, the *rne* genes are located in the nucleus.

In land plants and algae, the prokaryotic *fts* and *min* genes involved in plastid division and cpDNA replication have been discovered. The products of the *min* genes-MinD and MinE-interact with the tubulin-like FtsZ protein, which participates in plastid division.[50] FtsZ is encoded by a nuclear gene family. The *minD* and *minE* genes in *C. vulgaris* and *G. theta* are located in cpDNA.

In the plastid genomes of cryptomonads[45] and primitive red microalga *Cyanidioschyzon merolae,*[51,52] the *hlpA* and *dnaB* genes also related to organelle division have been identified. The *hlpA* product in cryptomonads, HlpA, is a functional homolog of the *E. coli* HU- and HMG1-like proteins. The location of genes responsible for plastid division (*minDE*) and DNA packaging (*hlpA*) in the same genome is the unique trait of the *G. theta* cpDNA outside bacteria.

The *dnaB* gene participating in *E. coli* DNA replication and coding for the DNA helicase was found in cpDNA of most nongreen algae. It is of particular interest that the chloroplast *dnaB* genes are very similar to their cyanobacterial homologs; however, no DnaB proteins have so far been discovered in chloroplasts. Meanwhile, the *A. thaliana dnaB* gene is nuclear, its transcript is formed in the cytoplasm, and its product appears to be mitochondrially targeted.[36]

Genes homologous to those of apicomplexan parasites have been found in algal cpDNAs. Thus, for example, a widespread bacterial *ycf24* gene detected in cpDNAs of all nongreen algae is highly homologous to ORF470 of *Plasmodium falciparum.* The data on *Synechocystis* sp. PCC 6803[37] indicate that *ycf24/ORF470* is probably involved in plastid division, and is thus a constitutive gene.

The cpDNA of all nongreen algae encode the metalloprotease FtsH. In land plants, this enzyme is nuclear-encoded. The FtsH proteins play an important, though not completely understood, role in chloroplast biogenesis.

In some algae, cpDNA encodes tmRNA, small RNA molecules that combine the functions of tRNA and mRNA and are widespread in bacteria. These tmRNAs encode proteolysis-inducing tag-peptides, which bind to the 3' end of damaged mRNA molecules during translation, initiating proteolysis of the defective polypeptide. Such a fascinating mode of proteolysis is unknown elsewhere in eukaryotes.[36]

Finally, plastid-containing eukaryotes carry the *hsp70* gene coding for the 70-kDa heat shock protein. In different algal lineages (*O. sinensis, G. theta, C. paradoxa, P. purpurea, C. caldarium, Pavlova lutherii,* and *Ch. reinhardtii*), this gene is chloroplast-encoded and highly homologous to the cyanobacterial *dnaK.* In land plants (*Spinacia oleracea, Citrullus lanatus, Cucumis sativus, Pisum sativum,* and *A. thaliana*), it is nuclear-encoded.

Sequencing of plastid genomes of land plants and algae, complete or partial sequencing of the cyanobacterial genomes, as well as progress in determination of the primary structure of nuclear genomes of plants provide convincing evidence in favor of the endosymbiotic origin of plastids. Present-day chloroplasts arose from a cyanobacterial endosymbiont.[2,52,53] Phylogenetic analysis of homologous genes in plastids and cyanobacteria and the comparison of gene cluster arrangement in cyanobacteria and plastids of the photosynthetic eukaryotes indicate that plastids have a monophyletic origin, being derived from a single cyanobacterial ancestor. The observation that plastid genomes of different land plants and algae contain similar gene complements also supports the monophyletic origin of plastids. Among the photosynthetic eukaryotes examined thus far, the multicellular red alga *P. purpurea* has the most ancient genome: the gene content in its cpDNA is the highest (252), and gene clusters are most close to those of cyanobacteria. During the conversion of a free-living cyanobacterial ancestor into highly specialized organelles, chloroplasts lost their autonomy and a considerable part of their genetic material. They became semiautonomous and preserved, according to different data, only 1 to 10% of genes found in the genomes of free-living cyanobacteria (for example, the *Synechocystis* PCC6803 genome contains 3,168 genes).[37] This means that many genes of the endosymbiont were lost during evolution and/or were transferred to the nucleus. In all likelihood mainly duplicated genes, which were already present in the eukaryotic host before the symbiotic event or were useless for the existence of the endosymbiont, were lost.

Phylogenetic analysis of nine taxa, including land plants and some algal species, showed that during evolution simultaneous gene losses in multiple independent lineages outnumber the phylogenetically unique losses by more than 4:1. Studies on nine chloroplast genomes demonstrated that 44 genes were lost by chloroplasts twice independently, 43 genes have undergone three parallel losses, and 14 genes were lost four times in independent lineages. In some instances, whole gene blocks (for example, *apc*, *cpc*, and *cpe* encoding phycobilisome components) were lost in algal lineages Chlorophyta and Stramenopiles.[54]

During evolution, some genes were transferred from the chloroplasts to the nucleus.[55-57] The products of these genes are reimported by chloroplasts with the help of transit peptides and a specific protein import mechanism. As a result, the proteins are retained in the chloroplasts, while the corresponding genes are concentrated in the nuclear chromosomes. The presence of only two membranes that surround chloroplasts allows these organelles to take up cytosolic protein precursors. This process is accompanied by the cleavage of transit peptides and the release of the processed polypeptides into the stroma. Similar events take place in mitochondria, which are also of endosymbiotic origin. The first convincing evidence for the endosymbiotic gene transfer was obtained about 10 year ago.[58] Plastids contain much more proteins than cpDNA encodes. The estimates of the total number of proteins in plastids differ considerably. According to some data, differentiated plastids of different types contain about 2000-2500 proteins.[59] Other authors (engaged in *Arabidopsis* studies) assume that their number may amount to 5000.[58] From the discrepancy between the number of genes in cpDNA and the number of proteins necessary for chloroplast functioning, it follows that plastids import a great number of proteins.

How many nuclear genes descend from cyanobacterial genomes? A comparative phylogenetic analysis of the genomes of *A. thaliana* and three cyanobacterial species (*Synechocystis* PCC6803, *Prochlorococcus marinus*, and *Nostoc punctiforme*) showed that of 24,990 proteins encoded by the nuclear genome in *Arabidopsis*, 4500 (18%) are encoded by their cyanobacterial homologs. These proteins belong to all functional classes, and most of them are imported from the nucleus to different cell compartments but usually not to chloroplasts. Furthermore, a great number of proteins of noncyanobacterial origin are imported from the nucleus to the chloroplasts. Therefore, there is no strict correlation between gene origin and protein compartmentation. Nuclear genes derived from the plastid ancestor have evolutionarily acquired new functions unusual for cyanobacteria (for example, pathogen resistance, protein secretion, etc.).[60] Genes for 117 proteins encoded by the nuclear genome of *Arabidopsis* are still found in the chloroplast genomes of land plants and algae. There are 44 chloroplast genes whose homologs are in the nuclear genome of at least one higher plant species, which supports endosymbiotic gene transfer to the nucleus.[54] Gene transfer is an ongoing process. It cannot be excluded that in millions of years, the genomes of cell organelles will contain even less genes or their genomes will be completely assimilated by the nucleus. Hydrogenosomes, ATP-producing organelles surrounded by a double membrane, have been discovered in amitochondriate protists. They originate from the same symbionts as mitochondria; however, they lack their own genome, which was completely assimilated by the nuclear genome. Thus, hydrogenosomes provide an example of a completed endosymbiotic gene transfer.

The genes for the regulatory proteins are supposed to be the first transferred to the nucleus, the genes for enzymes and structural proteins followed.[58] Multisubunit chloroplast proteins are the examples. Thus, the γ-subunit of the chloroplast ATPase (atpC) plays a regulatory role in the complex. The genes for the structural subunits (*atpA, atpB, atpE, atpF,* and *atpH*) are found in all cpDNAs of land plants, but the regulatory subunit atpC is nuclear-encoded (*atpC*) in all plants examined. The *atpC* gene is therefore supposed to be the first gene of this complex to have been transferred. The same phenomenon is observed in the chloroplast ATP-dependent protease Clp. The catalytic subunit (ClpP) is cpDNA-encoded in land plants, whereas the regulatory subunit (ClpC) is nuclear-encoded. The Rubisco catalytic subunit (rbcL) is encoded by the chloroplast genome, while the regulatory subunit rbcS is nuclear-encoded. This tendency holds for the plastid RNA polymerase involved in transcription of plastid genes whose σ-factors are nuclear-encoded. The examples mentioned above point to the general phenomenon: the regulatory genes are concentrated in the nucleus.[58]

Structural and functional studies on the chloroplast genomes, started about 50 years ago, progressed considerably owing to the development of such technologies as gene cloning and DNA sequencing. At first, the efforts were aimed at determining the complete primary structure of cpDNA from land plants. Further analysis of algal cpDNA, especially of nongreen algae, revealed several interesting features of cpDNA evolution.

Plastid genome organization indicates that all plastids are descendants of a single cyanobacterial ancestor. However, the details of endosymbiotic events remain obscure. In addition to morphological and biochemical markers, diverse molecular markers help one to unravel the evolutionary pathways leading to the establishment of plastids and to elucidate the early evolutionary history of these organelles. Sequencing of the plastid genomes of photosynthetic eukaryotes with special emphasis on certain algal species, since they retained more cyanobacterial traits than plastid genomes of land plants, seems most promising. Comparative analysis of the structure and composition of genes and gene clusters, gene order, the size of intergenic spacers, etc. is a useful means of determining the phylogenetic relationships between photosynthetic organisms, reconstructing the evolutionary events that led to present-day algal and land plant lineages, and obtaining more accurate estimates of the divergence times for the major land plant species.

In this review we focused on the recent data on the structure and functions of the chloroplast genome of land plants and algae, and only shortly discussed the problems of the origin and evolution of plastids. These problems deserve special analysis.

Acknowledgements

This work was supported by the Russian Foundation for Basic Research (project no. 03-04-49051).

References

1. Leister D. Chloroplast research in the genomic age. Trends in Genetics 2003; 19:47-56.
2. Raven JA, Allen JF. Genomics and chloroplast evolution: What did cyanobacteria do for plants? Genome Biology 2003; 4:209.1-209.5.
3. Wakasugi T, Tsudzuki T, Sugiura M. The genomics of land plant chloroplasts: Gene content and alteration of genomic information by RNA editing. Photosynthesis Res 2001; 70:107-118.
4. Cosner ME, Jansen RK, Palmer JD et al. The highly rearranged chloroplast genome of Trachelium caeruleum (Campanulaceae): Multiple inversions, inverted repeat expansion and contraction, transposition, insertions/deletions, and several repeat families. Curr Genet 1997; 31:419-429.
5. Lilly JW, Havey MJ, Jackson SA et al. Cytogenomic analyses reveal the structural plasticity of the chloroplast genome in land plants. The Plant Cell 2001; 13:245-254.
6. Shinozaki K, Ohme M, Tanaka M et al. The complete nucleotide sequence of the tobacco chloroplast genome: Its gene organization and expression. EMBO J 1986; 5:2043-2049.
7. Ohyama K, Fukuzawa H, Kohchi T et al. Chloroplast gene organization deduced from complete sequence of liverwort Marchantia polymorpha chloroplast DNA. Nature 1986; 322:572-574.

8. Hiratsuka J, Shimada H, Whittier R et al. The complete sequence of the rice (Oryza sativa) chloroplast genome: Intermolecular recombination between distinct tRNA genes accounts for a major plastid DNA inversion during the evolution of the cereals. Mol Gen Genet 1989; 217:185-194.
9. Ogihara Y, Isono K, Kojima T et al. Structural features of a wheat plastome as revealed by complete sequencing of chloroplast DNA. Mol Genet Genomics 2002; 266:740-746.
10. Sugiura M, Hirose T, Sugita M. Evolution and mechanism of translation in chloroplasts. Annu Rev Genetics 1998; 32:437-459.
11. Sato S, Nakamura Y, Kaneko T et al. Complete structure of the chloroplast genome of Arabidopsis thaliana. DNA Res 1999; 6:283-290.
12. Kato T, Kaneko T, Sato S et al. Complete structure of the chloroplast genome of a legume, Lotus japonicus. DNA Res 2000; 7:323-330.
13. Knox EB, Palmer JD. The chloroplast genome arrangement of Lobelia thuliniana (Lobeliaceae): Expansion of the inverted repeat in an ancestor of the Campanulales. Pl Syst Evol 1999; 214:49-64.
14. Hupfer H, Swiatek M, Hornung S et al. Complete nucleotide sequence of the Oenothera elata plastid chromosome, representing plastome I of the five distinguishable Euoenothera plastomes. Mol Gen Genet 2000; 263:581-585.
15. Schmitz-Linneweber CH, Maier R, Alcaraz JP et al. The plastid chromosome of spinach (Spinacia oleracea): Complete nucleotide sequence and gene organization. Plant Mol Biol 2001; 45:307-315.
16. Katayama H, Ogihara Y. Structural alterations of the chloroplast genome found in grasses are not common in monocots. Curr Genet 1993; 23:160-165.
17. Maier RM, Neckermann K, Igloi GL et al. Complete sequence of the maize chloroplast genome: Gene content, hotspots of divergence and fine tuning of genetic information by transcript editing. J Mol Biol 1995; 251:614-628.
18. Douglas SE. Chloroplast origins and evolution. In: Bryant A, ed. The Molecular Biology of Cyanobacteria. Amsterdam: Kluwer, 1994:91-118.
19. Wakasugi T, Sugita M, Tsudzuki T et al. Updated gene map of tobacco chloroplast DNA. Plant Mol Biol Rep 1998; 16:231-241.
20. Kowallik KV. Origin and evolution of chloroplasts: Current status and future perspectives. In: Schenk H, Herrmann RG, Jeon KW et al, eds. Eukaryotism and Symbiosis. Intertaxonic Combination versus Symbiotic Adaptation. Berlin: Springer, 1997:3-23.
21. Wolfe KH, Morden CW, Palmer JD. Function and evolution of a minimal plastid genome from a nonphotosynthetic parasitic plant. Proc Natl Acad Sci USA 1992; 89:10648-10652.
22. Bommer D, Haberhausen G, Zetsche K. A large deletion in the plastid DNA of the holoparasitic flowering plant Cuscuta reflexa concerning two ribosomal proteins (rpl2, rpl23), one transfer RNA (trnI) and an ORF2280 homologue. Curr Genet 1993; 24:171-176.
23. Odintsova MS, Yurina NP. RNA editing in chloroplasts and plant mitochondria. Genetika 2000; 37:307-320.
24. Schmitz-Linneweber C, Regel R, Du TG et al. The plastid chromosome of Atropa belladonna and its comparison with that of Nicotiana tabacum: The role of RNA editing in generating divergence in the process of plant speciation. Mol Biol Evol 2002; 19:1602-1612.
25. Wakasugi T, Nagai M, Kapoor M et al. Complete nucleotide sequence of the chloroplast genome from the green alga Chlorella vulgaris: The existence of genes possibly involved in chloroplast division. Proc Natl Acad Sci USA 1997; 94:5967-5972.
26. Turmel M, Otis CH, Lemieux C. The complete chloroplast DNA sequence of the green alga Nephroselmis olivacea: Insights into the architecture of ancestral chloroplast genomes. Proc Natl Acad Sci USA 1999; 96:10248-10253.
27. Sugiura M. The discovery of the complete sequence of tobacco and rice chloroplast genomes. In: Kung S-D, Yang S-F, eds. Discoveries in Plant Biology, Vol. 2. Singapore: World Scientific, 1998:45-60.
28. Freyer R, Neckermann K, Maier RM et al. Structural and functional analysis of plastid genomes from parasitic plants: Loss of an intron within the genus Cuscuta. Curr Genet 1995; 27:580-586.
29. Wakasugi T, Tsudzuki J, Ito S et al. Loss of all ndh genes as determined by sequencing the entire chloroplast genome of the black pine Pinus thunbergii. Proc Natl Acad Sci USA 1994; 91:9794-9798.
30. Millen RS, Olmstead RG, Adams KL et al. Many parallel losses of infA from chloroplast DNA during angiosperm evolution with multiple independent transfers to the nucleus. Plant Cell 2001; 13:645-658.
31. Guera A, De Nova PG, Sabater B. Identification of the Ndh(NAD(P)H-plastoquinone-oxidoreductase) complex in etioplast membranes of barley: Changes during photomorphogenesis of chloroplasts. Plant Cell Physiol 2000; 41:49-59.
32. Stoebe B, Martin W, Kowallik KV. Distribution and nomenclature of protein-coding genes in 12 sequenced chloroplast genomes. Plant Mol Biol Rep 1998; 16:243-255.

33. Bhattacharya D, Medlin L. Algal phylogeny and the origin of land plants. Plant Physiol 1998; 116:9-15.
34. Douglas SE. Plastid evolution: Origins, diversity, trends. Curr Opin Genet Dev 1998; 8:655-661.
35. Palmer JD. A single birth of all plastids? Nature 2000; 405:32-33.
36. Simpson CL, Stern DB. The treasure trove of algal chloroplast genomes. Surprises in architecture and gene content, and their functional implications. Plant Physiol 2002; 129:957-966.
37. Kaneko T, Sato S, Kotani H et al. Sequence analysis of the genome of the unicellular cyanobacterium Synechocystis sp. strain PCC 6803. II. Sequence determination of the entire genome and assignment of potential protein-coding regions. DNA Res 1996; 3:109-136.
38. Maul JE, Lilly JW, Liying Cui et al. The Chlamydomonas reinhardtii plastid chromosome. Islands of genes in a sea of repeats. The Plant Cell 2002; 14:2659-2679.
39. Hallick RB, Hong L, Drager RG et al. Complete sequence of Euglena gracilis chloroplast DNA. Nucl Acids Res 1993; 21:3537-3544.
40. Gockel G, Hachtel W. Complete gene map of the plastid genome of the nonphotosynthetic euglenoid flagellate Astasia longa. Protist 2000; 151:347-351.
41. Zhang Z, Green BR, Cavalier-Smith T. Single gene circles in dinoflagellate chloroplast genomes. Nature 1999; 400:155-159.
42. Glockner G, Rosenthal A, Valentin K. The structure and gene repertoire of an ancient red algal plastid genome. J Mol Evol 2000; 51:382-390.
43. Lemieux CX, Otis CH, Turmel M. Ancestral chloroplast genome in Mesostigma viride reveals an early branch of green plant evolution. Nature 2000; 403:649-652.
44. Turmel M, Otis C, Lemieux C. The chloroplast and mitochondrial genome sequences of the charophyte Chaetosphaeridium globosum: Insights into the timing of the events that restructured organelle DNAs within the green algal lineage that led to land plants. Proc Natl Acad Sci USA 2002; 99(17):11275-11280.
45. Douglas SE, Penny SL. The plastid genome of the cryptophyte alga, Guillardia theta: Complete sequence and conserved synteny groups confirm its common ancestry with red algae. J Mol Evol 1999; 48:236-244.
46. Reith M, Munholland J. Complete nucleotide sequence of the Porphyra purpurea chloroplast genome. Plant Mol Biol Rep 1995; 13:333-335.
47. Baum M, Cordier A, Schon A. RNase P from a photosynthetic organelle contains an RNA homologous to the cyanobacterial counterpart. J Mol Biol 1996; 257:43-52.
48. Stirewalt VL, Michalowski CB, Löffelhardt W et al. Nucleotide sequence of the cyanelle genome from Cyanophora paradoxa. Plant Mol Biol Rep 1995; 13:327-332.
49. Ebel C, Frantz C, Paulus F et al. Trans-splicing and cis-splicing in the colourless euglenoid, Entosiphon sulcatum. Curr Genet 1999; 35:542-550.
50. Osteryoung KW. Organelle fission. Crossing the evolutionary divide. Plant Physiol 2000; 123:1213-1216.
51. Ohta N, Sato N, Kuroiwa T. Analysis of the plastid genome of protoflorideophyceous algae Cyanidioschizon merolae. Plant Cell Physiol 1998; 39(Suppl 01):54.
52. Martin W, Stoebe B, Goremykin V et al. Gene transfer to the nucleus and the evolution of chloroplasts. Nature 1998; 393:162-165.
53. Zerges W. Does complexity constrain organelle evolution? Trends in Plant Sci 2002; 7(4):175-181.
54. Douglas AE, Raven JA. Genomes at the interface between bacteria and organelles. Phil Trans R Soc Lond B 2003; 358:5-18.
55. Martin W, Herrmann RG. Gene transfer from organelles to the nucleus: How much, what happens, and why? Plant Physiol 1998; 118:9-17.
56. Doolittle WF, Boucher Y, Nesbo CL et al. How big is the iceberg of which organellar genes in nuclear genomes are but the tip? Phil Trans. R Soc Lond B 2003; 358:39-57.
57. Stegemann S, Hartmann S, Ruf S et al. High-frequency gene transfer from the chloroplast genome to the nucleus. Proc Natl Acad Sci USA 2003; 100:8828-8833.
58. Allen JF. The function of genomes in bioenergetic organelles. Phil Trans R Soc Lond B 2003; 358:19-37.
59. Steiner JM, Löffelhardt W. Protein import into cyanelles. Trends in Plant Sci 2002; 7:72-77.
60. Martin W, Rujan T, Richly E et al. Evolutionary analysis of Arabidopsis, cyanobacterial, and chloroplast genomes reveals plastid phylogeny and thousands of cyanobacterial genes in the nucleus. Proc Natl Acad Sci USA 2002; 99:12246-12251.

CHAPTER 7

Comparison of the Immobilization Techniques for Photosystem II

Régis Rouillon,* Sergey A. Piletsky, Elena V. Piletska, Pierre Euzet and Robert Carpentier

Abstract

The main methods of immobilization employed to stabilize the life time of photosynthetic material are studied. Various parameters and properties concerning the immobilization procedures are evaluated: method, biological material, techniques to measure the photosynthetic activity, storage and operational stabilities. A comparison between two methods of immobilization (chemical and physical) to measure the effect of herbicides which inhibited photosynthesis is discussed. The practical implication of photosystem II is emphasized.

Introduction

Absorption and conversion of solar energy to chemical energy occur with an extremely high degree of efficiency within the photosynthetic membrane. The process begins with absorption of light by the antenna pigments, leads to charge separation in the reaction centres and produces reducing equivalents that are used to drive the biochemical processes of the organism.[1] The photosystem II complex (PSII) is essential to the regulation of photosynthesis since it catalyses the oxidation of water into oxygen and supports electron transport.[2] The above properties confer to photosynthetic materials and in particular to PSII a great potential for various biotechnological applications. Among the agricultural pollutants, herbicides inhibiting photosynthesis via targeting PSII function still represent the basic means of weed control.[3] This group consists of several classes of chemicals such as triazines (e.g., atrazine), phenylureas (e.g., diuron), or phenols (e.g., bromoxynil). In 45 references cited in this chapter, immobilized photosynthetic materials were used 21 times to detect herbicides in samples. In more of the substances which form part of the classes mentioned previously, it was possible also to detect other herbicides which have other modes of action (alachlor,[4] glyphosate[4,5]) and even insecticides.[6] The industrial pollution causes the release into the environment of toxic substances such as metals (Pb, Cd, Cr, Cu, Ni, Hg, Zn).[7] Although toxic metal cations may cause a wide variety of deleterious effects when taken up by plants, the photosynthetic apparatus appears to be especially sensitive to their toxicity. Photosystem II is particularly affected by these contaminants.[8,9] In view of these findings, the use of photosystem II as the biological receptor in biosensors provided an excellent tool for the detection of toxic metal cations.[10-16] Methanol vapors can be also detected with whole microalgae.[17] Among other interesting biotechnological applications using photosynthetic materials, one can find the nitrite uptake by *Chlamydomonas reinhardtii* cells growing in airlift reactors,[18] or the production of hydrogen gas through biophotolysis of water by chloroplasts membranes associated with native or

*Corresponding Author: Régis Rouillon—Université de Perpignan, Centre de Phytopharmacie, 52 Avenue Paul Alduy 66860 Perpignan, France. Email: rouillon@univ-perp.fr

Biotechnological Applications of Photosynthetic Proteins: Biochips, Biosensors and Biodevices, edited by Maria Teresa Giardi and Elena V. Piletska.

immobilized hydrogenase.[19] However, the isolated photosynthetic materials are affected by a relatively short active life time that limits their use. To circumvent this limitation, various immobilization techniques have been designed to improve the stability of biological functions. The objective of this study is to provide an overview of the main immobilization procedures used to stabilize photosynthetic materials. In this review we will discuss: (i) the main methods of immobilization; (ii) the choice of the biological material; (iii) the various techniques used to measure the photochemical activity, and (iv) the storage and operational activities after immobilization. A comparative study of thylakoid membranes sensitivity for herbicide detection after physical or chemical immobilization will be developed.

Main Methods of Immobilization

The immobilization methods are identified either as chemical or physical procedures depending on whether covalent bonds are established or not.[20]

Physical methods involve adsorption of the photosynthetic material on a support or the inclusion in a natural or synthetic gel (Table 1). Various supports were used: filter paper discs,[21] alumina filter discs,[11,22-24] glass microfibre filters,[4,6,17] or a column containing diethylaminoethyl (DEAE)-cellulose.[25] The immobilization was obtained by filtration on the support. After adsorption, the filter containing the biological material can be protected with a thin alginate layer that is hardened with $CaCl_2$[21] or with another filter disc.[11] The immobilization of photosynthetic material by adsorption on an inert or ion exchanger support is a simple, economic, soft technique that preserves the native activity of the material. However, the interaction forces which intervene are weak, hence desorption of the biocatalyst may be observed.

The immobilization by inclusion in a polymeric network, a technique rarely employed for enzymes because of the problems of salting out, was also developed for the immobilization of photosynthetic materials. The biocatalyst is distributed in the three-dimensional network of the gel. Natural or synthetic gels can be also employed to carry out a physical immobilization of the photosynthetic

Table 1. Main procedures to immobilize photosynthetic materials

Physical	Chemical
Adsorption	**Reticulation**
• filter paper disk	• glutaraldehyde
• alumina filter disk	
• glass microfibre filter	**Coreticulation: glutaraldehyde - different proteins**
• DEAE cellulose	
Gel inclusion	• gelatin
polysaccharide gel	• collagen
• agar	• bovine serum albumine
• agarose	
• carrageenan	
• alginate	
protein gel	
• gelatin	
synthetic gel	
• polyacrylamide	
• polyurethane	
• photocrosslinkable resin	
• vinyl	
• poly(vinylalcohol)	
• poly(vinyalcohol) bearing styrylpyridinum groups	

material. Among the natural gels, polysaccarides were used; seldom agar,[14] agarose,[3] carrageenan,[26] and in many works alginate.[3,18,23,26-31] This latter gel was often employed with a $CaCl_2$ solution to solidify the mixture. The addition of the mixture photosynthetic material-alginate in the shape of droplets in a solution of $CaCl_2$ leads to formation of gel beads. The alginate-biological material mixture can be done at low temperature whereas other gels like carrageenan, agar, or agarose must be heated. Gelatin is a protein gel which has interesting properties for the entrapment of biological materials.[3] Many synthetic gels were also employed with different results on the storage stability and the residual activity after entrapment. Among these synthetic products, one can find polyacrylamide gel,[32] polyurethane prepolymer (PU-3),[26,27,33] photo-crosslinkable resin prepolymer (ENT-4000),[26] vinyl monomers (HEA or M-23G),[34] poly(vinylalcohol) polymers,[35-37] poly(vinylalcohol) bearing styrylpyridinium polymers (PVA-SbQ).[13,15,16,31,38-42] To realize the entrapment, the polymerisation of some synthetic gels was initiated by specific chemicals (polyacrylamide gel) or by irradiation with UV for PVA-SbQ, white light for photo-crosslinkable resin prepolymer, γ-rays for vinyl monomers. The synthetic gels can be toxic. Bovine serum albumin (BSA) was used as a protectant and added to the mixture before immobilization with urethane prepolymer, vinyl monomers, poly(vinylalcohol), and PVA-SbQ polymers. BSA was shown to be very effective in preserving the activity of isolated chloroplasts stored in vitro.[43] An addition of BSA largely prevented the decrease of electron transport activity. For a maximal protective effect the albumin molecules have to be present during the homogenisation step and continuously in the photosynthetic preparation.[26] The concentration of BSA is important to obtain the best results. The effect of different BSA concentrations (0.1, 5 and 10%) on the storage stability at + 4°C in the dark was shown for a mixture of thylakoids and chloroplasts entrapped in PVA-SbQ. After a 7-day storage, there was no significant difference between the various concentrations. After a 60-day storage, the optimal activity was achieved with 1% BSA.[39]

The method of immobilization in a polyurethane gel is at the same time a physical method of immobilization by inclusion and a chemical one. Indeed, the isocyanate groups of the urethane prepolymer react with-OH functions of water and with particular-NH_2 functions of proteins located in the photosynthetic membrane structures. This chemical reaction is still reinforced by the addition of BSA to the prepolymer.

Covalent fixing on a support is a technique which was used often for the immobilization of enzymes but remained scarcely used during a long period for photosynthetic materials because of the loss of activity due to the denaturing effect of the bonding agent. But, glutaraldehyde, a common fixative for electron microscopy, was shown to preserve some Hill reaction activity in chloroplasts.[44] This product was often used as a chemical method to immobilize photosynthetic materials. In the case of reticulation, glutaraldehyde can be employed directly to provide the immobilization of the biological material.[45-49] In the case of the coreticulation, the denaturing effect of the reticulating agent can be decreased by the addition of proteins during the polymerization: gelatin,[26] collagen,[50] or bovine serum albumin (BSA-Glu).[3,5,10,12,26,27,31,33,42,51-54] Glutaraldehyde builds a network of covalent bonding mainly with the free-NH_2 groups of both photosynthetic material and added proteins. The role of the exogenous proteins is mainly to protect the photosynthetic material against too many bonding so that an adequate biological function is retained.

Various Immobilized Photosynthetic Materials

Figure 1, shows the distribution of 45 references[3-6,10-18,21-42,45-54] cited in this chapter according to the biological material used for the immobilization. The isolation of photosystem II submembrane fractions is time consuming. In consequence, this material was used in only three of the cited reports: spinach PSII particles immobilized in glutaraldehyde-bovine serum albumin[52] and in poly(vinyalcohol) bearing styrylpyridinum groups[16] or PSII particles prepared from the thermophilic cyanobacterium *Synechocococcus elongatus* stabilized with various procedures (gelatin, agarose, alginate and BSA-glu).[3]

On the contrary, the immobilization of whole cells can be performed in a relatively straightforward way. 14 references on 45 used the intact cells: 4 employed cyanobacteria only,[15,23,29,41] 8 immobilized eukaryotic microalgae,[4,11,14,17,18,21,24,28] and 2 both cyanobacteria and microalgae.[6,22] Among

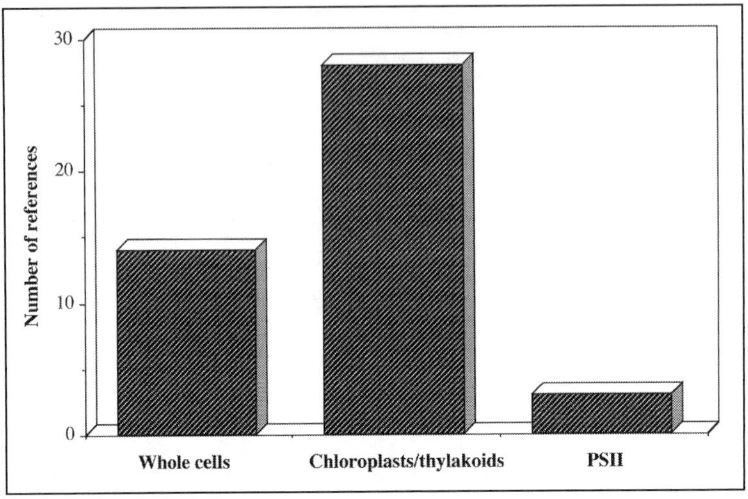

Figure 1. Distribution of 45 references cited in this chapter according to the photosynthetic material used for the immobilization.

cyanobacteria, *Synechococcus* strains were often immobilized: PCC6301,[22,23,29] or PCC 7942.[15] Among immobilized microalgae, *Chlorella vulgaris*[4,6,11,17,24] *Scenedesmus subspicatus*,[11,21] and *Selenastrum capricornutu*[11,14] were used. In all case the whole cells were immobilized with physical procedures.

The immobilization of intact cells is easy, although some disadvantages exist owing to the impermeability of the cell envelope to electrolytes, and the possibility of side reactions which may consume reactants and products. Partial or total enzymatic hydrolysis with lysozyme of the peptidoglycan polymer that forms the rigid cell wall of cyanobacteria yield cells that are permeable to ions. This enzymatic method of cyanobacteria cell wall disruption is useful where it is possible to preserve the photosynthetic activity.[20] It was shown that such permeabilization strongly increased the inhibition of photosynthetic activity by $HgCl_2$.[41] These cells of microalgae and cyanobacteria come from cultures which are not always realizable in a laboratory.

Between photosystem II submembrane fractions which must be isolated and whole cells which must be cultivated, many immobilization works (28 on 45 references cited) employed chloroplasts and thylakoids: intact chloroplasts, mixtures of chloroplasts and photosynthetic membranes, or specially, thylakoids alone. To obtain these photosynthetic membranes, the chloroplasts were subjected to osmotic shock by briefly placing them in a hypotonic medium. This procedure ruptures the chloroplast envelope and releases the stroma. The stripped chloroplasts were then returned to an isotonic medium. The interest of this photosynthetic material is justified by a direct contact between the reaction sites and the operation medium. The photosynthetic preparations were often obtained from spinach leaves.

Measure of PSII Activity after Immobilization

The PSII complex is known to be particularly sensitive to many substances and its activity is studied firstly. Different methods were used to detect the activity of PSII after immobilization of photosynthetic material (Fig. 2). There is no relation between the method of immobilization and the measurement technique of the photosynthetic activity.

The measurement of oxygen evolution using a Clark-type oxygen electrode is a standard procedure in the field of photosynthesis research. The measure is simple and the oxygen electrode constitutes an interesting method to evaluate the photosynthetic activity. Several types of photosystem II specific artificial electron acceptors can be used as electroactive mediators to maximize the photosynthetic activity. The more used mediator is potassium ferricyanide[26,34,37,46-49] but this one

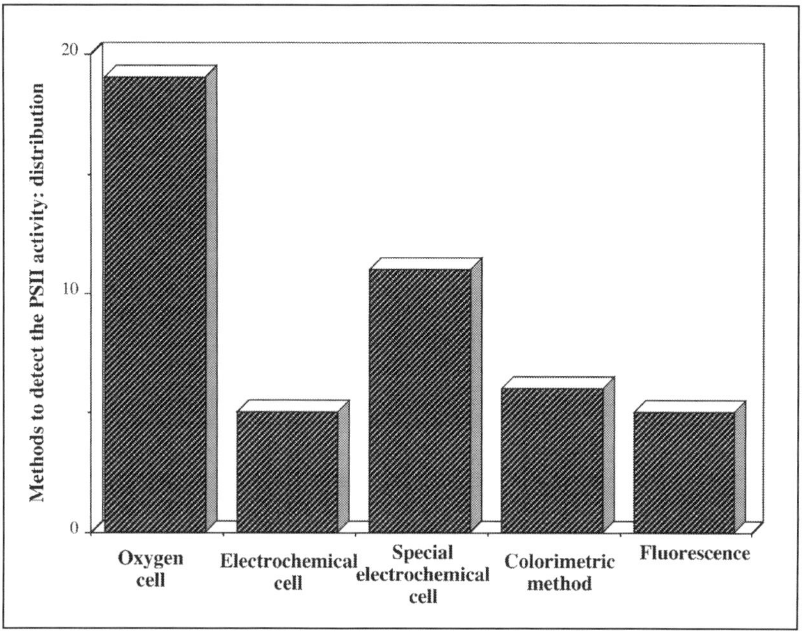

Figure 2. Distribution of methods* used in the references cited to detect PSII activity after immobilization. *Several methods are sometimes used in the same publication.

presents two principal disadvantages. Thermodynamically, potassium ferricyanide has a capacity to accept electrons directly from PSII and indeed it does so efficiently in subchloroplasts particles. However it was shown that in unfragmented thylakoid membranes ferricyanide acts more as a PSI electron acceptor than as a PSII acceptor.[55] In eukaryotic photosynthetic organisms, the photosynthetic electron transport sites are located on thylakoids stacks within the membrane-bound chloroplasts and access to these sites is restricted to lipophilic mediators. So potassium ferricyanide is not effective for whole cells of eukaryotic microalgae and intact chloroplasts. Substituted benzoquinones such as parabenzoquinone[27,38,39] or 2,5-dichlorobenzoquinone[12] can be employed to improve the PSII activity. The presence of ammonium chloride as uncoupling agent in the medium also makes it possible to increase the activity.[26,27,37-39] The presence of CdCl2 increases the activity of photosystem II at the level of electron transfer and of fluorescence.[1]

The measurement of the PSII activity can be based on amperometric detection of the reduced form of a specific artificial electron acceptor in an electrochemical cell. The mediator must be present in its oxidized form in the measuring buffer. Under illumination, this artificial electron acceptor is reduced and then reoxidized on the surface of a working electrode. Different mediators were used: potassium ferricyanide,[3,22,23] parabenzoquinone,[11,22] duroquinone,[3] 2,6 dimethylbenzoquinone.[11] With whole cells of eukaryotic algae, potassium ferrycianide, methyl-viologen, and DCIP were unable to access the photosynthetic electron transfer chain.[11]

A special electrochemical cell was often employed to detect the photosynthetic activity of immobilized thylakoids.[5,10,13,40,42,52-54] With this electrochemical micro-cell, it was possible to detect a photocurrent of free or immobilized thylakoid membranes in the absence of exogenous artificial electron acceptor. The photosynthetic origin of the photocurrent was demonstrated by an inhibition in the presence of the photosynthetic inhibitor diuron.[53] The inhibition of the photocurrent by catalase, an enzyme that degrades hydrogen peroxide, demonstrated that dissolved oxygen was involved.[10] The ability of oxygen to act as an electron acceptor for the photosynthesis membrane was initially demonstrated by Mehler.[56] This process, known as pseudocyclic electron transport or Mehler reaction is characterized by the reduction of oxygen on the reducing side of photosystem I to form superoxide ions. The latter dismute either enzymatically or spontaneously to hydrogen peroxide. Hydrogen peroxide is therefore oxidized at the working electrode. To use this electrochemical micro-cell for the measurement of photosynthetic activity in whole cyanobacteria cells[15,41] or photosystem II submembrane fractions[16,52] an artificial electron acceptor must be present in the electrolytic solution for charge transfer between the photosynthetic material and the working electrode. In these last works, the photosynthetic activity was evaluated with the artificial electron acceptor 2,5-dichlorobenzoquinone (DCBQ) which was reduced by photosystem II.

The difference in color between the oxidized and reduced forms of some mediators enable their use in colorimetric measurements of the photosynthetic activity. For example, under illumination, thylakoids after immobilization can reduce the blue 2,6 dichorophenolindophenol to the colorless leuco form.[25,30,31,35,36]

Chlorophyll a fluorescence induction is a widespread method to evaluate the photosynthetic activity.[57] This method is noninvasive, highly sensitive, fast, and easily measured. When chlorophyll molecules in photosystem II absorb light, that light may be assimilated into the light reactions of photosynthesis or may be released as fluorescence or heat energy. In vivo fluorescence increases when photosynthesis declines or is inhibited. Numerous environmental factors can affect the rate of electron transport between photosystem II and photosystem I due to interference with electron carriers between the two photosystems.[58] For example, when the diuron is added in the measured sample, electron transport from photosystem II to photosystem I is blocked resulting in maximum fluorescence. This method was often employed to detect the photosynthetic activity of immobilized photosynthetic material.[4,6,21,28,37]

Photosystem II Activity after Immobilization

The method of immobilization must answer several closely interdependent criteria: to preserve the photosynthetic activity, to improve the stability of this activity, and to preserve the morphological structure.

It was observed that the specific activity of PSII of chloroplasts immobilized in DEAE-cellulose was higher than those of the free chloroplats.[25] The activities of PSI and PSII catalyzed by free or immobilized chloroplasts were both insensitive to uncouplers after adsorption indicating that the immobilized chloroplasts had already been uncoupled.

Examples of PSII activities (oxygen production) after immobilization of thylakoids according to the procedure are shown in Table 2. As the uncoupling effect of each method varies, an uncoupler (NH_4Cl) was added to the reaction medium in order to obtain a well defined activity yield,

Table 2. PSII activity (oxygen production) after immobilization of thylakoids according to the procedure and after 400h of storage in the dark at 4°C[26]

Immobilization Procedure	PSII Activity (%) after Immobilization[a]	PSII Activity (%) after 400h of Storage[b]
BSA-glutaraldehyde	68	15.2
Urethane polymer-BSA	50	9.4
Gelatin-glutaraldehyde	45	0
Urethane polymer	28	4.5
Alginate gel	22	0
Photo-crosslinkable resin	19	0
Carrageenan gel	3	0

[a]Activity is given relative to the value before immobilization. [b]Activity is given relative to the initial value after immobilization.

Table 3. PS II activity (oxygen production) after entrapment[a] of a mixture of chloroplasts and thylakoids in different PVA-SbQ and after 90-day storage[b] in the dark at 4°C[39]

PVA-SbQ polymers	600	1200	1700	2300	3500	3500*
PSII activity (%) after entrapment[a]	25	25	30	36	58	60
PS II activity (%) after 90-day storage at 4°C[b]	0	0	20	10	12	20

[a]Activity is given relative to the value before entrapment. [b]Activity is given relative to the initial value after entrapment. 3500*: betaine form.

expressing the integrity of the electron transfer chain. The oxygen production was measured using potassium ferricyanide as electron acceptor.[26]

The activity after immobilization is depending on the method used. The weakest loss of activity after immobilization was obtained with the coreticulation by the BSA-glutaraldehyde method. Heating of the carrageenan gel can explain the important loss of photosynthetic activity. The presence of BSA in the mixture with the urethane polymer produced an increase in the activity after immobilization. This result confirms the protective role of proteins.

The characteristics of the material used for the immobilization are also important. There are several polymers of PVA-SbQ which differ mainly by the degree of polymerisation.[39] The residual PSII activity obtained after entrapment of a mixture of chloroplasts and thylakoids in the various PVA-SbQ polymers is shown in Table 3. The values were obtained in the presence of NH_4Cl and parabenzoquinone as electron acceptor. The 3500* gel is a betaine form which has fewer styrylpyridinium groups. It was showed that styrylpyridinium groups were toxic for some enzymes whereas betaine groups stabilize them. The residual photosystem II activity increases with the degree of polymerization: from 25% with polymer 600 to 60% with polymer 3500. There was no difference between polymers 3500 with or without betaine. The PVA-SbQ polymers require light to carry out the polymerization. In spite of operating conditions unfavorable to the preservation of a good photosynthetic activity, the thylakoides kept a good activity after entrapment.

Photosystem II submembrane fractions immobilized both in BSA-glutaraldehyde and in PVA-SbQ (polymer 3500) showed a similar residual activity (60%) after immobilization.[16,52]

Storage and Operational Stabilities after Immobilization

In vivo, the cellular environment contributes to the maintenance of the structures and photosynthetic activities. The cell is able to generate the factors necessary to the survival of the metabolic activities and to preserve the structural and enzymatic integrity thus allowing resistance to many external factors like the variations of temperature or light. After isolation of the photosynthetic material, the activity decreases quickly. The stability of the activity depends particularly on the conditions imposed on the biological material (temperature, ionic force, pH, light, utilization). Two forms of stability must be considered:

- The stability of the activity according to the storage period of the photosynthetic material. The biological preparations are maintained in dark at fixed temperatures.
- The functional stability of the photosynthetic material. The preparations are exposed to operating light, temperature conditions and often acceptor of electrons.

The storage stability varies according to various parameters such as the technique of immobilization, the characteristics of products employed for the stabilization, the biological material used and the conditions of conservation.

The activity of thylakoids immobilized according to various procedures after 400h of storage at 4°C in dark are shown in Table 2. Only polyurethane, polyurethane-BSA, and BSA-glutaraldehyde kept a residual activity. Other works confirmed that the immobilization by coreticulation in an albumin-glutaraldehyde matrix gave a good preservation of the thylakoids activity.[12] The best storage conditions were found at 4°C in dark. Under these storage conditions, the immobilized thylakoids remained stable for about 200h before they started to lose their oxygen evolving activity. This phenomenon was also encountered in preparations kept at 22°C in dark though the period of stability was decreased by a factor of 10. When the samples were stored at 22°C in light, even the immobilized preparations started to decline immediately after their preparation. Both free and immobilized photosynthetic membranes degraded at a faster rate in light in comparison with the samples kept in dark.[12]

The PVA-SbQ allowed a good preservation of activity but the characteristics of the polymer used for the immobilization are important. A mixture of chloroplasts and thylakoids maintained free in rinsing medium at + 4°C in dark lost its activity (measurements of the oxygen production) after one week. The immobilization of the same biological material in PVA-SbQ increased considerably the storage stability (Table 3). After a storage of 90 days, at 4°C in dark and in dry state, the polymers 1700 and 3500* retained 20% of their initial activity, polymers 2300 and 3500, respectively, 10% and 12%, whereas the polymers 600 and 1200 were not active.[39] The stability of the thylakoids entrapped in PVA-SbQ 2300 was checked under storage at-18°C after 24h drying at 4°C before freezing. After 90-day storage, the activity of the photosynthetic membranes stored at-18°C was five times greater than at 4°C. After a 427-day storage the membranes retained 20% of their activity.[39] At-20°C thylakoid preparations immobilized in BSA-glutaraldehyde kept 40% of their activity after a 40-day storage and 50% after a 20-day storage in alginate gel.[31]

The life time of whole cells of cyanobacteria and microalgae is improved after immobilization. Whole cells of *Scenedesmus susbspicatus* immobilized on filter paper disks and then covered with a thin alginate layer can be stored at 4°C over a period of about 6 months without significant lost in fluorescence properties.[21] Cyanobacteria cells (*Synechococcus* sp. PCC6301) entrapped in dehydrated alginate beads had a useful shelf life of between 5 and 10 weeks.[23] After entrapment in PVA-SbQ, the photosynthetic activity of *Synechoccus* sp. PCC7942 progressively decreased for 2 weeks. However, the values obtained at 2 weeks were still recorded after 60 days of storage with 45% of initial activity at 4°C and 90% of the initial activity at-18°C.[15]

The immobilization of the photosynthetic material also increases the operational activity. Oxygen production by native and immobilized thylakoids was continuously monitored at 20°C under saturating illumination.[26] After 50 min, the native thylakoids are completely inactive while the thylakoids immobilized according to different procedures keep some residual activity. If these results are compared with those of storage stability in dark, it becomes obvious that the continuous use under illumination accelerated the inactivation rate by a factor of about 300. Moreover, among the immobilization procedures used, the BSA-glutaraldehyde method presented the best protection against photoinactivation.[26]

After entrapment in DEAE-cellulose, the PSII activity of a chloroplast suspension was assayed as DCIP photoreduction with water as the electron donor. Concentration of reduced DCIP in the effluent increased rapidly after the onset of illumination, reaching a plateau after 10 min. The maximum level was sustained for 2h. Then, the activity decreased during continuous operation, reaching 50% of the initial activity after 4.5h. This decline seemed to be mainly due to a decrease in the activity of the chloroplasts, not the desorption of chloroplasts during operation.[25]

In order to measure the long-term stability, the activity of whole cells of *Chlorella vulgaris* strain immobilized on a glass microfiber filter on the tip of an optical fiber was tested in the flow mode for 2 weeks. The fluorescence activity was stable for the first week. A decrease in the second week was attributed to the loss of biological activity.[4]

The method used to measure the activity of immobilized biological materials can interfere with the operational activity. PSII complexes isolated from the thermophilic cyanobacterium *Synechococcus elongatus* and immobilized in BSA-glutaraldehyde on the surface of a screen-printed sensor composed of a graphite working electrode and Ag/AgCl reference electrode shown a weak operational half-life of about 8h if the electrodes were coupled and a good half-life (24h) with separated electrodes. The low stability with the coupled electrodes was caused by Ag^{+} ions released from the reference electrode, which were toxic for PSII activity.[3] In the same way, the operational life of whole cells of *Synechococcus* sp. PCC6301 immobilized in alginate gels was greater (up to 1 week) when potassium ferricyanide was used as mediator compared with less than 1 day when a benzoquinone mediator was employed. Both types of mediators can be employed to monitor cyanobacterial photosynthetic activity; however nonpenetrating mediators such as potassium ferricyanide are less hostile to the microbial biocatalyst than lipophilic mediators such as a benzoquinone.[23]

Physical or Chemical Immobilization Comparative Study

In the physical procedure using poly(vinylalcohol) bearing styrylpyridinium (PVA-SbQ), the biological material is stabilized without chemical bonding and it is confined into a network established by the polymerized PVA-SbQ. In the chemical method where the photosynthetic material is immobilized by coreticulation in an albumin-glutaraldehyde crosslinked matrix (BSA-Glu), the glutaraldehyde builds a network of covalent bonding mainly with the free NH_2 of both photosynthetic material and albumin. The albumin's role is mainly to protect the photosynthetic material against too many bonding so that adequate biological function is retained. Further, the spongious material that is formed is appropriate for migration and diffusion of electroactive species.[59] An application of chloroplasts immobilized in a poly(vinylalcohol) film was reported and the authors mentioned that although the immobilized material offered a longer active lifetime to the electrochemical cell than native chloroplasts, the use of the poly(vinylalcohol) film was limited by its resistance to diffusion of added chemicals.[60] Consequently, the two immobilization procedures (PVA-SbQ and BSA-Glu) were compared upon the sensitivity of the thylakoid membranes to detect nine herbicides targeting photosystem II.[42] Despite the largely differing mode of immobilization, the procedures led to strikingly similar detection capabilities for herbicides. The two methods showed a similar sensitivity for five herbicides: atrazine, diuron, tebuthiuron, propanil, bentazone. Cyanazine and hexazinone were more easily detected by the BSA-Glu immobilized samples whereas the PVA-SbQ entrapped samples were more sensitive to metribuzin and bromoxynil. These herbicides were all triazines except bromoxynil, a phenol derivative. On a practical point of view, this comparative study showed that both immobilization approaches were equivalent in their performances of detection toward the herbicides tested. However, the physicochemical properties of the materials differ. The PVA-SbQ immobilized samples are hydrophilic and they dislocate during prolonged exposure to aqueous solutions. In contrast, the BSA-glu immobilization procedure produces an insoluble matrix that is well suited for longer incubation periods in aqueous solution potentially improving the detection of chemicals with a slower rate of inhibition. On the other hand, the dry state preservation of the PVA-SbQ immobilized material is advantageous because it prevents microbial contamination of the samples. Such contamination could be observed with the wet BSA-Glu immobilized samples if they were kept more than 2 weeks even at 4°C. The use of antimicrobial agents should be avoided as they can possibly interfere with the herbicide detection.

Conclusion

The possibilities of using the photosynthetic systems immobilized in many biotechnological applications can explain the great interest carried to the various methods of stabilization. The photosystems are enzymatic complexes which are subjected to the effects of various pollutants. Among others, the toxic substances found in agricultural, industrial, and municipal effluents can deteriorate PSII in the photosynthetic organisms of aquatic and terrestrial biomass. The use of photosynthetic materials as biological receptors in a biosensor is an excellent tool for the detection of many pollutants.

References

1. Franco E, Alessandrelli S, Masojidek J et al. Modulation of D1 protein turnover under cadmium and heat stresses monitored by methionine incorporation. Plant Sci 1999; 144:53-61.
2. Goetze DC, Carpentier R. Monitoring oxygen reduction by photosystem I in whole thylakoid membranes using a photoelectrochemical cell. J Photochem Photobiolo 1990; 8:17-26.
3. Koblizek M, Maly J, Masojidek J et al. A biosensor for the detection of triazine and phenylurea herbicides designed using photosystem II coupled to a screen-printed electrode Biotechnol Bioeng 2002; 78:110-116.

4. Naessens M, Leclerc JC, Tranh-Minh C. Fiber optic biosensor using Chlorella vulgaris for determination of toxic compounds. Ecotox Environ Safe 2000; 46:181-185.
5. Laberge D, Chartrand J, Rouillon R et al. In vitro phytotoxicity screening test using immobilized spinach thylakoids. Environ Toxicol Chem 1999; 18:2851-2858.
6. Sanders CA, Rodriguez M, Greenbaum E. Stand-off tissue-based biosensors for the detection of chemical warfare agents using photosynthetic fluorescence induction. Biosens Bioelectron 2001; 16:439-446.
7. Bodzek D, Janoszka B, Dobosz C et al. Determination of polycyclic aromatic compounds and heavy metals in sludges from biological sewage treatment plants. J Chromatogr 1997; 774:177-192.
8. Krupa Z, Baszynski T. Some aspects of heavy metals toxicity towards photosynthetic apparatus-direct and indirect effects on light and dark reactions. Acta Physiol Plant 1995; 17:177-190.
9. Carpentier R. The negative action of toxic divalent cations on the photosynthetic apparatus. In: Pessarakli M, ed. Handbook of Plant and Crop Physiology. New York: Marcel Dekker Inc., 2002:763-772.
10. Carpentier R, Loranger C, Chartrand J et al. Photoelectrochemical cell containing chloroplasts membranes as a biosensor for phytotoxicity measurements. Anal Chim Acta 1991; 249:55-60.
11. Pandard P, Vasseur P, Rawson DM. Comparison of two types of sensors using eukaryotic algae to monitor pollution of aquatic systems. Wat Res 1993; 27:427-431.
12. Loranger C, Carpentier R. A fast bioassay for phytotoxicity measurements using immobilized photosynthetic membranes. Biotechnol Bioeng 1994; 44:178-183.
13. Rouillon R, Gingras Y, Carpentier R et al. Detection of heavy metals using thylakoids entrapped in polyvinylalcohol bearing styrylpyridinium groups. In: Mathis P, ed. Photosynthesis from Light to Biosphere. Dordrecht, Boston, London: Kluwer Academic Press, 1995:933-936.
14. Lukavsky J, Marsalk B. The evaluation of toxicity by a biosensor with immobilized algae. Arch Hydrobiol Suppl 1997; 119:147-155.
15. Rouillon R, Tocabens M, Carpentier R. A photoelectrochemical cell for detecting pollutant-induced effects on the activity of immobilized cyanobacterium Synechococcus sp. PCC 7942. Enzyme Microb Tech 1999; 25:230-235.
16. Rouillon R, Boucher N, Gingras Y et al. Potential for the use of photosystem II submembrane fractions immobilised in poly(vinylalcohol) to detect heavy metals in solution or in sewage sludges. J Chem Technol Biot 2000; 75:1003-1007.
17. Naessens M, Tranh-Minh C. Whole-cell biosensor for direct determination of solvent vapours. Biosens Bioelectron 1998; 13:341-346.
18. Vilchez C, Vegas M. Nitrite uptake by immobilized Chlamydomonas reinhardtii cells growing in airlift reactors. Enzyme Microb Technol 1995; 17:386-390.
19. Cocquempot MF, Aguirre R, Lissolo T et al. Coimmobilization effect on H_2 production by a chloroplasts membranes-hydrogenase system. Biotechnol Lett 1982; 4:313-318.
20. Papageorgiou GC. Immobilized photosynthetic microorganisms. Photosynthetica 1987; 21:367-383.
21. Frense D, Muller A, Beckmann D. Detection of environmental pollutants using optical biosensor with immobilized algae cells. Senso Actuat B 1998; 51:256-260.
22. Rawson DM, Allison JW, Cardosi MF. The development of whole cell biosensors for on-line screening of herbicide pollution of surface waters. Toxic Assessment: An internationaly quarterly 1987; 2:325-340.
23. Rawson DM, Willmer AJ, Turner APF. Whole-cell biosensors for environmental monitoring. Biosens 1989; 4:299-311.
24. Pandard P, Rawson DM. An amperometric algal biosensor for herbicide detection employing a carbon cathode oxygen electrode. Environ Toxicol and Water Quality 1993; 8:323-333.
25. Shioi Y, Sasa T. Immobilization of photochemically-active chloroplasts onto diethylaminoethyl-cellulose. FEBS Lett 1979; 101:311-315.
26. Cocquempot MF, Thomasset B, Barbotin JN et al. Comparative stabilization of biological photosystems by several immobilization procedures. 2. Storage and functional stability of immobilized thylakoids. Eur J Appl Microbiol Biotechnol 1981; 11:193-198.
27. Thomasset B, Barbotin JN, Thomas D. Fluorescence and photoacoustic spectroscopy of immobilized thylakoids. Biotechnol Bioeng 1983; 25:2453-2468.
28. Hertzberg S, Jensen A. Studies of alginate-immobilized marine microalgae. Bot Mar 1989; 32:267-273.
29. Preuss M, Hall EAH. Mediated herbicide inhibition in a pet biosensor. Anal Chem 1995; 67:1940-1949.
30. Brewster JD, Lightfield AR, Bermel PL. Storage and immobilization of photosystem II reaction centers used in a assay for herbicides. Anal Chem 1995; 67:1296-1299.

31. Piletskaya EV, Piletsky SA, Sergeyeva TA et al. Thylakoid membranes-based test-system for detecting of trace quantities of the photosynthesis-inhibiting herbicides in drinking water. Anal Chim Acta 1999; 391:1-7.
32. Karube I, Aizawa K, Ikeda S et al. Carbon dioxide fixation by immobilized chloroplasts. Biotechnol Bioeng 1979; 21:253-260.
33. Thomasset B, Friboulet A, Barbotin JN et al. Modulation by a high citrate concentration of kinetic parameters and of functional stability of two immobilized thylakoid systems. Biotechnol Bioeng 1986; 28:1200-1205.
34. Fujimura T, Yoshii F, Kaetsu I. Stabilization of photosystem II (O_2 evolution) of spinach chloroplasts by radiation-induced immobilization. Plant Physiol 1981; 67:351-354.
35. Ochiai H, Shibata H, Matsuo T et al. Immobilization of chloroplasts photosystems with polyvinyl alcohols. Agric Biol Chem 1978; 42:683-685.
36. Ochiai H, Shibata H, Fujishima A et al. Photocurrent by immobilized chloroplast film electrode. Agric Biol Chem 1979; 43:881-883.
37. Park IH, Seo SH, Lee HJ et al. Photosynthetic characteristics of polyvinylalcohol-immobilized spinach chloroplasts. Korean J Bot 1991; 34:215-221.
38. Rouillon R, Tocabens M, Marty JL. Stabilization of chloroplasts by entrapment in polyvinylalcohol bearing styrylpyridinium groups. Anal Lett 1994; 27:2239-2248.
39. Rouillon R, Mestres JJ, Marty JL. Entrapment of chloroplasts and thylakoids in polyvinylalcohol-SbQ. Optimization of membrane preparation and storage conditions. Anal Chim Acta 1995; 311:437-442.
40. Rouillon R, Sole M, Carpentier R et al. Immobilization of thylakoids in polyvinylalcohol for the detection of herbicides. Sensor Actuat B 1995; 26-27:477-479.
41. Avramescu A, Rouillon R, Carpentier R. Potential for use of a cyanobacterium Synechocystis sp. immobilized in poly(vinylalcohol): Application to the detection of pollutants. Biotechnol Tech 1999; 13:559-562.
42. Laberge D, Rouillon R, Carpentier R. Comparative study of thylakoid membranes sensitivity for herbicide detection after physical or chemical immobilization. Enzyme Microb Tech 2000; 26:332-336.
43. Friedlander M, Neuman J. Stimulation of photoreactions of isolated chloroplasts by serum albumin. Plant Physiol 1968; 43:1249-1254.
44. Park RB, Kelly J, Drury S et al. The Hill reaction of chloroplasts isolated from glutaraldehyde-fixed spinach leaves. Proc Natl Acad Sci USA 1966; 55:1056-62.
45. Packer L, Allen JM, Starks M. Light-induced ion transport in glutaraldehyde-fixed chloroplasts: Studies with nigericin. Arch Biochem Biophys 1968; 128:142-152.
46. West J, Packer L. The effect of glutaraldehyde on light-induced H^+ changes, electron transport, and phosphorylation in pea chloroplasts. Bioenergetics 1970; 1:405-412.
47. Hardt H, Kok B. Stabilization by glutaraldehyde of high-rate electron transport in isolated chloroplasts. Biochim Biophys Acta 1976; 449:125-135.
48. Papageorgiou GC, Isaakidou J. The pH dependence of the photosynthetic electron transport in glutaraldehyde-treated thylakoids. FEBS Lett 1982; 138:19-24.
49. Mishra SR, Sabat SC. Effect of calcium ion on Hydrilla verticillata thylakoid membrane O_2 evolution. Indian J Biochem Bio 1995; 32:94-99.
50. Howell JM, Vieth WR. Biophotolytic membranes: Simplified kinetic model of photosynthetic electron transport. J Mol Catal 1982; 16:245-298.
51. Papageorgiou GC. Immobilization of photosynthetically active intact chloroplasts in a crosslinked albumin matrix. Biotechnol Lett 1983; 5:819-824.
52. Carpentier R, Lemieux S. Immobilization of a photosystem II submembrane fraction in a glutaraldehyde cross-linked matrix. Appl Biochem Biotechnol 1987; 15:107-117.
53. Carpentier R, Lemieux S, Mimeault M et al. A photoelectrochemical cell using immobilized photosynthetic membranes. Bioelectrochem Bioenerg 1989; 22:391-401.
54. Purcell M, Carpentier R, Belanger D et al. Immobilized plant thylakoid membranes as a biosensor for herbicides. Biotechnol Tech 1990; 4:363-368.
55. Izawa S. Acceptors and donors for electron transport. Method Enzymol 1980; 24:413-434.
56. Mehler A. Studies on reactions of illuminated chloroplasts. I. Mechanism of the reduction of oxygen and other Hill reagents. Arch Biochem Biophys 1951; 33:65-77.
57. Lazar D. Chlorophyll a induction. Biochim Biophys Acta 1999; 1412:1-28.
58. Thomson JA. Cellular fluorescence capacity as an endpoint in algal toxicity testing. Chemosphere 1997; 35:2027-2037.
59. Carpentier R, Lemieux S, Mimeault M. Photocurrent generation by thylakoid membranes immobilized in an albumin-glutaraldehyde cross-linked matrix. Biotechnol Lett 1988; 10:133-136.
60. Ochia H, Shibata H, SawaY et al. Properties of the chloroplast film electrode immobilized on a tin dioxide-coated glass plate. Photochem Photobiol 1982; 35:149-155.

CHAPTER 8

Comparison of Photosynthetic Organisms at Various Evolutionary Stages for Protein Biochips

Maria Teresa Giardi,* Dania Esposito and Giuseppe Torzillo

Abstract

Many chromophore molecules, such as bacteriochlorophylls, bacteriopheophytins and quinones, are arranged in Reaction Centers with a relevant distance and energy status such as to ensure unidirectional electron transfer. Therefore even a single Reaction Center is a sophisticated molecular device suitable for technological approaches.

The structures and functions of the photosynthetic proteins differ in photosynthetic organisms at various evolutionary stages, allowing their exploitation in various technological applications.

Introduction

Photosynthesis was well-established on the Earth by approximately 3.5 billion years ago, and it is widely believed that the ancient photosynthetic organisms had metabolic capabilities similar to those of modern cyanobacteria. All photosystems are built on a general theme but with some variations. They are composed of three major components: the Reaction Center (RC) that carries out photochemical charge separation and electron transport, the inner antenna consisting of pigment proteins tightly associated with RC, and the peripheral or outer antenna.[1]

The development of the two photosystems, PSII and PSI, and the ability to evolve oxygen was acquired by the ancestral algae and higher plants by an endosymbiotic event that can have turned a primordial organism cyanobacterium like into a cell organelle, the chloroplast. Hence, the common features of the photosystems of all classes of extant organisms came from their ancient origin, and their evolutionary relationships have been analysed using sequence analysis and biophysical measurements. The results indicate that all RCs fall into two basic categories: those with pheophytin and a pair of prenylquinones as early acceptors, and those with iron sulphur clusters as early acceptors. No simple linear branching evolutionary scheme can account for the distribution patterns of RC in existing photosynthetic organisms, and lateral transfer of genetic information is considered as a likely possibility.[2,3]

The consistency of the RC indicates that structural or assembly needs dictate a particular packing arrangement of the transmembrane proteins that embed the cofactors. The latter consists of a well conserved tetrapyrrole dimer, as the primary electron donor, suggesting that this is an essential requirement for any functional photosystem.[4] Within this constraint, the individual characteristics of the donor can be modified to adjust the changes in the photosynthetic pathway due to different light conditions, as well as the presence of UV and ionising radiation, that characterized

*Corresponding Author: Maria Teresa Giardi—Group on Photosynthetic-Based Biosensors National Council of Research-IC, Via Salaria km 29.3, Area of Research of Rome, Rome, Italy. Email: giardi@mlib.cnr.it

Biotechnological Applications of Photosynthetic Proteins: Biochips, Biosensors and Biodevices, edited by Maria Teresa Giardi and Elena V. Piletska.

a primordial environment. For example, in order to gain the functional capability over oxidize water, the primary donor needs both to become highly oxidizing and to coordinate electron and proton transfer with a metal complex.[5] Comparison of the biosynthetic pathways of the tetrapyrroles found in various RCs was also used to track the evolutionary path. Hence, consideration of both the energetic requirements and the pigment composition, needed for photosynthetic capability, has led to specific scenarios for the stages in the evolution of photosynthesis.[6] Information on phylogenetic distribution and evolution of bioenergetic pathways has been gathered from complete bacterial and archaeal genome sequences obtained from a wide range of evolutionary lines. Although there is general agreement that cyanobacteria gave rise to plastids, one of the unresolved questions remains whether the primary endosymbiotic event involved a single (monophyletic) or multiple (polyphyletic) cyanobacterial ancestors.[7]

The photosynthetic apparatus represents a crucial contact point between the organism and its environment and its flexibility on the one hand and stability on the other are instrumental to survival. In this context, great importance is assumed by the turnover ability of the D1 protein of RC. The D1 protein is the product of the *psbA* gene, and its synthesis and degradation are known to be regulated by light through redox state modification of the quinone PQ. Environmental extremes that negatively impact photosynthesis seem to act primarily by damaging RC at the level of D1 protein metabolism.[8] Previous experiments have shown that degradation of the D1 protein is involved in the functional damage. Thus, in an intact system with active protein synthesis, the activity of RC not only depends on physiological light intensity but also on the presence of interfering radiation, the probability of damage to PSII and the capacity of the organism to increase D1 protein turnover.[8-10]

The unstable character of the D1 protein has been conserved throughout evolution among oxygen-evolving species. This fact is physiologically significant to the survival of photosynthetic oxygenic organisms under the extreme conditions of primitive Earth.

Reaction Centers and Photosynthetic Proteins

All organisms rely on chemiosmotic membrane systems for energy transduction; the great variety of participating proteins and pathways can be reduced to a few universal principles of operation. This chemical basis of bioenergetics can be understood with relationship to the origin and early evolution of life.

Among the five domains of life on Earth, three of them, Eucarya, Bacteria and Archaea, are capable of using solar energy to grow, but there are large differences in the way they convert energy.

In Eucarya there are two main groups of photosynthetic organisms: plants and algae. Both of them perform oxygenic photosynthesis, that is, they evolve oxygen, using water as a source of electrons. For this purpose, they are equipped with two photosystems PS I and PS II, acting in series in a very efficient way, creating a light-driven flux of electrons from a redox couple with an high redox potential, H_2O/O_2, to a low potential redox couple $NADH/NAD^+$. The electron flux is also coupled to the generation of a proton gradient that drives the phosphorylation of ADP to ATP. In a second stage, in which light energy is not necessary, the "high energy" compounds generated in the first phase are used for the biosynthesis of glucose starting from CO_2 as carbon source.

Halobacteria, belonging to the Archaea, use light to generate a proton gradient across their plasmatic membrane by the bacteriorhodopsin (bR), a transmembrane protein. The quantum of light absorbed promotes a cis-trans isomerisation of the retinal chromophore cofactor which in turn generates a conformational change in the bR promoting the translocation of protons. The protein changes its colour during the isomerization process (as is described in the section *Applications of Bacteriorhodopsin*). Since there is no oxidation/reduction chemistry and no fixation of CO_2 some biologists do not consider halobacteria as photosynthetic.

Bacteria exhibit a large variety of photosynthetic organisms, performing oxygenic as well as anoxygenic photosynthesis. Among oxygenic photosynthesising prokaryotes cyanobacteria and prochlorophytes are capable of extracting electrons from water, possess two photosystems, but they lack organelles. Prochlorophytes posses key characteristics similar to those of plant chloroplasts, which have led scientists to believe that nonextant prochlorophytes were the organisms that formed symbiotic relationships with eukaryotes and developed into plant chloroplasts.[11] The anoxygenic photosynthetic bacteria possess only one photosystem and the biosynthesis of photosynthetic pigments is inhibited by the presence of oxygen. They are divided into four main groups: (i) purple bacteria, which in turn are subdivided into sulphur and non sulphur (depending on whether they produce or not elemental sulphur as a final product of their metabolism); both of them have RC similar to the PS II; (ii) green sulphur bacteria (*Chlorobiaceae*), whose RCs are similar to PS I; (iii) green gliding bacteria or green nonsulphur bacteria (*Chloroflexaceae*); (iv) gram positive bacteria (*Heliobacteriaceae*). The green and red sulphur bacteria are characterised by photoautotrophic growth, while green and red non sulphur bacteria grow photoheterotrophically.

The photosystems of green plants and photosynthetic bacteria appear to function with basically the same sort of mechanisms of energy transfer, primary charge separation, electron transfer, charge stabilization, but the molecular constituents of the reaction center are quite different. Photosystem I contains iron-sulphur proteins as electron acceptors so can be called "iron-sulphur (FeS) type" reaction center, while photosystem II contains pheophytin as the primary electron acceptor and quinone as the secondary acceptors so it can be called "pheophytin-quinone (Φ·Q) type".[12] The reaction center of purple bacteria, green nonsulphur bacteria, and PSII are (Φ·Q) type. Green sulphur bacteria, heliobacteria, and PSI have (FeS) type reaction centers.[4]

Pigment molecules, in the photosynthetic systems, are bound to polypeptides forming pigment-protein complexes. All photosynthetic systems contain bacteriochlorophyll or chlorophyll (tetrapyrroles) that absorb light over a wide range of solar spectrum and focus that energy on the reaction center. The reaction center of purple bacteria contains four bacteriochlorophylls, two bacteriopheophytins, two quinones, one carotenoid, and one nonheme iron arranged into two symmetrically-related branches. Photosystem II of algae and higher plants has a large number of cofactors and proteins, but can be isolated pigment-protein complexes containing six chlorophylls, two pheophytins, two carotenoids, two prenylquinones and three proteins (reaction center II).[13-15]

While most of the tetrapyrrole pigments in the RC (FeS) type function as light harvesting, six central tetrapyrroles and three iron-sulphur complexes appear to constitute the photochemical heart of the reaction center.

The relationship of the various organisms with respect to the type of Reaction Center is exemplified by their electron-transport diagrams, called Z-scheme (Fig. 1).[4]

RCII of plants, algae and cyanobacteria and RC of purple bacteria show significant amino acid sequence homologies in their reaction center proteins, that are the D1 and D2 proteins for the former and the L and M proteins for the latter. Moreover, they contain analogous electron carriers, but the Cyt *b559* in PSII has no comparable carrier in purple bacteria.

The most remarable distinction, however is the unique ability of PSII to oxidize water. Photosystem II constantly replenishes atmospheric oxygen and supports aerobic life on our planet through the highly oxidizing water-splitting reaction triggered by solar energy. This reaction is catalysed by a cluster of four manganese atoms bound to the luminal surface of the complex. PSII consists of over 25 polypeptides, which make up the oxygen evolving complex, a light-harvesting chlorophyll protein complex capturing light, and a reaction center involved in primary charge separation. Light energy is captured by the chlorophylls bound to the proteins and transferred to the reaction center pigments. Here, the excitation of the chlorophyll P680 leads to primary charge separation and formation of the radical pair $P680^+Pheo^-$; this radical pair is stabilised when $Pheo^-$ transfers its electron to QA, the primary quinone acceptor, and then to the secondary quinone acceptor, QB-bound to D1 protein. The electron is then transferred to the linked Photosystem I for the further reduction of NADP to NADH, necessary for the synthesis of basic organic compounds.

In conclusion, the general pattern of primary charge separation and hence the support of life, is remarkably conserved, although the specific pathway utilized for these processes varies depending upon the organism.[11-15]

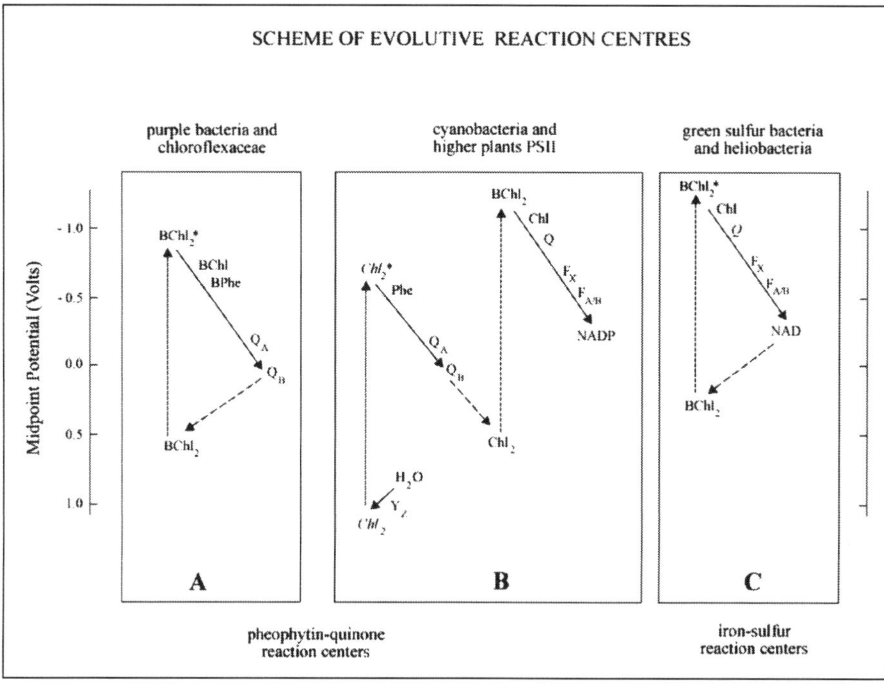

Figure 1. Pheophytin-quinone and iron-sulfur Reaction Centers. The dotted line represents the absorption of light by the primary electron donor (Chl2 or BChl2). The line shows the energy transfers in the Reaction Center, from the PSII tyrosine residue (Yz), through the monomer bacteriochlorophyll (BChl), A) the monomer bacterio-pheophytin (BPhe), or B) pheophytin (Phe) and quinone transfer components, QA and QB, in the pheophytin-quinone type of Reaction Center, and C) through the monomer chlorophyll (Chl), quinone (Q) and F components in the iron-sulfur Reaction Centers.

Technological Applications

In recent years, some progress on isolation of reaction centers and PSII have been obtained. It is now possible to isolate stable and pure preparations of photosystem II from plant thylakoid and cyanobacterial membranes by mild detergent solubilization. The most important fact, for nanotechnology purposes is that the isolated photosystem II is still able to transfer electrons, evolve oxygen and fluoresce.[16]

Moreover, technological applications can take advantage from recent advance in PSII molecular biology and in site-directed mutagenesis, which has produced a number of mutants of bacteria, algae and cyanobacteria (e.g., *Rhodobacter, Synechocystis, Chlamydomonas*) resistant to extreme conditions. The mutants show altered aminoacid composition of the D1 protein. Cysteines and Histidines, which can help immobilisation of the proteins, have also been introduced into the D1 protein by side-directed mutagenesis.[17]

The reaction center of photosynthetic bacteria can be easily isolated and it is particularly stable against denaturation. Therefore, its photosynthetic proteins are suitable for realization of devices, promoting a light-induced electron transfer across lipid membranes.[16]

Depending on its origin, RC technology can provide different applications. RC of bacteria has been utilised for building biochips; RCII of cyanobacteria, algae and higher plants has been applied for photonic-crystal bandgap materials, biosensors, biodevices and photoelectrical cells; bR of halobacteria has been applied for the reversible holographic memory, neural logic gates, nonlinear optical filters and pattern-recognition systems.

The advantage of using photosynthetic and light-dependent protein biodevices depends on the enzyme specificity in recognizing certain analytes or particular physical chemical conditions. Moreover, their biochemical activities can be easily monitored by amperometric, potenziometric and optical systems.

Some chemicals, such as triazines, phenyluree, phenolic compounds are able to bind PSII inhibiting its activity. The binding affinity of chemicals depends on the D1 amino acid composition and the structural conformation of the QB binding site, where they are bound. Molecular studies have succeeded in verifying the effect of a single amino acid modification on chemical binding affinity.[17] Other D1 modifications, such as phosphorylation and alteration of turnover rate, seem to be very important for its activity.[9]

An overview of the technological products from various evolutive photosynthetic organisms is reported.

Biochips of Reaction Centers

The building of biological devices based on RCs required a previous control of the molecular orientation of immobilised proteins or membrane fragments. The Langmuir-Blodgett (LB) technique allows an ordered immobilisation by the quantitative transfer of a compressed monolayer of aliphatic molecules to a solid support.[18] Repeated immobilizations of the chromatophore membranes have been often used to prepare multilayers of oriented membrane on a solid surface.[19] Bacterial RC of *Rhodopseudomonas sphaeroides* has been spread on an air/aqueous interface, compressed and transferred quantitatively to either glass or transparent, tin oxide-coated slides.[20] These assemblies permit the concomitant measurement of both optical and electrical activities of the deposited RC by connecting the electrodes to a voltage-clamp. Measurements of the light-induced electrical transients show charging and discharging currents, which are correlated to the RC photochemical activity. Though, the measured electron transfer was at best only 10-12% of that measured by optical assay, thus suggesting the orientation of proteins films on the air/aqueous interface. The reaction centers must be predominately oriented with opposing directions of electron transfer, having only a slight, variable tendency to align with the ubiquinone directed toward the aqueous phase.[20] Linear dichroism techniques reveal the angular inclination of the RC with respect to the solid support. The spectra indicate a close correlation between the structure of the RC in the coating with that in the membrane plane in vivo. Finally the optical properties of the monolayer are essentially unaltered from those in solution, suggesting that the RC are quantitatively transfer onto the slide.[20] Other analytical methods have been developed for a detailed characterization of the LB films and to optimise the construction of stable and photo-active films of isolated reaction centers. In particular, compressibility and stability properties of the films at the air/water interface are examinated through compression isotherms and the spectral quality, the orientation of the deposited films and their photo-activity are analysed by spectrophotometric and redox potentiometric techniques.[21] Highly packed RCs in the film have been obtained using the interaction with quinonylphospholipid.[22] It was shown that the secondary structure of the protein was not affected in LB films by heating up to 200°C, while in solution it was completely lost at 55°C.[23] The improvements in LB technique have made it successfully applied. Moreover, using LB multilayers composed of two types of bacteria RC mixture (*Rhodobacter sphaeroides* mixed with *Rhodopseudomonas viridis*), the spectral region of the light/solar energy absorbed by the system was extended.[24]

An attempt to produce films of chromatophores was made by Clayton by drying aqueous suspensions onto solid supports.[25] Dehydration of RC from *Rhodopseudomonas sphaeroides* alters their adsorption spectra, photochemical efficiency, fluorescence yield and reaction kinetics of electrons that return from reduced ubiquinone to oxidized bacteriochlorophyll. The properties of the air-dried films were restored by exposure to water vapour. It suggests that water helps to preserve an optimal configuration for photochemical efficiency.

The development of photochemical active films of chromatophores has also involved the dialysis method.[26] The photoactive sheet derived from *Rhodopseudomonas viridis* had characteristic properties of a three-dimensional thin crystal, showing linear dichroism in the absorption spectrum. That indicates a correct orientation in the arrangement of the chromatophores.

The orientation and conformation of protein at the surface of electrodes represent key factors, as they influence the interfacial electron speed. A direct reversible fast electron transfer was observed for RC of *Rhodobacter sphaeroides* reconstituted in polycation sandwiched monolayer film.[27] This layer-by-layer technique allows a such ordered orientation of RC that favours the electron transfer in film. Furthermore, RC in the film features a photo-induced redox-peak fluctuation, suggesting an intact and functional state; redox peaks were also found dependent on pH, implying a proton-coupled electron transfer.[27]

The characterization of the topological structure of the chromatophore membranes is an important feature as the orientation of the photosynthetic system with respect to the electrode can have implications for the electrical measurements. In particular, must be investigated whether the layer formed on a lipid-coated support consists of a continuous structure or of unsealed vesicles, the size of the vesicles, and whether an interior aqueous phase is preserved. Fusion of chromatohores from photosynthetic bacteria *Rhodobacter sphaeroides* with a supported lipid layer was characterized by Keller et al.[28] The type and the capacitance of the support and the vectorial transfer of protons were investigated by comparing different support such as a Mylar film or a planar modified gold electrode, and by adding the ionophore gramicidin, an inhibitor of proton flow through ATPase. The results demonstrated the fusion of individual chromatophores to form independent blisters which preserve an interior aqueous compartment. Under current-clamp conditions the photovoltage is independent of the capacitance of the support and the kinetics is directly related to the membrane potential of the individual blisters.

Hara et al characterized topological orientation of RC from *Rhodopseudomonas viridis* using antibodies against H- and C-subunits.[29] About 40% of the membranes were unsealed vesicles in the inside-out orientation exposing the cytoplasmic side whereas 20% of those were unsealed vesicles in the right side-out orientation exposing the periplasmic side. The residual 20% was in the form of a flat sheet structure. The RC electro-deposited in film generated photocurrent and photovoltage between indium-tin-oxide and gold electrodes. Addition of buffer salts such as sodium phosphate enhanced very much the photoelectrical response in the film.[29]

Photodevices

The conversion of visible light energy to electrical or chemical energy, via the isolated photosynthetic apparatus, is an interesting and advancing field of solar energy bioconversion research. Experiments have been carried out to test the possibility of light to electrical energy conversion by photochemical cells containing photosynthetic complexes. Photosystem II enriched particles and thylakoid membranes from plants, and chromatophores of photosynthetic bacteria, have all been used as constituents of photoelectrochemical cells.

A photocell based on chromatophores was built by Tamura et al.[30] The photocell comprising two sheets of SnO electrodes, ensured a sharp gradient of light intensity by using a chromatophore suspension of high optical density to generate an asymmetric distribution of charge for electron flow. An oriented flow of electrons was achieved.

Thermal stability on a photoelecrode made from dried film of a RC isolated from *Rhodopseudomonas viridis* was examined by Miyake et al.[31] Below 60°C, neither the optical absorbance of the film nor its photoelectric response were affected when heated for 1 h. When the temperature was raised over 120°C the electronic response was lost. X-ray diffraction of the film showed that heating did not destroy the secondary structure but the bundles of α-helices of the proteins were distorted. Light-induced electrical response of dried RC films was positively effected by the addition of cytochrome c.[32]

The orientation of the RCs immobilised as LB films on the electrode surface of photoelectric device was controlled using various substrates with different surface wettabilities and the degree of alignment was evaluated by measuring the polarities of light-induced electric responses. The orientation was also measured by the enzyme-linked immunosorbent assay that uses antibodies to distinguish opposite sites of the RCs. The resulting photocells showed a steady-state photocurrent with the current flow direction regulated by altering the orientation of the RCs.[33]

When TiO_2 semiconductor electrodes coated with sensitizer dyes were illuminated by visible light, the dye molecules become excited and inject electrons into the conduction band of TiO_2. When the electrons from the TiO_2 (as anodes) were channelled into a counterelectrode (a platinum cathode), a photocurrent was generated. The continuous generation of photocurrent was obtained using a photoelectrochemical cell with a dye-coated TiO_2 anode covered with photosystem II enriched particles and a platinum cathode immersed in a buffered electrolyte, containing benzoquinone as electron mediator (Fig. 2). This device accomplishes the direct photocleavage of water using the oxygen-evolving technique for the conversion of light to both electrical and chemical energy. By comparing the oxygen evolution activity and the amount of photocurrent generated from a particular PSII preparation, it was found that approximately 20% of the photosynthetic electron transfer activity of PSII measured as oxygen evolution rate was manifested as photocurrent.[34]

This suggests that PSII could greatly improve the efficiency of light conversion in photoelectrochemical cells: their use is now limited by the low efficiency of light conversion due to

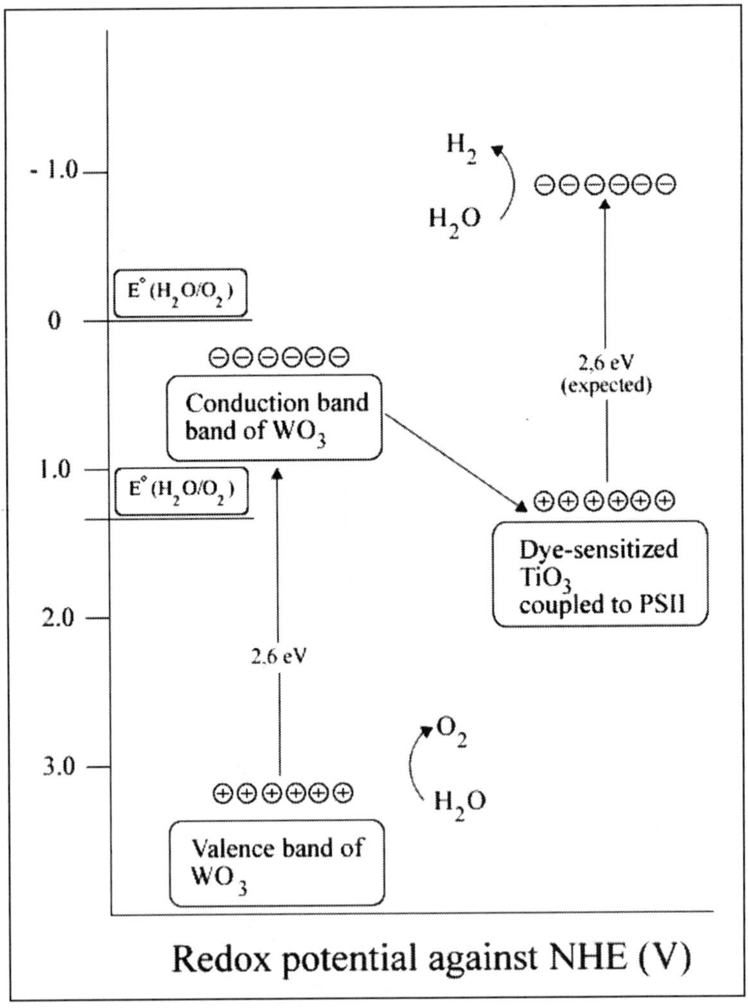

Figure 2. Dye-sensitized TiO_2 coupled to PSII.

the use of dyes with a tight wavelength range of capturing light. The utilization of an array of various RC chromatophores could enlarge the wavelength range of capturing light.

Applications of Bacteriorhodopsin

Bacteriorhodopsin is an excellent molecule for photonics. The protein is found in the purple membrane of the bacteria *Halobacterium salinarium*, that grows in salt marshes, and has evolved to exist in half-a-dozen stable states within a convenient, reversible photocycle. This robust protein, coupled with the emergence of genetic engineering, forms the basis of a variety of applications. Unlike many other photosynthetic biomolecules, bR is protected against photo-induced breakdown which is caused by reactive oxygen, singlet oxygen and free radicals. Its robustness results from bR's evolution in the tough salt marsh environment. It has learned to cope with extreme variations in light and heat and its natural function is to provide energy in low-oxygen anaerobic conditions.

Vectorial proton translocation through membranes is a fundamental energy conversion process in biological cells. The bR is the 26 kDa transmembrane protein that acts as a light-driven proton pump, converting light energy into a proton gradient. bR is the only protein constituent of the purple membrane, which also contain ten haloarcheal lipids per protein unit that have identical saturated side chains, but differ in the nature of their polar heads. bR is made up of seven alpha-helices that enclose an all-trans retinal chromophore linked via a protonated Schiff base to residue Lys216. Upon light absorption, retinal undergoes an isomerization process that results, under normal conditions, in the translocation of a proton from the cytoplasmic side to the extracellular side of the membrane.[35] The native photocycle has several spectroscopically unique steps which occur in a roughly linear order. The important photochemical event in this cycle is a trans to cis photoisomerization around the thirteenth carbon atom to the fourteenth carbon double bond in the chromophore.[36] At around the temperature of 80 K, the native protein undergoes this photocycle and switches between a green absorbing state to and a red absorbing state. At approximately room temperature, the protein switches between a green absorbing state and a blue absorbing state. In both the ground (green) and excited (red or blue) states, the chromophore displays several metastable configurations.[37] The main event follows two steps: a change in the shape of the conformational potential energy surface resulting from electron excitation and a conformational change. The time scales of the processes of interest (photoisomerization and proton transfer) are very short, ranging from tens to hundreds of femtoseconds. The molecule is the ultimate chameleon and its ability to change state is the basis of several photochromic applications, such as data storage and associative memories.

The range of potential applications for which bacteriorhodopsin has been investigated is remarkable. It includes reversible holographic memory, neural logic gates, spatial light modulation, nonlinear optical filters, photonic-crystal bandgap materials, pattern-recognition systems, optical switches and picosecond photodetectors. Several studies on the application of bR as molecular memory are patented by Japanese.

The need to explore different forms of computer architecture in order to improve the overall intelligence of a computer is especially important. Another important determinant in pursuing computer performance is miniaturisation. The increasing degree of miniaturisation of the individual components (integrated circuits) results in increasing the capabilities of each device component, and its speed of operation, thereby, decreasing the consumption of energy, size and weight. It may would seem that continuing miniaturisation could result in the implementation of better computer architecture, but physicists and engineers see the limit of the present technology due to the growth of production costs: every time the chip size lowers by a factor two, the production costs rise by a factor five. The recognition of these ultimate limits has inspired scientists to seek inspiration from biology. This is because a living organism operates with functional elements which are of molecular dimensions and which actually exploit quantum and thermal fluctuation phenomena. The hope of breaking the barrier of miniaturisation seems to lie in the utilisation of organic and biological materials and the exploitation of their chemistry, and the utilisation of radically different computer architecture.

The application of bR as protein biochip offered a new possibility in the molecular memory research. As an example of current work, consider the molecular optical memory research underway by Birge and coworkers.[38,39] Using the purple membrane from the bacterium *Halobacterium halobium*, a working optical bistable switch, fabricated in a monolayer by self-assembly, that reliably stores data with 10,000 molecules per bit was realised. The molecule switches in 500 femtoseconds and the actual speed of the memory is currently limited by how fast you can steer a laser beam to the correct spot on the memory.

Conclusion

These results show that the photosynthetic proteins are powerful promising technological systems for the conversion of solar energy into electrical energy.

This work was supported by the EU contract QLK3-CT-2001-01629 and by the Italian Space Agency.

References

1. Ke B. Photosynthesis: Photobiochemistry and photobiophysics. Kluwer Academic Publishers, 2001:1-41.
2. Turner PE, Chao L. Sex and the evolution of intrahost competition in RNA virus phi6. Genetics 1998; 50:523-532.
3. Woese C. The universal ancestor. Proc Natl Acad Sci USA 1998; 95:6854-6868.
4. Allen JP, Williams JC. A review. Photosynthetic reaction centers. FEBS Lett 1998; 438:5-9.
5. Hoganson CW, Babcock GT. A metalloradical mechanism for the generation of oxygen from water in photosynthesis. Science 1997; 277(5334):1953-1956.
6. Blankenship RE, Hartman H. The origin and evolution of oxygenic photosynthesis. Trends Biochem Sci 1998; 23:94-97.
7. Giovannoni SJ, Turner S, Olsen GJ et al. Evolutionary relationships among cyanobacteria and green chloroplasts. J Bacteriol 1988; 170(8):3584-3592.
8. Giardi MT, Masojidek J, Godde D. Effects of abiotic stresses on the turnover of the D1 reaction centre II protein. Physiol Plant 1997; 101:635-642.
9. Giardi MT. Phosphorylation and disassembly of photosystem II as an early stage of photoinhibition. Planta 1993; 190:107-113.
10. Mattoo A, Giardi MT, Raskind A et al. Dynamic metabolism of photosystem II reaction center proteins and pigments. A review. Physiol Plant 1999; 107:454-461.
11. Madigan M, Martiko J, Parker J. Brock biology of microorganisms. 9th ed. Prentice-Hall Inc. New Jersey: Upper Saddle River, 2000.
12. Blankenship RE. A review. Origin and early evolution of photosynthesis. Photosynth Res 1992; 33:91-111.
13. Deisenhofer J, Michel H. Nobel Lecture. The photosynthetic reaction centre from the purple bacterium Rhodopseudomonas viridis. EMBO J 1989; 8:2149-2170.
14. Feher G, Allen JP, Okamura MY et al. Structure and function of bacterial photosynthetic reaction centres. Nature 1989; 339:111-116.
15. Lancaster RD, Ermler U, Michel H. In: Blankenship RE, Madigan MT, Bauer CE, eds. Anoxygenic Photosynthetic Bacteria. Dordrecht: Kluwer, 1995:503-526.
16. Koblizek M, Masojidek J, Komenda J et al. A sensitive photosystemII-based biosensor for detection of a class of herbicides. Biotechnol Bioeng 1998; 60:664-669.
17. Johanningmeier U, Sopp G, Brauner M et al. Herbicide resistance and super-sensitivity in Ala (250) or Ala (251) mutants of the D1 protein in Chlamydomonas reinhardtii. Pesticide Biochem Physiol 2000; 66:9-19.
18. Blodgett KB, Langmuir I. Built-up films of barium stearate and their optical properties. Phys rev 1937; 51:964-982.
19. Hara M, Majima T, Ajiki SI et al. Multilayer preparation of bacterial photosynthetic membrane with a certain orientation immobilized on the solid surface by avidin-biotin interaction. Bioelectrochem Bioenerg 1996; 41:127-129.
20. Tiede DM, Mueller P, Dutton PL. Spectrophotometric and voltage clamp characterization of monolayers of bacterial photosynthetic reaction centers. Biochim Biophys Acta 1982; 681:191-201.
21. Alegria G, Dutton PL. Langmuir-Blodgett monolayer films of bacterial photosynthetic membranes and isolated reaction centers: Preparation, spectrophotometric and electrochemical characterization. Biochim Biophys Acta 1991; 1057:239-257.

22. Miyake J, Hara M. Protein-based nanotechnology: Molecular contruction of proteins. Material Science and Engeneering 1997; C4:213-219.
23. Nicolini C, Erokhin V, Antolini F et al. thermal stability of protein secondary structure in Langmuir-Blodgett films. Biochim Biophys Acta 1993; 1158:273-278.
24. Miyake J, Hara M, Goc J et al. Deactivation of excitation energy in bacterial photosynthetic reaction centres in Langmuir-Blodgett films. Spectrochimica Acta Part A 1997; 53:1485-1493.
25. Clayton RK. Effects of dehydration on reaction centers from Rhodospeudomonas sphaeroides. Biochim Biophys Acta 1978; 504:255-264.
26. Hara M, Asada Y, Miyake J. Photoreaction unit sheet of Rhodopseudomonas viridis. Biosci Biotech Biochem 1993; 57(6):871-874.
27. Kong J, Sun W, Wu X et al. Fast reversible electron transfer for photosynthetic reaction center from wild type Rhodobacter sphaeroides reconstituted in polycation sandwiched monolayer film. Bioelectrochem Bioenerg 1999; 48:101-107.
28. Keller S, Riou Y, Laval JM et al. Fusion of chromatophores from photosynthetic bacteria with a supported lipid layer: Characterization of the electric units. FEBS Letters 2000; 478:213-218.
29. Hara M, Ajiki S, Miyake J. Topological characterization and immobilization of a chromatophore membrane from Rhodopseudomonas viridis or application as a photoelectrical device. Supramolecular Science 1998; 5:717-721.
30. Tamura T, Sato A, Ajiki SI et al. A photocell based on a high concentration of chromatophore. Bioelectrochem Bioenerg 1991; 26:117-122.
31. Miyake J, Majima T, Namba K et al. Thermal stability of dried photosynthetic membrane film for photoelectrodes. Materials Science and Engineering 1994; Cl:63-67.
32. Majima T, Miyake J, Hara M et al. Light-induced electrical responses of dried chromatophore film: Effect of the addition of cytochrome c. Thin Solid Films 1989; 180:85-88.
33. Yasuda Y, Sugino H, Toyotama H et al. Control of protein orientation in molecular photoelectric devices using Langmuir-Blodgett films of photosynthetic reaction centers from Rhodopseudomonas viridis. Bioelectrochem Bioenerg 1994; 34:135-139.
34. Rao KK, Gratzel M, Evans MCW et al. Photocurrent generation from water via PSII membranes immobilized on dye-derivatized TiO_2 electrodes. In: Baltscheffsky M, ed. Current Research in Photosynthesis, Vol I. 1990:619-622.
35. Brown LS, Váró G, Hatanaka M et al. The complex extracellular domain regulates the transient deprotonation and reprotonation of the retinal Schiff base during the bacteriorhodopsin photocycle. Biochemistry 1995; 34:12903-12911.
36. Yamazaki Y, Tuzi S, Saitô H et al. Hydrogen-bonds of water and C = O groups coordinate long-range structural changes in the L photointermediate of Bacteriorhodopsin. Biochemistry 1996; 35:4063-4068.
37. Zhou F, Windemuth A, Schulten K. Molecular dynamics investigation of the proton pump cycle of Bacteriorhodopsin. Biochemistry 1993; 32(9):2291-2306.
38. Birge RR. Protein based computers. Scientific American 1995; 272(3):90-95.
39. American Chemical Society. In: Birge RR, ed. Molecular and Biomolecular Electronics. Washington DC: 1994:131-133, (491-510).

Chapter 9

Signal Transduction Techniques for Photosynthetic Proteins

Pinalysa Cosma, Francesco Longobardi and Angela Agostiano*

Introduction

The red-ox processes of biomacromolecules play an essential role in living systems: a series of electron-transfer reactions between donor and acceptor substances, immobilized in the membrane or dissolved in the aqueous phase, are involved in the conversion of solar energy in photosynthetic systems or in the cell respiration process. The catalytic role of proteins in signal transduction pathways mediating substrate metabolism in living process is well known.

The red-ox character of the components from the most relevant biological apparatus suggests the use of electrochemical methods to follow the reactions that occur during the metabolic processes. Although the interferences of metal electrodes with complex biological systems have often discouraged their use, recent improvements in electrochemical devices (properly designed cell, modified electrodes etc.) and in biochemical preparation methods, together with the formulation of new theoretical models, makes the electrochemical methods of analysis highly sensitive to the red-ox path of biological systems and very competitive to supply thermodynamic and kinetic information on the species involved.

In photosynthetic proteins, the primary charge separation and the sequence of electron transfer reactions can be utilized in the photosignal generation in electrochemical cells. The signal is the result of a combination of photophysical, photochemical and electrochemical events. The first is related to electronic excitation followed by charge separation, the second deals with reactions of excited molecules and finally, the electrochemical step involves a charge transfer at the interface between the electrolyte and the electrode.

Electrodes can be directly involved in exchanging electrons with the cofactors of the proteins. Two general strategic approaches have been taken in the literature.

One is to adsorb photochemically competent samples directly onto electrodes thereby establishing a direct electrochemical contact between the two,[1-35] while the other involves suspensions of membrane samples in conjunction with exogenous acceptors and/or donors.[36-50] In both cases, the electrons arriving at the acceptor side of the photosystems are intercepted by the working electrode of the cell instead of the physiological acceptors.

A brief description of most electrochemical techniques used in photosynthetic studies and some of their applications are reviewed below, with the aim of illustrating the potential of these techniques to obtain information on the nature and dynamics of the red-ox species involved within biological processes.

*Corresponding Author: Angela Agostiano—Dipartimento di Chimica, Università di Bari and CNR-IPCF sez. Bari, Via Orabona 4, 70126 Bari, Italy. Email: agostiano@chimica.uniba.it

Biotechnological Applications of Photosynthetic Proteins: Biochips, Biosensors and Biodevices, edited by Maria Teresa Giardi and Elena V. Piletska. ©2006 Landes Bioscience.

Cyclic Voltammetric Experiment

In cyclic voltammetric (CV) experiments[51] the voltage is applied to a stationary electrode (working electrode) immersed in an unstirred solution. It is scanned linearly from the initial value (E_i) to a limit value identified as switch potential (E_s), using triangular scan rates ranging from a few millivolts to hundreds of volts per second. The current response (obtained under diffusion controlled mass transfer conditions at the electrode) is plotted as a function of the applied potential. The complete resulting voltammogram displays a cathodic and anodic peak shaped waveform corresponding respectively to the reduction and oxidation of electroactive species in solution. The peak separation and height in the direct and reverse scan can be used as reversibility criteria and can provide information on the number of electrons involved in the electron transfer.

The possibility of variation of the scan rate over a wide range allows the study of the electron transfer involving transient species (up to half-life time in the order of milliseconds) and make this technique particularly useful in the investigation of stepwise reactions, as those involved in the photosynthetic and respiratory processes, often allowing the direct observation of reactive intermediates.

It is well known, that in chloroplast thylakoid membrane the absorption of light induces a vectorial electron transfer through the two photosystems in a sequential process involving the formation of red-ox species.[52] The use of a one compartment three electrode photoelectrochemical cell, first designed by Allen and Crane[36] and successively modified by Carpentier et al,[30] can allow these species to be revealed, if they are dissolved in the aqueous phase. The CV is, indeed, a powerful technique for detection of reactions proceeding, following or interposed between electron transfer to the electrode.

A very detailed theory on the effect of coupled chemical reactions on the potential, current and shape of voltammetric peak has been developed in the works of Saveant and Vianello[53] and Nicholson and Shain.[54] Based on these theories, the O_2 uptake by PS I following a Mehler-type reaction can be followed voltammetrically.[31] It is known that this reaction produces superoxide ion (O^{2-}) that quickly and spontaneously dismutates to H_2O_2, whose oxidation to give oxygen can be revealed at a platinum electrode under irradiation (see Fig. 1). In the dark (curve A) the figure shows the peaks corresponding to the oxygen evolution from water and the relative reduction in the reverse scan. Under saturating light condition (curve B) the voltammogram no longer shows a peak in the region of oxygen formation, but a flat wave with a half-peak potential, corresponding to the H_2O_2 oxidation potential (see curve C). The continuous supply of H_2O_2 at the electrode by the O_2 uptake from the photosystem I, accounts for the curve shape and the potential independence from the scan rate, as predicted by the theory on the preceding chemical reaction coupled with an irreversible electron transfer to the electrode. Evidences of the increased O_2 production under light is in addition provided by the increased value of the current corresponding to the O_2 reduction in the reverse scan.

In the absence of intervening exogenous acceptors or donors, the samples of whole thylakoid membranes generate photocurrent by the cooperative functioning of both PS II and PS I operating in series. Alternatively, their individual reactions can generate photocurrent if PS I is provided with an artificial electron donor system or PS II complemented with an artificial electron acceptor. Under irradiation, the change in the red-ox state of these mediators as a consequence of their interaction with specific sites of the electron transfer chain can be monitored by the appearing of new peaks (the technique can reveal species up to a concentration limit of 10^{-5} M), or by the variation in the current intensities in the cyclic voltammogram of thylakoids or sub-membrane particles.[31,32]

Accordingly, Figure 2, shows the cyclic voltammograms of PS II enriched membranes in the dark, and illuminated with and without a common PS II acceptor, DCBQ.[32] In the absence of a suitable acceptor, PS II does not evidence any significant photocurrent (Fig. 2, solid line), indicating that the direct electron exchange between the photosystem and the electrode does not take place.

The presence of DCBQ in the samples (Fig. 2, dotted line) results in a broad photocurrent peak with an onset potential of about 50 mV and is positioned at the potential corresponding to the oxidation of DCBQ, whose voltammogram in a buffer solution is reported in the inset of the same figure. A huge increase of the current corresponding to the reduction of oxygen is also observed.

The combined evidence of an earlier photocurrent onset potential relative to the whole thylakoid membranes and its well defined peak in addition to the voltammogram's negative slope at the potential

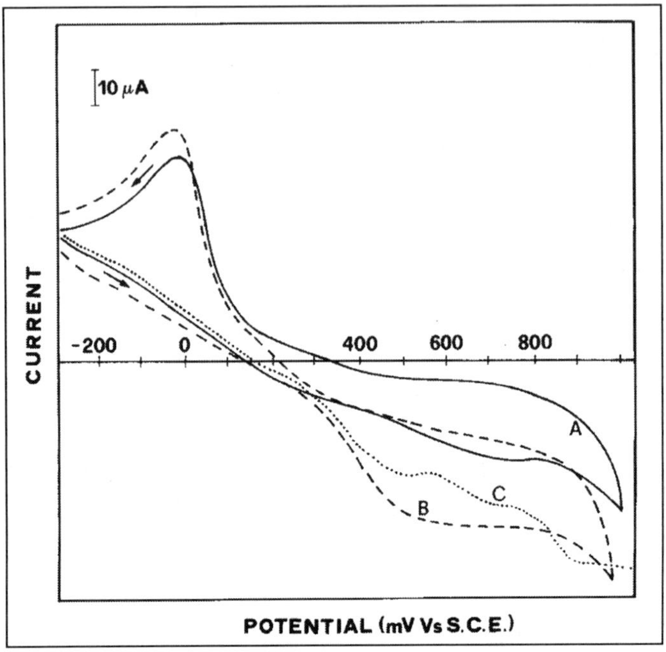

Figure l. Cyclic voltammogram of thylakoid membrane at a platinum electrode (A) in the dark, (B) in light, (C) in the dark in the presence of 0.01% H_2O_2 (scan rate 5 mV s^{-1}) as in ref.31.

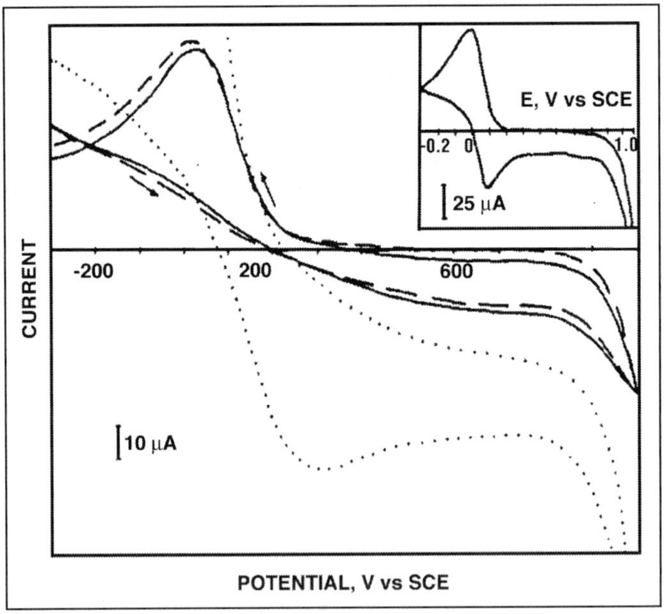

Figure 2. Current-potential curves of PS II enriched membranes (Chl conc. = 250 μg ml^{-1}): (-----) in the dark; (——) illuminated; (······) illuminated in the presence of 0.6 mM DCBQ as in reference 32. Inset shows the cyclic voltammogram of 0.6 mM DCBQ alone in buffer solution. Scan speeds in all cases = 10 mV s^{-1}.

corresponding to H_2O_2 oxidation, bar the involvement of the Mehler reaction in the photocurrent generation by PS II enriched membranes and indicates that DCBQ is the electroactive species.

Chronoamperometric Experiment

In chronoamperometric (CA) experiments[51] (see Fig. 3) the working electrode potential is changed instantaneously from the *initial potential* to the *first step potential*, and it is held at this value for the *first step time*. This is a single potential step experiment. In a double potential step experiment, the potential is changed to the *second step potential* after the first step time, and it is then held at this value for the *second step time*. The current is monitored as a function of time.

The analysis of chronoamperometry data is based on the Cottrell equation, which defines the current-time dependence for linear diffusion control:

$$i = nFACD^{1/2}\pi^{-1/2}t^{-1/2}$$

where:

n = number of electrons transferred/molecule
F = Faraday's constant (96,500 C mol^{-1})
A = electrode area (cm^2)
C = concentration (mol cm^{-3})
D = diffusion coefficient ($cm^2\ s^{-1}$)

This indicates that, under these conditions, there is a linear relationship between the current and the 1/square root of time. A plot of i vs. $t^{-1/2}$ is often referred to as the Cottrell plot.

The slope and intercept of the Cottrell plot can be measured. Since the slope is determined by n, A, C, and D, one of these parameters can be calculated, provided the other three are known.

Double potential step techniques can be used to investigate the kinetics of chemical reactions following electron transfer. In fact, considering the reaction R→O + ne^-, only a fraction of the molecules of O that are formed as a result of the first potential step are reduced again during the second step. Therefore, the current due to the second step, i_r, is less than that due to the first step, i_f. If O undergoes a chemical reaction to a molecule that is not reduced after the second step, then even fewer molecules of O are available for reduction, and i_r shows a corresponding decrease. The rate of the chemical reaction (k) can be calculated by investigating the effect of changing the *step time* on the i_r/i_f ratio and comparing these values to published working curves.

The CA is also used to measure photocurrents, which is the current that flows through a photosensitive device as the result of absorption of light. The photocurrents can be used to record an action spectrum, which is a parameter that describes the relative effectiveness of energy at different wavelengths in producing a particular biological response (as photocurrent). An action spectrum is used as a "weighting factor" for the UV spectrum to find the actual biologically effective dose (BED) for a given effect.

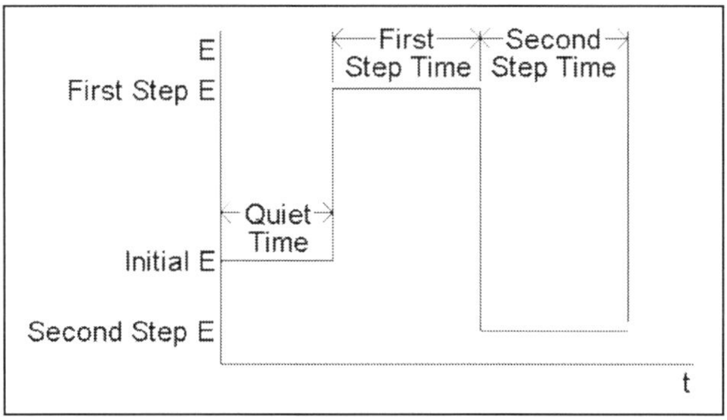

Figure 3. Potential wave form for chronoamperometry.

Several papers have been published on the use of a photosynthetic electrochemical micro-cell[28-30] and suitable mediators, to study the generation of photo-currents and photo-potentials by photosynthetic membranes and submembrane fractions enriched with PS I and PS II. The suitability of the cell to study the specific behaviour of the two photosystems and their synchronized activity was verified successfully. The same technique has also been applied to measure the photoelectrochemical response of samples illuminated by single wavelength light (action spectra).[32]

The Figure 4, shows the absorption spectra measured with a conventional spectrophotometer of three types of membrane samples along with the corresponding action spectra. The action spectrum of illuminated whole thylakoid membranes is shown in Figure 4A, evidencing no noticeable

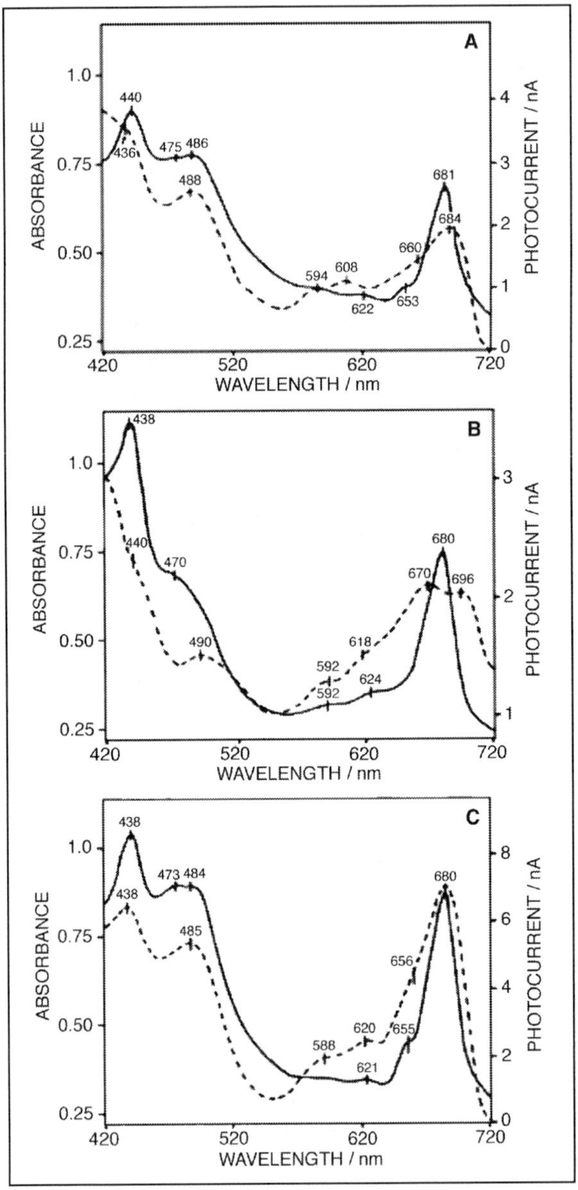

Figure 4. Action (-----) and absorption spectra (——) of a thylakoid membrane (A), of a PS I enriched membrane (B), of a PS II enriched membrane (C) from reference 32.

differences from the absorption counterpart reported in the same figure. In PS II-enriched membranes (Fig. 4B), the cooperative functioning of PS I and PS II is manifested particularly in the broadness of the long wavelength Chl *a* action peak relative to its absorption spectrum. However, the absorption spectrum of PS I-enriched membranes (Fig. 4C) does not show identifiable Chl *b* bands and shows only Chl *a* and carotenoid peaks and the relative action spectrum reveals a PS I unique splitting of the long wavelength Chl *a* action peak (696 and 670 nm). This may reasonably indicate a contribution to the photocurrent not only by the reaction center, but also by antennae Chl *a* of the light harvesting complex I (LHCI), thus confirming an increased light harvesting capacity relative to PS II antennae and further, a very strongly coupled energy transfer to the reaction center. In addition, the relatively small photocurrent contribution of carotenoids in PS I-enriched membranes may be due to the smaller set of this pigment in PS I compared to PS II (especially in the LHCII antennae complex)[31] and possibly reflects a greater role of these pigments in photoprotection than in light harvesting. For terrestrial higher plants, this is an especially important advantage as carotenoids are more efficient in harvesting light than chlorophyll (Siefermann-Harms, 1987). Furthermore, the spectra reflect these pigments having a well coupled energy transfer to the reaction centers.

Considered collectively, the spectra presented show sufficient evidence that the photoelectrochemical cell allows both the donor and acceptor sides of photosynthetically competent membrane samples to be investigated. Action spectroscopy can consequently show if the major photosynthetic pigment components of the membrane preparations are intact and functioning in a cooperative, coupled manner, such as would be expected in vivo. Thus, the usefulness of the photoelectrochemical cell as a tool for studying photosynthesis extends to investigating agents or circumstances that affect the intramembrane (or intraphotosystem) electron transfer events.

The photogalvanic measurements can be carried out using an experimental apparatus, such that reported in Figure 5, in which the compartments containing the electrode on which the submembrane

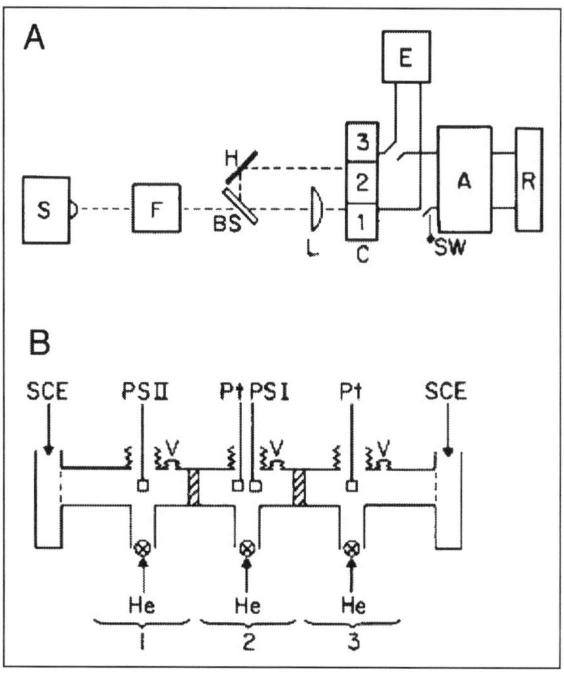

Figure 5. A) Experimental apparatus for photogalvanic measurements: S = source; F = water filter; BS = beam splitter; L = lens; C = cell; A = current amplifier; R = recorder, E = electrometer. B) Details of three-compartment photoelectrochemical cell from reference 55.

particles have been directly deposited are filled with a buffer solution and physically separated by that containing the a platinum electrode in the presence of a mediator.[55]

The Figure 6A, shows a photocurrent-time curve for the PS II electrode against the platinum electrode in the presence of $[Fe(CN)_6]^{3-}/[Fe(CN)_6]^{4-}$. An anodic photocurrent was observed upon illumination. This current attained a maximum value, I_{max} ~ 7 μA, after 15 s, then slowly decayed at longer periods of time to a steady value of ~4 μA. On turning off the light, the photocurrent reverts to a negative (cathodic) signal before returning to the steady values of the dark current level. The PS II electrode potential was shifted in the light to +49 mV versus calomel. In Figure 6B, the effects of changing the nature of the mediator is shown, for example, when a 1.1×10^{-3} M solution of $DCPIH_2$ and dehydroascorbate was used instead of $[Fe(CN)_6]^{3-}/[Fe(CN)_6]^{4-}$. The dehydroascorbate acted as the acceptor to PS II and $DCPIH_2$ the donor to PS I. A similar time behaviour for the generation and decay of the photoanodic current was observed, except for a decrease in the maximum value of the photocurrent, *i.e.*, I_{max} = 3.7 μA. The corresponding electrode potential was +33 mV.

The experimental effects reported above are readily understood in terms of the known photochemical properties of PS II and PS I particles. The light reaction of PS II particles results in the oxygen evolution from water splitting and the reduction of an electron acceptor such as ferricyanide. The light reaction of PS I particles leads to $NADP^+$ reduction in the presence of a donor such as $DCPIH_2$.

In the cell, the acceptors for PS II and PS I are physically separated from the compartments containing the PS II and PS I electrodes (see Fig. 5). The passage of electrons from the donors PS II and PS I particles to their respective acceptors can take place only via the Fermi levels of the Pt metal.

From Figures 6 and 7, we observe that the ratio of the PS II anodic photocurrent involving $[Fe(CN)_6]^{3-}/[Fe(CN)_6]^{4-}$ couple (E_0 = 360 mV) to that involving the dehydroascorbate/$DCPIH_2$ couple (E_0 = 120 mV) is 1.59. These observations are consistent with the expected behaviour shown

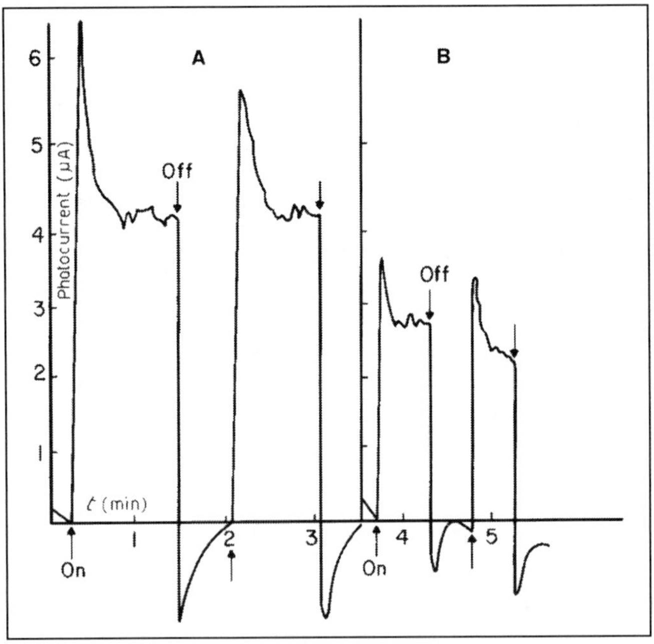

Figure 6. A) Photocurrent *versus* time behaviour for PS II-covered electrode in pH 7 phosphate buffer solution *versus* Pt electrode in compartment (2) containing 1.0×10^{-3} M 1:1 $[Fe(CN)_6]^{3-}/[Fe(CN)_6]^{4-}$ in pH 7 phosphate buffer. B) Same as in (A), with 1.0×10^{-3} M 1:1 dehydroascorbate $DCPIH_2$ in pH 7 phosphate buffer solution in compartment (2) from reference 56.

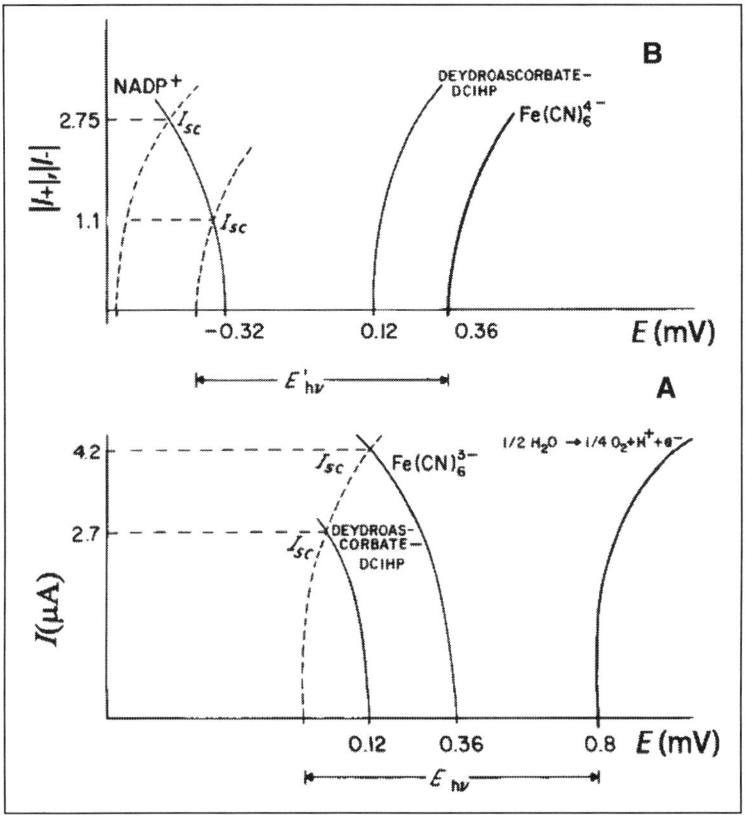

Figure 7. A) Current-potential behaviour for red-ox reactions at PS II electrode in compartment (1) and Pt in compartment (2). B) Current-potential behaviour for red-ox reactions at PS I electrode in compartment (2) and Pt in compartment (3). (———) Dark; (— — —) light; I_+ = anodic current; I_- = cathodic current; I_{sc} = short circuit photocurrent; E_h = photopotential of chlorophyll photoexcitation as in reference 55.

in Figure 7, in which the current-potential curves for the cathodic and anodic branches of the relevant photoelectrochemical reactions are shown. The Figure 7A, shows the current potential (*I-E*) behaviour of the transfer of electron from water oxidation to the acceptors, $[Fe(CN)_6]^{3-}$ and dehydroascorbate/ $DCPIH_2$, in the PS II light reaction. The short-circuit currents, I_{sc}, are obtained at the intersections of the *I-E* plots for the donor and acceptor. In dark conditions no I_{sc} is possible due to mismatches in the donor and acceptor red-ox potentials. In the light the potential for water oxidation is effectively shifted by the PS II photopotential, resulting in the crossing of the donor and acceptor *I-E* curves. The schematic representation in Figure 7B of the electron transfer from the donors, $[Fe(CN)_6]^{4-}$ and dehydroascorbate/$DCPIH_2$ to $NADP^+$ in the PS I reaction is similarly explained.

Most part of the studies of red-ox processes involving a series of vectorial electron-transfer reactions between cofactors immobilized in membrane protein often make use of substances soluble in solution, which are able to either mediate the electron transfer to an electrode or undergo selective electron transfer with one of the components of the chain in order to isolate single steps of the series.[31,32,36,56-59] The choice of the more suitable mediators to use in the different systems is based on several considerations: the red-ox potential, the capability of a fast equilibration with the protein and (eventually) with the electrode, the aptitude to diffuse both in the aqueous phase or in the protein environment and not to chemically react with the biological red-ox component.[60]

A preliminary investigation of kinetics and equilibrium of the reactions of electron transfer between mediators and proteins is therefore useful in order to evaluate the efficiency of the mediation and the quality of the deduced information on the red-ox properties of the protein itself.

In a recent paper[46] the chronoamperometry has been used to study the competitive occurrence of an electrochemical reaction of the mediators at the electrode and their chemical reaction with the cofactors of the Reaction Center of the photosynthetic bacterium *Rhodobacter Sphaeroides*. The overall process is modeled by a set of differential equations that allow the calculation of k_{ch} and k_{el}, the kinetic constants of the chemical and electrochemical reactions respectively.

The reaction center (RC) represents the minimum structural unit capable of the primary event in photosynthesis. In this pigment-protein complex the absorbed light energy is used to transfer an electron against a red-ox potential difference, thus determining a charge separation across the photosynthetic membrane.[21,22]

Absorption of a photon promotes the primary electron donor, a bacteriochlorophyll dimer (P), to its excited state. An electron is consequently transferred through a molecule of bacteriopheophytin to the primary ubiquinone-10 acceptor (Q_A). The stabilization of the primary charge-separated state $P^+Q_A^-$ is achieved by the electron transfer to a second molecule of quinone (Q_B) and the replacement of the electron on P^+ by reduced cytochrome (cyt) c_2.

The chronoamperometric technique was used to study the interaction between the reaction center and both potassium ferrocyanide and the 2,3-dimethoxy-5-methyl-p-benzoquinone (UQ-0), reductants of the photoxidized dimer and of the quinones in the site A (Q_A) and B (Q_B) of the protein respectively.

The response of the electrode, kept under illumination, to the addition of RC to a stirred solution of $K_4[Fe(CN)_6]$, Figure 8A, showed an increase of the current, followed by the gradual restoring of the initial value. Figure 8B reports the data relative to the interaction between the reaction center and the UQ-0. In the absence of RC a reduction current is recorded, whose intensity decreases exponentially in time. When the electrolysis is carried out in a solution of ubiquinone containing the protein, the reduction current of UQ-0 does not show any remarkable variations for almost 1h, thereafter starting to decrease slowly with time. The potential was set at -0.2 V, a value at which the mediator UQ-0 is reduced electrochemically and then successively chemically reduces the Q_B acceptor of the RC.

The data reported in Figure 8 can be interpreted by considering the competitive occurrence of two reactions, the chemical oxidation of the mediators (M) by the cofactors (C) of the RC, and their reduction at the electrode.

The overall process can be described by the following reactions:

$$M_{Red} + C_{Red} \xrightarrow{k_{ch}} M_{Ox} + C_{Red}$$

$$M_{Ox} + e^- \xrightarrow{k_{el}} M_{Red}$$

k_{ch} (M^{-1} s^{-1}) is the kinetic constant for the reaction between either the ferrocyanide and the photoxidized dimer of the bacteriochlorophyll incorporated into the RC or between the UQ-0 and the Q_B, while k_{el} (s^{-1}) is the rate constant of the electrochemical reduction of the mediators.

The results of the modeling procedure are reported in the inset of Figure 8. In the case of potassium ferrocyanide (inset of Fig. 8A), the simulation reproduces quite well the magnitude of the current jump and the time necessary to reach the maximum. The optimal fit has been obtained for a value of the k'_{ch} equal to 3.1×10^2 M^{-1} s^{-1} indicative of an equilibration process of the mediator with the RC on a time scale comparable with that occurring at the electrode.

In the case of the quinone, (inset of Fig. 8B), the calculated current gives a value of $k'_{ch} = 1.4 \times 10^4$ $M^{-1}s^{-1}$. This value is ~50 times greater than that found using the ferrocyanide, showing a noticeably faster equilibration of the UQ-0 with the protein. This information is of relevant interest when mediators are used in time resolved experiments, where, aside from their red-ox potential, a critical role is played by their ability to undergo a fast reaction with the acceptor or donor side of the reaction center.

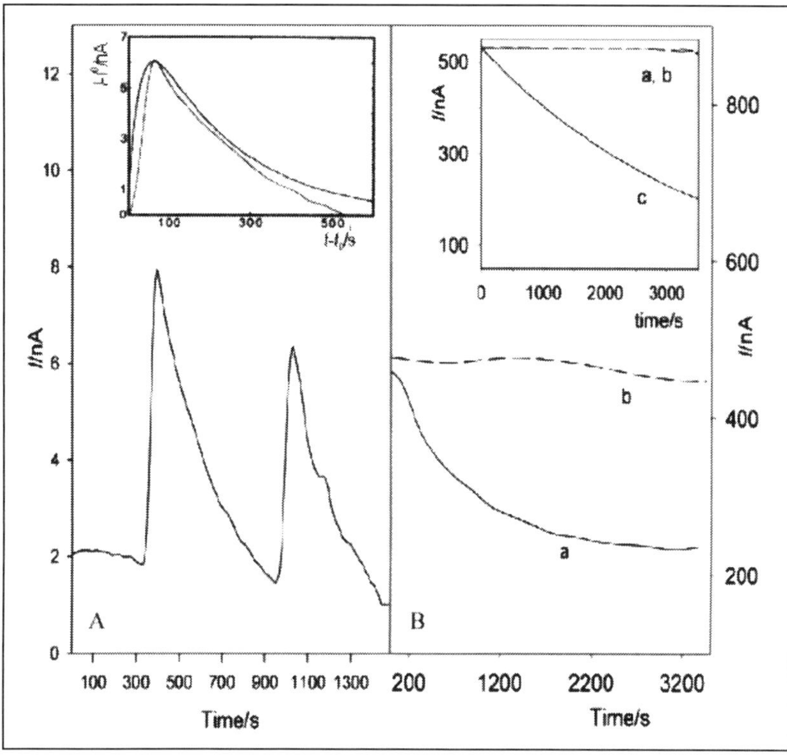

Figure 8. A) Chronoamperometry recorded for a solution of 10-5 M [$K_4Fe(CN)_6$] in the presence of two sequential additions of RC as in reference 46. Inset: Enlargement of the first addition of RC (curve b) and its optimal fit (curve a) as in reference 46. B) Chronoamperometry recorded for a solution 10^{-5} M UQ-0 at different RC concentrations as in reference 46. Inset: Calculated current.

Overview of Recent Applications

An attempt to follow by direct electrochemistry the red-ox reactions involving the cofactors of the RC embedded in lipid films on pyrolytic graphite electrodes has been recently carried out,[23] allowing the evaluation of the peaks relative to quinones and the primary donor. Direct electrochemistry of cofactors was also realized for RC in a lipid film on graphite and ITO or sandwiched between polycation layers on gold, permitting the determination of their midpoint potentials by cyclic and square wave voltammetry. In this case evidence of the presence of peaks relative to the bacteriopheophytin was reported for the first time.[24]

By using several mediators in a ultra-thin-layer electrochemical cell the determination of midpoint potentials and the investigation of red-ox-poised electron transfer reactions in isolated reaction centers from *Rhodobacter sphaeroides* was carried out by chronoamperometry and by spectroscopy in the visible/near-IR region.[37]

The oxidation of bacteriochlorophyll (BChl) in the RC was also investigated by spectroelectrochemistry, with the aim of obtaining an order of susceptibility to oxidation related to BChl-Bchl and Bchl-protein interactions.[61] Red-ox potentiometric techniques were used on Langmuir-Blodgett highly oriented films of chromatophores and isolated reaction centers deposited on glass, giving evidence of properties of the pigments and cofactors quite similar to those exhibited in vivo by the photosynthetic apparatus.[27] Exogenous quinones immobilized on electrodes or dissolved in solution were used as electron transfer mediators to study the photoelectrochemical activity of the RC for the development of photobioelectrodes for energy conversion.[38-40]

Recently new methods, based on perturbations on the linear sweep voltammetry response of the mediator in the presence of the protein,[41] a mediated thin-layer voltammetry technique,[42] cyclic voltammetric simulation applied to an electrochemically mediated enzyme reaction[44] have been settled to gain information on the protein-mediator interactions. More recently the Scanning Electrochemical Microscopy (SECM) was used to probe the red-ox activity of individual cells of purple bacteria, by using two groups of mediators (hydrophilic and hydrophobic species) in order to gain information on the dependence of measured rate constant on the formal potential of the mediator in solution. By this technique an evaluation of the intracellular potential was also performed.[45]

Several papers in the last decade have dealt with the electrochemical response of thylakoid and subthylakoid membrane particles of green plants, solubilized or adsorbed on electrode. Very recently[27] well defined and reversible voltammetric signals for a Photosystem I reaction center in a lipid film on a graphite electrode have been detected and assigned to a phylloquinone and iron-sulfur cluster, allowing also the determination of the respective electrochemical constant rate. A catalytic process of injection of electrons to ferrodoxin in solution was also observed electrochemically, mimicking the in vivo electron shuttle during photosynthesis.[27] The signal relative to the P700, the chlorophyll dimer of PS I, was also obtained by direct electrochemistry in solution.[62] The combined action of electrochemistry and Fourier Transform Infrared difference Spectroscopy allowed the investigation of two aggregation states, monomeric and trimeric, of the PS I, through the red-ox titration of the 700 nm absorption band.[47,48] The number of the electrons involved in the charge transfer, equal to 1, shown as the protein-protein interaction does not affect the red-ox properties of P700. Photoelectrochemical measurements of PS I complexes in the presence of red-ox dyes[49] were performed and the electrogenic phase in the time domain of ms was assigned to the electron transfer from the dye to the chlorophyll of P700, giving insight in to the dispute of whether the in vivo reduction of the $P700^+$ is due to an electron transfer within the RC of within the RC-cytochrome c_6 complex.

The electrochemical response from the thylakoid membrane in the micro-cell has also been used to detect impairment of the photoinduced electron transport due to the presence of herbicides, thus envisaging a possible use of the system as an environmental sensor.[33] The possibility of using the suppression of the photoelectrochemical signal to monitor the presence of inhibitors of the electron transport was also tested by using intact cells of cyanobacteria Synechoccus and a series of exogenous electron acceptors.[34] The application of electrochemical techniques to gain information from intact cells has been also verified in green bacteria, determining the photopotential action spectra of samples immobilized between two transparent electrodes.[35] The charge separation in the reaction centers and the diffusion of charges from the antenna pigments through the membrane was identified as being responsible for the signal generation. Another successful example of electrochemistry performed on cells is represented by the measurements of the in vivo topology and electron transport of individual guard cell in *Tradescantia fluminensis* carried out by scanning electrochemical microscopy.[50]

In conclusion, the electrochemical technique has proven to be a useful tool for investigating the physicochemical processes in biologically relevant systems, providing information on the nature of the red-ox species participating to the electron transfer and their reactivity.

References

1. Bedja I, Kamat PV, Hotchandani S. Electrochemical induced fluorescence quenching and photocelectrochemical behavior of chlorophyll a- modified SnO2 films. J Appl Phys 1996; 80:4637-4643.
2. Yang Y, Zhou R, Han Y et al. Electrochemical study of chlorophyll a adsorbed on SnO_2. J Electroanal Chem 1994; 370:269-271.
3. Naser NS, Planner A, Franckowiak D. Action spectra of the photopotential generation for pigment and dye solutions in nematic liquid crystals located in the electrochemical cell. J Photochem Photobiol A: Chemistry 1998; 113:279-282.
4. Ptak A, Der A, Toth-Boconadi R et al. Photocurrent kinetics (in the microsecond time range) of chlorophyll a, chlorophyll b and stilbazolium merocyanine solutions in a nematic liquid crystal located in an electrochemical cell. J Photochem Photobiol A: Chemistry 1997; 104:133-139.

5. Ptak A, Chrzumnicka E, Dudkowiak A et al. Electrochemical cell with bacteriochlorophyll c and chlorophylls a and b in nematic liquid crystal. J Photochem Photobiol A: Chemistry 1996; 98:159-163.
6. Moncelli MR, Becucci L, Dolfi A et al. Monolayers and multilayers of chlorophyl a on a mercury electrode. Bioelectrochem 2002; 56:159-162.
7. Tadini Buoninsegni F, Becucci L, Moncelli MR et al. Electrochemical and photoelectrochemical behavior of chlorophyll a films adsorbed on mercury. J Electroanal Chem 2003; 550-551:229-240.
8. Guidelli R, Becucci L, Dolfi A et al. Some bioelectrochemical applications of self-assembled films on mercury. Solid State Ionics 2002; 150:13-26.
9. Kalyasundaram K, Grätzel M. Applications of functionalized transition metal complexes in photonic and optoelectronic devices. Coord Chem Rev 1998; 77:347-414.
10. Moser JE, Bonnote P, Grätzel M. Molecular photovoltaics. Coord Chem Rev 1998; 171:245-250.
11. Hagfeldt A, Grätzel M. Light-induced redox reactions in nanocrystalline systems. Chem Rev 1995; 95:49-68.
12. Curri ML, Petrella A, Striccoli M et al. Photochemical sensitisation process at photosynthetic pigments/Q-sized colloidal semiconductor hetero-junctions. Synth Met 2003; 139:593-596.
13. Witt HT. Energy conversion in the functional membrane of photosynthesis. Analysis by light pulse and electric pulse methods: The central role of the electric field. Biochim Biophys Acta 1979; 505:355-427.
14. Volkov AG, Gugeshashvili MI, Zelent B et al. Light energy conversion with chlorophyll a and pheophytin a monolayers at the optically transparent SnO_2 electrode: Artificial photosynthesis. Bioelectrochem Bioenerg 1995; 38:333-342.
15. Kay A, Grätzel M. Artificial photosynthesis. 1. Photosensitization of titania solar cells with chlorophyll derivatives and related natural porphyrins. J Phys Chem 1993; 97:6272-6277.
16. Hotchandani S, Kamat P. Modification of electrode surface with semiconductor colloids and its sensitization with chlorophyll a. Chem Phys Lett 1992; 191:320-326.
17. Grätzel M. Photoelectrochemical cells. Nature 2001; 414:338-344.
18. Gao FG, Bard AJ, Kispert LD. Photocurrent generated on a carotenoid-sensitized TiO2 nanocrystalline mesoporous electrode. J Photochem Photobiol A 2000; 130:49-56.
19. Pan J, Benko G, Xu Y et al. Photoinduced electron transfer between a carotenoid and TiO_2 nanoparticle. J Am Chem Soc 2002; 124:13949-13957.
20. Konovalova TA, Kispert LD, Konovalov VV. Surface modification of TiO_2 nanoparticles with carotenoids. EPR study. J Phys Chem B 1999; 103:4672-4677.
21. Allen JP, Feher G, Yeats TO et al. Structure of the reaction center from rhodobacter sphaeroides. In: Breton J, Vermeglio A, eds. The Photosynthetic Bacterial Reaction Center, Structure and Dynamics. New York: Plenum, 1988:5-12.
22. Tiede DM, Budil DE, Tang J et al. Symmetry breaking structures involved in the docking of cytochrome c and primary electron transfer in reaction centers of rhodobacter sphaeroides. In: Breton J, Vermeglio A, eds. The Photosynthetic Bacterial Reaction Center, Structure and Dynamics. New York: Plenum, 1988:13-20.
23. Munge B, Pendon Z, Frank HA et al. Electrochemical reactions of redox cofactors in Rhodobacter sphaeroides reaction center proteins in lipid films. Bioelectrochem 2001; 54:145-150.
24. Kong J, Lu Z, Lvov Y et al. Direct electrochemistry of cofactor redox sites in a bacterial photosynthetic reaction center protein. J Am Chem Soc 1998; 120:7371-7372.
25. Zou Y, Zhao J, Chen Z et al. Influence of pigment substitution on the electrochemical properties of Rhodobacter sphaeroides 601 reaction centers. Series C: Life Science 2001; 44:524-532.
26. Alegria G, PL, Dutton I. Langmuir-blodgett monolayer films of bacterial photosynthetic membranes and isolated reaction centers: Preparation, spectrophotometric and electrochemical characterization. Biochim Biophys Acta 1991; 1057:239-257.
27. Munge B, Das SK, Ilagan R et al. Electron transfer reactions of redox cofactors in spinach photosystem I reaction center protein in lipid films on electrodes. J Am Chem Soc 2003; 125:12457-12463.
28. Carpentier R. A photosynthetic electrochemical micro-cell. Curr Top in Electrochem 1997; 4:173-181.
29. Mimeault M, Carpentier R. Electrochemical monitoring of electron transfer in thylakoid membranes. Enz Microb Technol 1988; 10:691-694.
30. Lemieux S, Carpentier R. Properties of a photosystem II preparation in a photoelectrochemical cell. J Photochem Phobiol B: Biology 1988; 2:221-231.
31. Agostiano A, Goetze DC, Carpentier R. Photoelectrochemistry of thylakoid and sub-thylakoid membrane preparations: Cyclic voltammetry and action spectra Electrochim. Acta 1993; 38:757-762.

32. Agostiano A, Goetze DC, Carpentier R. Cyclic voltammetry measurements of the photoelectrogenic reactions of thylakoid membranes. Photochem Photobiol 1992; 55:449-455.
33. Laberge D, Rouillon R, Carpentier R. Comparative study of thylakoid membranes sensitivity for herbicide detection after physical or chemical immobilization. Enz Microb Technol 2000; 26:332-336.
34. Croisetiere L, Rouillon R, Carpentier R. A simple mediatorless amperometric method using the cyanobacterium Synechococcus leopoliensis for the detection of phytotoxic pollutants. Appl Microbiol Biotechnol 2001; 56:261-264.
35. Ptak A, Dudkowiak A, Franckowiak D. Photoelectrical properties of green bacteria cells and cell fragments located in electrochemical cell. J Photochem Photobiol A: Chemistry 1998; 115:63-68.
36. Allen MJ, Crane AE. Null potential voltammetry - an approach to the study of Plant photosystems. Bioelectrochem Bioenerg 1976; 3:84-91.
37. Moss DA, Leonhard M, Bauscher M et al. Electrochemical redox titration of cofactors in the reaction center from Rhodobacter sphaeroides. FEBS Lett 1991; 283:33-36.
38. Katz EY, Solov'ev AA, Photobioelectrodes on the basis of photosynthetic reaction centers. Study of exogenous quinones as possible electron transfer mediators. Anal Chim Acta 1992; 266:97-106.
39. Katz EY, Shkuropatov AY, Shuvalov VA. Electrochemical approach to the development of a photoelectrode on the basis of photosynthetic reaction centers. Bioelectrochem Bioenerg 1990; 23:239-247.
40. Solov'ev AA, Katz EY, Shuvalov VA et al. Improving the electrochemical reversibility of quinones on a platinum surface to enhance the characteristics of photobioelectrodes based on photosynthetic reaction centers. Elektrokhimiya (USSR) 1992; 28:1762-1771.
41. Parker VD, Roddik A, Seefeld LC et al. Determination of rate and equilibrium constants for the reactions between electron transfer mediators and proteins by linear sweep voltammetry. Anal Biochem 1997; 249:212-218.
42. Parker VD, Seefeld LC. A Mediated thin-layer voltammetry method for the study of redox protein electrochemistry. Anal Biochem 1997; 247:152-157.
43. Yokoyama Y, Kayanuma K. Cyclic voltammetric simulation for electrochemically mediated enzyme reaction and determination of enzyme kinetic constants. Anal Chem 1998; 70:3368-3376.
44. Liu Y, Seefeld LC, Parker VD. Entropies of redox reactions between proteins and mediators: The temperature dependence of reversible electrode potentials in aqueous buffers. Anal Biochem 1997; 250:196-202.
45. Cai C, Liu B, Mirkin MV et al. Scanning electrochemical microscopy of living cells. 3. Rhodobacter sphaeroides. Anal Chem 2002; 74:114-119.
46. Agostiano A, Caselli M, Cosma P et al. Electrochemical investigation of the intercation of different mediators with the photosynthetic reaction center from rhodobacter sphaeroides. Electrochim Acta 2000; 45:1821-1828.
47. Hamacher E, Kruip J, Roegner M et al. Characterization of the primary electron donor of photosystem I, P700, by electrochemistry and Fourier transform infrared (FTIR) difference spectroscopy. Spectroch. Acta Part A: Molecular and Biomolecular Spectroscopy 1996; 52A:107-121.
48. Hamacher E, Kruip J, Roegner M et al. Characterization of the primary electron donor of photosystem I, P700, by electrochemistry and Fourier transform infrared (FTIR) difference spectroscopy. In: Mathis P, ed. Photosynthesis: From Light to Biosphere, 10th Proc. Int. Phosynthesis Congr., Montpellier, Vol. 2. Dordrecht: Kluwer, 1995:95-98.
49. Gourovskaya KN, Mamedov MD, Vassiliev IR et al. Electrogenic reduction of the primary electron donor $P700^{+}$ in photosystem I by redox dyes. FEBS Lett 1997; 414:193-196.
50. Tsionsky M, Cardon ZG, Bard AJ et al. Photosynthetic electron transport in single guard cells as measured by scanning electrochemical microscopy. Plant Physiol 1997; 113:895.
51. Bard AJ, Faulkner LR. Controlled potential microelectrode techniques. Potential Sweep Methods. In Electrochemical Methods: Fundamentals and Applications. New York: Wiley and Sons, 1980:213-248.
52. Forti G. Energy conversion in higher plants and algae. In: Ametz J, ed. Photosynthesis. Amsterdam: Elsevier, 1987:1-20.
53. Saveant SM, Vianello E. Potential-sweep chronoamperometry theory of kinetic currents in the case of a first order chemical reaction preceding the electron-transfer process. Electrochim Acta 1967; 8:629-646.
54. Nicholson RS, Shain I. Theory of stationary electrode polarography. Single Scan and Cyclic Methods Applied to Reversible, Irreversible, and Kinetic Systems. Anal Chem 1964; 37:706-723.
55. Agostiano A, Fong FK. In vitro photoelectrochemical model of the Z scheme in green plant photosynthesis. Bioelectrochem Bioenerg 1987; 17:325-337.

56. Glaser EG, Crofts AR. A new electrogenic step in the ubiquinol: Cytochrome c_2 oxidoreductase complex of Rhodopseudomonas sphaeroides. Biochim Biophys Acta 1984; 766:322-333.
57. Pan RL, Fan I-Ji, Bhardwaj R et al. A photosynthetic photoelectrochemical cell using flavin mononucleotide as the electron acceptor. Photochem Photobiol 1982; 35:655-664.
58. Gross EL, Youngman DR, Winemiller SL. An FMN-photosystem I photovoltaic cell. Photochem Photobiol 1978; 28:249-256.
59. Agostiano A, Caselli M. Photoelectrochemistry of thylakoid membranes. Bioelectrochem Bioenerg 1997; 42:255-262.
60. Dutton PL. Redox potentiometry: Determination of midpoint potentials of oxidation-reduction components. In Biological Electron Transfer Systems. In Methods in Enzymology, Biomembranes. Academic Press, 1978:411-435.
61. Kropacheva TN, Hoff AJ. Electrochemical oxidation of bacteriochlorophyll a in reaction centers and antenna complexes of photosynthetic bacteria. J Phys Chem B 2001; 105:5536-5545.
62. Kievit O, Brudvig GW. Direct electrochemistry of photosystem I. J Electroanal Chem 2001; 497:139-149.

CHAPTER 10

Biotechnological and Computational Approaches for the Development of Biosensors

Giulio Testone,* Donato Giannino, Domenico Mariotti, Prashant Katiyar, Mayank Garg, Emanuela Pace and Maria Teresa Giardi

Introduction

For ages, humans have developed technologies to exploit living organisms and their metabolism to produce food (e.g., bread, cheese and wine). Modern biotechnology implies the industrial use of scientific knowledge of cellular and molecular processes to make or modify products, to improve plants and animals, or to develop microorganism for specific uses.[1]

Nowadays, the term biotechnology is associated with techniques such as genetic engineering, cell fusion, novel bioprocessing, bioremediation etc. The introduction of genetic manipulations in large-scale processing has raised several questions regarding health risks for consumers and environment. Moreover, the scenario regarding biotechnology perception and acceptance is rather complex (for a review: Plant Genetic Engineering. Towards the Third Millennium. Proceedings of the International Symposium on Plant Genetic Engineering edited by A.D. Arencibia, Elsevier Press, 2000). Among Europeans, there has been great suspicion and reluctance to accept genetically modified food on a wide scale, though medical and environmental applications of biotechnology have been recognised as useful and subsequently have been favoured.[2] Over the past ten years, biotech devices (biosensors) have been developed to monitor polluting molecules. Biosensors employ natural or genetically engineered living organisms, or part of them, whose use is tightly controlled so as to prevent their release into the environment. Moreover, the lack of ethical and moral implications have contributed to generate their approval from the public.[3]

The development of biosensors for the monitoring of pesticides responds to the demand for in-situ, rapid, simple, selective tools able to replace those based on sophisticated and expensive technologies (e.g., gas chromatography, high performance liquid chromatography and mass spectrometry). A biosensor incorporates biological or biomimetic molecules, associated to a transducer system, which identifies a pollutant and provides selectivity. The recognition of a specific or a class of pesticides causes physical and chemical changes of the sensing element which are converted into signals that are measured by the transducer system. In biosensor assembly, the choice of a sensing component can range from natural or lab-engineered biomolecules (e.g., nucleic acids and proteins) to artificial receptors such as molecular imprinting polymers.[4] The sensing elements are often selected on the basis of interactions between harmful and target molecules occurring in the cell. The advantages of using biological material is that biological molecules are highly specific.[5] Hence,

*Corresponding Author: Giulio Testone—Institute of Biology and Agricultural Biotechnology, CNR, via Salaria km 29,3, 00016, Monterotondo Scalo, Rome, Italy.
Email: giulio.testone@mlib.cnr.it

Biotechnological Applications of Photosynthetic Proteins: Biochips, Biosensors and Biodevices, edited by Maria Teresa Giardi and Elena V. Piletska.

biosensors may provide information on the potential toxicity of a pollutant. The photosynthesis machinery comprises several components which can be used as sensors to detect herbicides and determine their toxic effects.[6-9] Herbicides kill or damage infesting weeds by targeting a wide range of plant metabolic processes. For instance, photosynthetic herbicides mainly affect the photosystem II (PSII) by direct contact. PSII is a multi-enzymatic chlorophyll-protein complex located in thylakoid membranes of cyanobacteria, algae and plants. The PSII reaction center comprises D1 and D2 proteins which are responsible for the electron transfer to the plastoquinone (Q_B) during the light reactions of photosynthesis. A few domains of D1 protein form the Q_B niche, which is also the target of photosynthetic herbicides, so that they compete with the plastoquinone in affecting D1 protein functionality.[6] Amino acid mutations of D1 protein provide different degrees of sensitivity and resistance of plant species to herbicides.

In our laboratory, D1 protein of algae and plants was chosen as a biomediator and its overexpression in heterologous hosts was a mandatory step to construct new biosensor prototypes. This manuscript focuses on a few strategies useful to produce D1 protein based sensors. Subsequently, we describe some bioinformatic approaches to establish amino acid mutations, which can specifically enhance sensitivity and selectivity of sensing elements towards pollutants.

Synthesis of Biomediators in Bacterial Hosts

Several factors affect protein synthesis in heterologous organisms: the biology of host systems, the genetic stability of recombinant vectors, the strength of promoter driving gene transcription, and the properties of template protein (size, structure and stability, post-translational editing, correct folding and host toxicity). Hosts for in vitro protein synthesis range from eukaryotic to prokaryotic organisms, each with its pros and cons with respect to purposes.

The Gram-negative bacterium *Escherichia coli* is a versatile host and can be used both as a whole cell for biosensor purpose, if the sensing protein is made accessible to the pollutant by exposure on the bacterium surface, and as a factory to yield amounts of protein biomediator to be purified and assembled in a biosensor. The overexpressed proteins in *E. coli*, depending on their target signals, can be transported to the outer cell and secreted in the cultivation medium or localised in distinct cell compartments. We intend as the outer cell of *E. coli* the extracellular space and the cell envelope, which consists of the outer membrane, the periplasm space and the cytoplasmic membrane.

Outer Cell Driven Proteins

Proteins destined for export are engineered as chimerical polypeptides which contain cytoplasm-to-target signals. These signals usually derived from secretory proteins involved in the translocation of heterologous polypeptides across the inner membrane.[10] For example, the organophosphorous hydrolase (OPH) which breaks neurotoxic organophosphate compounds, has been engineered and displayed on the cell surface of *E. coli* to create whole cell pesticide biosensors. To display OPH, the chimerical construct harboured the Lpp-OmpA anchor system.[11] This contained the signal sequence of mature lipoprotein (Lpp) for proper localisation in the outer membrane and the signal sequence of the outer membrane protein A (OmpA) for the transport of passenger proteins across the outer membrane.[12] In cell-surface display systems recombinant enzyme activity can decrease compared to that of respective free forms due to a different working environment.[13] In fact, functional OPH was recently displayed on the surface of *Moraxella sp.* and the quantity achieved was much greater than that in recombinant *E. coli.*[14] Further improvement was obtained using the ice nucleation protein-based target system of *Pseudomonas syringae.*[14,15]

The isolation and purification of recombinant proteins produced by bacteria is greatly facilitated if they are secreted into broth. Protein secretion of Gram negative bacteria comprises three main pathways: type I, that is a signal sequence independent transport system mediated by an ATP-binding cassette transporter; type II, or general secretion pathway, that exploits the presence of N-terminal signals to target proteins to the cytoplasmic membrane; pathogenic bacteria use type III to release virulence proteins harbouring non canonical N-terminal signals which are not processed.[16,17] Nowadays type I is best engineered to secrete passenger proteins: the fusion of α-haemolysin C-terminal domain and β-lactamase and chloramphenicol acetiltransferase is a successful tool to secrete these proteins into the medium.[16] In addition, in some mutants of *E. coli*, the periplasm targeted proteins are directly transported into the medium, hence the use of strains with leakage properties which can also enhance protein recovery from the broth.[18]

A correct folding of recombinant proteins is regulated by the redox potential of the outer compartments and performed by foldases. These catalyse disulfide bond formation and peptidyl-proline isomerization,[19] prevent protein aggregation in inclusion bodies and hamper the proteolysis by cytoplasmic enzymes.[20,21] The coexpression of eukaryotic foldases leads to an increased yield of eukaryotic recombinant proteins in *E. coli*.[22] Moreover, the supply of reduced glutathione generates favourable environmental conditions for carrying out a correct folding in the periplasm.[22]

Finally, proteins targeted to a periplasm run proteolytic risks due to proteases densely distributed in the periplasm (e.g., DegP, Tsp, protease III and OmpT etc.). Heterologous proteins can be increased in the periplasm by protease inhibitors or by using engineered *E. coli* strains lacking periplasm proteases.[23-25] Alternatively, the heterologous protein can be protected from proteases by cleavage site mutagenesis or by fusion with peptides which hamper the recognition of cleavage sites.[26]

Intracellular Proteins

The over production of recombinant polypeptides in bacteria often leads to the accumulation of partially or misfolded proteins which aggregate into inclusion bodies. The enzyme acetylcholinesterase (AChE), that is used as a biomediator for herbicide biosensor, was overexpressed in *E. coli*.[27] Unfortunately, AChE accumulated in inclusion bodies and only a very small amount of active form was retrieved by refolding and oxidation.[27] Kinetic models suggest that the increase of a correctly folded protein can be achieved when the rate of protein synthesis decreases. In fact, the amount of native protein increased when the host was grown at sub-optimal temperature (e.g., 30°C) or supplied with nonmetabolizable carbon sources (e.g., desoxyglucose).[28] Consequently, the partial induction of strong promoters and/or the use of a weak promoter are also valid tools to modulate the protein expression rate.[20]

Chaperones are proteins essential for the correct folding of proteins in the cell under physiological and stress conditions. They prevent aggregation of unfolded polypeptides and have a role in the correct refolding of chaperone-bound denatured polypeptides. Several cytoplasmic chaperone complexes have been characterised in *E. coli*: the ribosome-associated trigger factor assists in the folding of de-novo nascent chains; the DnaK-DnaJ-GrpE system recognizes chains during extension; the GroEL-GroES complex acts on the folding of intermediates that expose hydrophobic surfaces; the Clp ATPases are regulatory components of the ATP-dependent Clp serine proteases, but they can also function as chaperònes independently of ClpP.[29,30] The co-expression of chaperones and the optimisation of *E. coli* growth conditions have been shown to enhance the production, assembly and solubility of correctly folded protein.[19] For instance, the overexpression of GroEL-GroES complex facilitated the assembly of Rubisco,[31] of plant ferredoxin-NADP reductase[32] and of cyclohexanone monoxygenase.[33]

Chimerical constructs that contain hydrophilic molecules such as thioredoxin (Trx), ubiquitin, glutathione-S-transferase (GST) or maltose-binding protein (MBP) increase the solubility of recombinant proteins.[26] The ability of MBP to inhibit aggregation and improve solubilisation of passenger polypeptides is higher than that of GST and Trx. Moreover, fusion with MBP performs efficient initiation of translation, stability and protection against proteolysis and represents a tag for purification processes.[34] In some cases, fusion with MBP improves the proper folding of passenger protein into its biologically active conformation. In fact, MBP seems to act as an molecular chaperone preventing interaction between hydrophobic sequences on adjacent nascent chains.[35]

The accumulation of proteins in inclusion bodies offers the advantage that aggregated polypeptides are biologically inactive, non toxic to the host and insensitive to degradation.[26] The amount of recombinant protein in inclusion bodies can represent 50% of the total content of cell protein under optimised conditions.[28] However, the recovery of the biologically active protein requires methods of solubilisation, renaturation and purification. Nowadays, there are simple and handy lab techniques to recover hydrophobic polypeptides, which mostly aggregate.[36,37] Thus, the formation of inclusion bodies is a desired event: proper environmental conditions, such as high temperatures and media acidification, and the use of strong promoters (T7, *trp* or *tac*) can fulfil the purpose.[26] Finally, gene constructs that contain codons which are rare in *E. coli* may be inefficiently expressed and hamper the production of recombinant proteins.[38] These problems can be solved either by site-directed mutagenesis to replace rare codons with the *E. coli*-preferred triplets or by coproduction of these rare tRNAs.[39]

In recent years, yeast expression systems have been developed as an alternative to bacterial host. Strains of *Pichia pastoris* and *Saccaromyces cerevisiae* offer the advantage of an easy microbial growth and gene manipulation and provide eukaryotic environment for post-translational processing, such as proteolytic processing, folding, disulfide bridge formation, and glycosylation. This approach results in a product that is often identical, or very similar, to the native protein.[40,41] The choice of host system depends on many factors, such as the size, structure and stability of the gene product, and the requirements for post-translational modifications for biological activity. Moreover, the amount required, convenient cost and quality of the final product, have to be considered.

Bioinformatics to Develop Protein Based Biosensors

Bioinformatics and Protein Modelling

Bioinformatics is a discipline developed on the knowledge of biology, statistics and informatics to manage and elaborate an enormous amount of data generated by biochemical, genetic and biomolecular technologies. Regards proteins, the number of sequenced cDNAs and their deduced products stored in databanks is greater than that of three-dimensional structures solved by X-ray crystallography or nuclear magnetic resonance (NMR). The reason lies in the technical difficulty of protein crystallization, the high costs and the length of time necessary for the elucidation of three-dimensional structures.[42] Consequently a lot of interest has been focused on the theoretical and predictive computational approaches aimed at the determination of protein structures.

These approaches include: (i) de novo methods, which directly predict a structure by physical-chemical principles; (ii) template-based methods, which develop protein modelling via homology (or comparative) analyses and fold recognition via "threading".[43] The homology based approach is a "high-throughput and low-resolution" technique,[42] but it represents a reliable way to yield reasonably accurate protein structures.[44] Four phases typically constitute the homology modelling: (i) the identification of polypeptide templates with well characterised structures, stored in data bank (e.g.: the Brookhaven protein, PDB); (ii) the alignment between the query sequence and its templates; (iii) the development of a rough model and (iv) its subsequent refinement.[45] The process of template search and the quality of the alignment are crucial to determine the accuracy of a three-dimensional model,[46] in fact, an incorrect alignment can impair the domains of the model, which will exhibit structural divergences from the real structure.[47] The choice of template structures, based on pair sequence alignment, should be limited to those sharing at least a 30% sequence identity; this value allows a reliable use of one structure as template for the query protein.[48] Analyses based on multiple sequence alignment can significantly increase the underlying signal by reducing the noise; hence a sequence identity of 15% is often sufficient to detect a remote homology in sequence databases.[44] The procedures to identify the best alignment between the target sequence and bank stored homologues are based on algorithms that assign a score for each possible alignment. Practically, homologues are recovered after plotting the target sequences with sequences stored in databases by heuristic algorithms such as FASTA and BLAST. The former matches the best alignment between the entire query sequence and a specific database (global alignment), whereas the latter identifies local sequence homologies (local alignment). Moreover, FASTA can compare one or a couple of amino acids, whereas BLAST can compare groups of amino acids enabling a faster procedure. In addition, a substitution score is calculated for each pair of amino acid that can be aligned (positive similarity) and for "gaps" and "mismatches" (negative similarity). The complete set of these scores composes a substitution matrix. Several matrices have been developed based on mutation rates of homologous proteins in database; among them, PAM and BLOSUM matrices, are the most popular.[44]

In order to construct a three-dimensional homology model two different methodologies can be pursued.[47] A modelling procedure based on fragment homology uses the alignment between the query sequence and known proteins to identify the regions with the highest level of sequence conservation.[47] Hence, the model is designed on highly conserved domains, whereas variable regions (usually represented by loops) are subsequently modelled with specific tools (SCWRL, MaxSprout, etc). On the other hand, the restraint methodology uses the alignment between the query sequence and its templates to derive geometrical restraints, such as limits of inter-atomic distance and dihedral angle ranges of backbones and side chains. MODELLER is one of the most commonly used software, which calculates restraints of the target sequence on the basis of empirical data collected from numerous alignments and performs a subsequent statistical elaboration.[49] Moreover, the software estimates physical force fields to prevent atoms from clashing against each other and constructs a three-dimensional model containing nonhydrogen atoms in main chains and side chains.

The initial three-dimensional models often require refinement of bond lengths, angles and torsion angles that may be imprecise, hence a process of energy minimization is performed.[50] Molecular potential energies are calculated via quantum (QM) and/or molecular mechanics (MM). With the MM method the molecules are handled as "objects" responding to the classic mechanic law, independently from both electron presence and quantum mechanic rules. On the contrary, the QM method takes in account the electronic structure of molecules providing more rigorous structures than those of MM. However, the QM method is limited to small molecules due to the demand for a complex computational analysis.[47] The structure refinement implies that the atoms of a molecule are moved from their starting position so that the potential energy of the new arrangement is calculated. The changes may involve one or multiple atoms, external or internal coordinates, and simple or complex algorithms until a new structure at a lower degree of potential energy is generated. This process is repeated several times until no further adjustment of the atomic position leads to a energy minimisation, hence the structure is "optimised".

A Bioinformatic Approach to Design D1 Protein Based Sensors

Three-dimensional models are useful to study and predict mutual matches between proteins or protein docking to varied molecules. This procedure is crucial for the identification of putative catalytic residues and the search for new inhibitor/substrate complexes. Moreover, the improvement of protein stability and the detection of amino acid targets by experiments of site-directed mutagenesis can be pursued.

Our research focuses on biosensors which employed the D1 protein as the sensing element for polluting herbicides. Thus, the D1 protein-herbicide interactions can be investigated by the combination of homology based protein modelling and virtual-mutagenesis. The ideal engineered D1 protein sensing should perform a higher herbicide binding affinity, maintain the electron transport capacity and exhibit enhanced stability of a high temperature as compared to wild type D1 protein. Up to now, an accurate crystallographic structure of the D1 protein from higher plants has not yet been achieved though three-dimensional models of PSII components have been developed.[51-53] These exploit the homology of D1/D2 polypeptides with the L/M subunits of the bacterial reaction center of purple nonsulphur bacteria *Rhodobacter sphaeroides* and *Rhodopseudomonas viridis*, for which high resolution crystal structures were determined.[52,54,55] Anyway, a model founded solely on the homology with prokaryotic systems may result incomplete, as specific PSII features could be ignored.

The PSII structure of the unicellular green alga *Chlamydomonas reinhardtii* was modelled by combining results from the experimental data on components of the PSII[55] and standard homology approaches. This approach leads to a refined model of the PSII reaction center comprising D1 and D2 proteins, accessory chlorophylls, a manganese cluster, two molecules of β-carotene and cytochrome b559.[53] To our aims, the D1 protein structure was extrapolated from this model with a few added refinements (e.g., addition of missing hydrogen atoms, removal of a few cofactors etc.). Structure validation was checked by WHAT IF and SYBYL programs and a structure with minimal conformational energy was produced. Current protein algorithms generally generate a large number of docked conformations with favourable surface complementarities. This process is followed by the reranking of the conformations by using a potential which approximates the free energy.[56] Hence, herbicide dockings in D1 binding pocket were estimated by the GRAMM program (global range molecular modelling) that scored the lowest energy state. Subsequently, a refinement of the D1 protein-herbicide docked state was achieved by a combination of literature data on mutated D1 proteins of *C. reinhardtii* with the data from PSII models.

In the eukaryotic PSII complex, the D1 protein binds the cofactor plastoquinone in the Q_B pocket that is the same target of the herbicide atrazine,[53] so the Q_B binding site was accurately characterised. The protein region spanning the 8 angstrom radius of the Q_B site was selected, four different orientations of atrazine were established and the resulting structures were energy minimised. The lowest energy structure was used to estimate and compare the energy changes resulting from amino acid substitutions tests in the Q_B site. All possible rotamers were established and visualized for each mutation. Key residues were defined as those which altered specific features of the D1 protein upon substitution. In detail, mutation targets did not include the residues His215 and His272 indispensable for the electron transport, Ser264 specific for the atrazine binding, and Gly256 since this mutation was predicted to cause steric hindrance. All the established mutations were than energy minimized using the SYBYL program. About fifty mutations were selected for molecular modelling since they exhibited comparable, or minor energy values than those calculated for the wild type D1 protein bound by atrazine. However, some mutations (e.g.: Tyr246), though energetically stable, were further ignored because they could hamper the correct entrance of atrazine into the binding pocket. The validation of the selected mutations was performed using the Ramachandran plot parameters, since the residues involved the D1 protein-atrazine complex were in bet helix conformation.[57]

The modelling described above represents a powerful tool to acquire fundamental knowledge of the structure and function of the D1 protein, and it could lead to the creation of mutated D1 proteins which can function as highly effective biomediators in recognising a target herbicide.

References

1. Davies WP. An historical perspective from the Green Revolution to the gene revolution. Nutr Rev 2003; 61(6 Pt 2):124-134.
2. Braun R. People's concerns about biotechnology: Some problems and some solutions. J Biotechnol 2002; 98(1):3-8.
3. Gaskell G, Allum N, Bauer M et al. Biotechnology and the European public. Nature Biotechnology 2000; 18:935-938.
4. Sergeyeva TA, Piletsky SA, Brovko AA et al. Selective recognition of atrazine by molecularly imprinted polymer membranes. Development of conductometric sensor for herbicides detection. Anal Chimica Acta 1999; 392:105-111.
5. Belkin S. Microbial whole-cell sensing systems of environmental pollutants. Curr Opin Microbiol 2003; 6(3):206-212.
6. Koblizek M, Masojidek J, Komenda J et al. A sensitive photosystem II based biosensor for detection of a class of herbicides. Biotechnol Bioengineering 1998; 60:664-669.
7. Koblizek M, Maly J, Masojídek J et al. A Sensitive photosystem II-based biosensor for detection of a class of herbicides. Screen printed electrodes as transduction devices. Biotechnol Bioeng 2002; 78:110-116.
8. Giardi MT, Koblizek M, Masojidek J. Photosystem II-based biosensors for the detection of pollutants. Biosens Bioelectron 2001; 16:1027-1033.
9. Vedrine C, Leclerc JC, Durrieu C et al. Optical whole-cell biosensor using Chlorella vulgaris designed for monitoring herbicides. Biosens Bioelectron 2003; 18(4):457-463.
10. Cornelis P. Expressing genes in different Escherichia coli compartments. Curr Opin Biotechnol 2000; 11:450-454.
11. Richins RD, Kaneva I, Mulchandani A et al. Biodegradation of organophosphorus pesticides by surface-expressed organophosphorus hydrolase. Nat Biotechnol 1997; 15(10):984-987.

12. Lee YS, Choi JH, Xu Z. Microbial cell-surface display. Trends Biotech 2003; 21(1):45-52.
13. Van der Vaart JM, te Biesebeke R, Chapman JW et al. Comparison of cell wall proteins of Saccharomyces cerevisiae as anchors for cell surface expression of heterologous proteins. Appl Environ Microbiol 1997; 63:615–620.
14. Shimazu M, Mulchandoni A, Chen W. Cell surface display of organophosphours hydrolase using ice nucleation protein. Biotechnol Prog 2001; 17:76–80.
15. Mulchandani P, Chen W, Mulchandani A et al. Amperometric microbial biosensor for direct determination of organophosphate pesticides using recombinant microorganism with surface expressed organophosphorus hydrolase. Biosens Bioelectron 2001; 16(7-8):433-437.
16. Sandkvist M, Bagdasariant M. Secretion of recombinant proteins by Gram-negative bacteria. Curr Opin Biotechnol 1996; 7:505-511.
17. Thanassi DG, Hultgren SJ. Multiple pathways allow protein secretion across the bacterial outer membrane. Curr Opin Cell Biol 2000; 12(4):420-430.
18. Gumpert J, Hoischen C. Use of cell wall-less bacteria (L-forms) for efficient expression and secretion of heterologous gene products. Curr Opin Biotechnol 1998; 9(5):506-509.
19. Georgiou G, Valaxt P. Expression of correctly folded proteins in Escherichia coli. Curr Opin Biotechnol 1996; 7:190-197.
20. Baneyx F. Recombinant protein expression in Escherichia coli. Curr Opin Biotechnol 1999; 10:411-421.
21. Mar Carriò M, Villaverde A. Role of molecular chaperones in inclusion body formation. FEBS Letters 2003; 537:215-221.
22. Ostermeier M, De Sutter K, Georgiou G. Eukaryotic protein disulfide isomerase complements Escherichia coli dsbA mutants and increases the yield of a heterologous secreted protein with disulfide bonds. J Biol Chem 1996; 271(18):10616-10622.
23. Meerman HJ, Georgiou G. Construction and characterization of a set of E. coli strains deficient in all known loci affecting the proteolytic stability of secreted recombinant proteins. Biotechnology 1994; 12(11):1107-1110.
24. Park SJ, Georgiou G, Lee SY. Secretory production of recombinant protein by a high cell density culture of a protease negative mutant Escherichia coli strain. Biotechnol Prog 1999; 15(2):164-167.
25. Kandilogiannaki M, Koutsoudakis G, Zafiropoulos A et al. Expression of a recombinant human anti-MUC1 scFv fragment in protease-deficient Escherichia coli mutants. Int J Mol Med 2001; 7(6):659-664.
26. Jonasson P, Liljeqvist S, Nygren PA et al. Genetic design for facilitated production and recovery of recombinant proteins in Escherichia coli. Biotechnol Appl Biochem 2002; 35:91-105.
27. Fischer M, Ittah A, Liefer I et al. Expression and reconstitution of biologically active human acetylcholinesterase from Escherichia coli. Cell Mol Neurobiol 1993; 13(1):25-38.
28. Lilie H, Schwarz E, Rudolph R. Advances in refolding of proteins produced in E. coli. Curr Opin Biotechnol 1998; 9:497-501.
29. Houry WA. Chaperone-assisted protein folding in the cell cytoplasm. Curr Protein Pept Sci 2001; 2(3):227-244.
30. Hoskins JR, Sharma S, Sathyanarayana BK et al. Clp ATPases and their role in protein unfolding and degradation. Adv Protein Chem 2001; 59:413-429.
31. Weber F, Keppel F, Georgopoulos G et al. The oligomeric structure of GroEL/GroES is required for biologically significant chaperonin function in protein folding. Nat Struct Biol 1998; 5(11):977-985.
32. Dionisi HM, Checa SK, Krapp AR et al. Cooperation of the DnaK and GroE chaperone systems in the folding pathway of plant ferredoxin-$NADP^+$ reductase expressed in Escherichia coli. Eur J Biochem 1998; 251(3):724-728.
33. Lee DH, Kim MD, Lee WH et al. Consortium of fold-catalyzing proteins increases soluble expression of cyclohexanone monooxygenase in recombinant Escherichia coli. Appl Microbiol Biotechnol published online 2003.
34. Kapust RB, Waugh DS. Escherichia coli maltose-binding protein is uncommonly effective at promoting the solubility of polypeptides to which it is fused. Protein Sci 1999; 8(8):1668-1674.
35. Bach H, Mazor Y, Shaky S et al. Escherichia coli maltose-binding protein as a molecular chaperone for recombinant intracellular cytoplasmic single-chain antibodies. J Mol Biol 2001; 312(1):79-93.
36. Middelberg PJA. Preparative protein refolding. Trends Biotechnol 2002; 20(10):437-443.
37. Tsumoto K, Ejima D, Kumagai I et al. Practical considerations in refolding proteins from inclusion bodies. Protein Expression Purification 2003; 28:1-8.
38. Zahn K. Overexpression of an mRNA dependent on rare codons inhibits protein synthesis and cell growth. J Bacteriol 1996; 178:2926-2933.

39. McNulty DE, Claffee BA, Huddleston MJ et al. Mistranslational errors associated with the rare arginine codon CGG in Escherichia coli. Protein Expr Purif 2003; 27(2):365-374.
40. Cereghino GP, Cregg JM. Applications of yeast in biotechnology: Protein production and genetic analysis. Curr Opin Biotechnol 1999; 10:422-427.
41. Rai M, Padh H. Expression systems for production of heterologous proteins. Current science 2001; 80(9):1121-1128.
42. Maggio ET, Ramnarayan K. Recent developments in computational proteomics. Drug Discov Today 2001; 6(19):996-1004.
43. Moult J. Predicting protein three-dimensional structure. Curr Opin Biotechnol 1999; 10(6):583-588.
44. Xu D, Xu Y, Uberbacher EC. Computational tools for protein modeling. Curr Protein Pept Sci 2000; 1(1):1-21.
45. Szklarz GD, Halpert JR. Use of homology modeling in conjunction with site-directed mutagenesis for analysis of structurefunction relationship of mammalian cytocromes P450. Life science 1997; 61(26):2507-2520.
46. Fiser A, Sanchez R, Melo F et al. Comparative protein structure modeling. Computational Biochemistry and Biophysics. New York: Marcel Dekker, 2001:275–312.
47. Forster MJ. Molecular modelling in structural biology. Micron 2002; 33:365-384.
48. Prasad JC, Comeau SR, Vajda S et al. Consensus alignment for reliable framework prediction in homology modeling. Bioinformatics 2003; 19(13):1682-1691.
49. Sali A, Overington JP. Derivation of rules for comparative protein modeling from a database of protein structure alignments. Protein Sci 1994; 3:1582-1596.
50. Murphy RB, Philipp DM, Friesner RA. A Mixed quantum mechanics/molecular mechanics (QM/MM) method for large-scale modeling of chemistry in protein environments. J Computational Chem 2000; 21(16):1442-1457.
51. Santini C, Tidu V, Tognon G et al. Three-dimensional structure of the higher-plant photosystem II reaction center and evidence for its dimeric organization in vivo. Eur J Biochem 1994; 221(1):307-315.
52. Svensson B, Etchebest C, Tuffery P et al. A model for the photosystem II reaction center core including the structure of the primary donor P680. Biochemistry 1996; 35(46):14486-14502.
53. Xiong J, Subramanian S, Govindjee. A knowledge-based three dimensional model of the Photosystem II reaction center of Chlamydomonas reinhardtii. Photosynth Res 1998; 56:229-254.
54. Svensson B, Vass I, Cedergren E et al. Structure of donor side components in photosystem II predicted by computer modelling. EMBO J 1990; 9(7):2051-2059.
55. Xiong J, Subramaniam S, Govindjee. Modeling of the D1:D2 proteins and cofactors of the photosystem II reaction center: Implications for herbicide and bicarbonate binding. Protein Sci 1996; 5:2054–2073.
56. Camacho CJ, Vajda S. Protein–protein association kinetics and protein docking. Curr Opin Struct Biol 2002; 12(1):36-40.
57. Lovell SC, Davis IW, Arendall IIIrd WB et al. Structure validation by Calpha geometry: Phi, psi and Cbeta deviation. Proteins 2003; 50(3):437-450.

CHAPTER 11

The Problem of Herbicide Water Monitoring in Europe

Licia Guzzella* and Fiorenzo Pozzoni

Abstract

Among human activities agriculture is one of the principal activities responsible for the damages that can affect water resources. The increase in food production achieved in last decades has been possible mainly because of massive use of fertilizers and pesticides. Data on pesticide pollution are rather scarce; monitoring data on water contamination is based on few investigations, focussing on a limited area and on few compounds of interest. The large number of compounds (approximately 600, which were approved for use) and confidentiality of manufacturers make it difficult to obtain accurate and current information on pesticide application. Herbicides are the second most important class of pesticides used in the European Union. A major difficulty in estimating water quality related to herbicide contamination is due to seasonal changes of field application and low maximum admissible concentration required by European policy. Unfortunately, systematic monitoring is not yet routine in the EU. That is why the data on maximum permissible concentration of herbicides significantly varies among European countries.

Introduction

Increase in food production has been a common policy objective throughout Europe for several decades. Farmers increased agricultural output significantly between the 1940s and the 1990s in response to such policies. This agricultural policy determined an overall mechanisation combined with the abandonment of traditional practices, reliance on nonrenewable inputs such as inorganic fertilizers and pesticides. Diffuse losses from agriculture represent the main source of nitrate pollution in European waters. For instance, more than fifty percent of nutrient load to the Danube River comes from agricultural input.[1]

According to the land use in Europe, agriculture accounts for more than 42 percent of the total land area, although the proportion varies from less than 10 per cent in Finland, Sweden and Norway to 70 per cent or more in Hungary, Ireland and UK. The use of agrochemical compounds (fertilizers and crop protection chemicals) has played a significant role in increasing the crop yields in the European agricultural land.

Community legislation distinguishes between "active substances" and "plant protection products" (PPPs). The active substance is considered the ingredient that can affect the plant biological activity while the PPP represents the commercial formulate. Other ingredients, the coadjutant compounds, can improve the activity, increasing the adherence to the plant surfaces or the homogeneity and stability of the commercial formulates.

PPPs are used to protect plants or plant products against harmful organism or to prevent the negative action of infesting organisms. They can act in a variety of modality but the main one is the

*Corresponding Author: Licia Guzzella—IRSA, CNR, via della Mornera 25, 20047 Brugherio, Milan, Italy. Email: guzzella@irsa.cnr.it

Biotechnological Applications of Photosynthetic Proteins: Biochips, Biosensors and Biodevices, edited by Maria Teresa Giardi and Elena V. Piletska.

pest killing, therefore they are considered as pesticides. The *pesticide* definition is used for plant protection products and it indicates any substances used in agriculture to control, destroy, repel or attract a plant pest. Among the pesticides there are plant protection products that act as *herbicides*, i.e., utilized to control undesirable or noxious plant growth, generally called weeds, in the crop production but also in noncrop areas where it is necessary to limit the plant growth.

A fundamental contribution to the Green Revolution has been the development and application of pesticides for the control of a wide variety of insects and weeds, which are responsible for decreasing quantity and quality of food production. The history of pesticide development can be resumed as followed: in 1800-1920—early organic formulates based on nitrophenols, chlorophenols, naphthalene (main characteristics: lack specificity and toxicity to nontarget organisms); in 1945-1955—organochlorinate formulates based on DDT, HCH (main characteristics: persistence, good selectivity, good agricultural properties, harmful ecological effects); in 1955-1970—cholinesterase inhibitor formulate based on organophosphorus compounds, carbamates (main characteristics: lower persistence, toxic for non target organisms, some environmental problems due to their high solubility); in 1970-1985—formulates based on synthetic pyrethroids and biological pesticides (main characteristics: lack of selectivity, resistance, costs and variable persistence); in 1985—development of genetically engineered organisms.[2]

The large use of pesticides in agriculture has began in 1950s that is indicated as the *chemical age*; in 1993 the American Chemical Society identified that 13 million of chemicals were sold in the world and 500.000 new compounds were added in the commerce annually.

Pesticide and Herbicide Use in Europe

Agriculture can be considered one of the few human activities in which chemicals are intentionally released into the environment with the aim of killing living organisms.

Statistical data on PPP sales divided into four main chemical categories (herbicides, fungicides, insecticides and other pesticides), which are expressed in tons of active ingredients, are gathered annually in most of the Member States and reported to Eurostat (Fig. 1). Data are available for the period 1990-1998. A statistical analysis of the data showed that countries with the largest agricultural areas (France, Italy, Spain and Germany) demonstrated the highest consumption of pesticides in term of sold volumes. In terms of pesticide usage per hectare, Italy, France, the Netherlands and Belgium emerged as the heaviest users, reflecting the intensive nature of agriculture in these countries and/or particular characteristics of the crops, requiring more intensive pesticide treatment than others (e.g., fruit and grapes).

A wide range of different chemical compounds are currently in use in the EU. More than 600 single, active compounds are applied by farmers on stable cultivated agricultural soil.[3] Herbicides

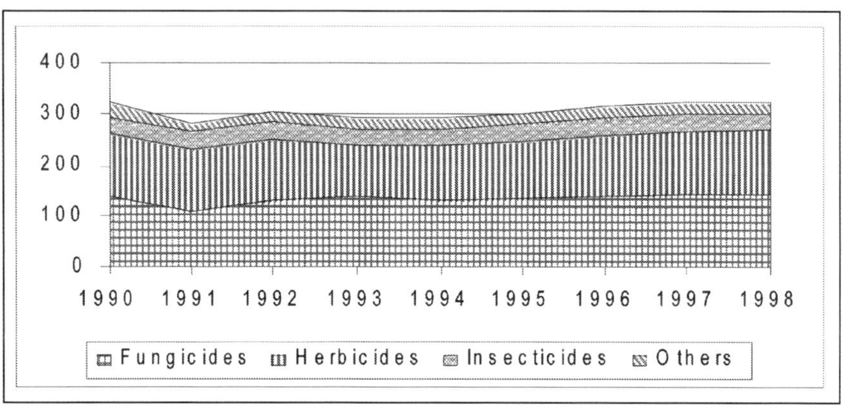

Figure 1. Total sales of pesticides in the EU-15 (in kilotons of active ingredients).

include chemical classes such as carbamates, ureas, sulphonylureas, triazines, chlorophenoxyacids and others, eg glyphosate. For example, the use of 525 active ingredients is allowed in France, 391 in Italy, 531 in Spain, 450 in UK.

Agriculture is by far the biggest PPP-using sector; non agricultural use is estimated as only 2% of the total pesticide use.[4]

The EU currently declares approximately 320.000 tons of active substances sold per year.[5] This amount represents 25% of the total world market of PPPs. The most sold products are: fungicides (43%), followed by herbicides (36%), insecticides (12%) and other pesticides (9%).

The crop protection market is estimated to represent more than 6 billion of euros for the European Union. PPP sales considered in monetary terms were increased annually in the years preceding 1999.[6]

In term of volumes of pesticide sold (Fig. 1) during 1991 and the period between 1993 and 1995 the use of pesticides decreased, while the trend reversed in the years thereafter with a rising tendency. An increase in the total volume of used pesticides does not necessarily mean an increase in the risks associated with their use but, on the other wise, the opposite can not be said for a reduction in the pesticide usage.

The chemical nature and the volume of PPP applied vary both on the basis of agricultural crops (for example the largest quantities of PPPs are used for grapes and cereal crops) and on a range of other factors, such as the weather trend, the seasonal factors and the prices of pesticides.

The rate of pesticide application per hectare also varies widely between the different European countries. In the last years, rate of application was lowest in the Northern European countries and highest in Southern and Western ones. Besides the highest application rates were observed in the Netherlands.

In the Northern and Central European countries, the use of herbicides is prevalent (Fig. 2),[7] while in the Southern and Western ones the use of insecticides and fungicide dominates. The rate range varies from less than 2 kg/ha in Denmark, Finland and Sweden to over than 10 kg/ha in Belgium, Ireland and Netherlands according to Wossink et al.[8] Intensive farming systems with a great crop yield per hectare favourite a higher usage of pesticides. That's why the pesticide usage in the Netherlands is so high, even if the volume of the pesticide applied per unit of crop production is low. This consideration allows concluding that the sale of pesticides in the Netherlands is equivalent to 1.56 kg/year per 1000 euros of crop production against a value of 2.19 kg /year calculated for Sweden.

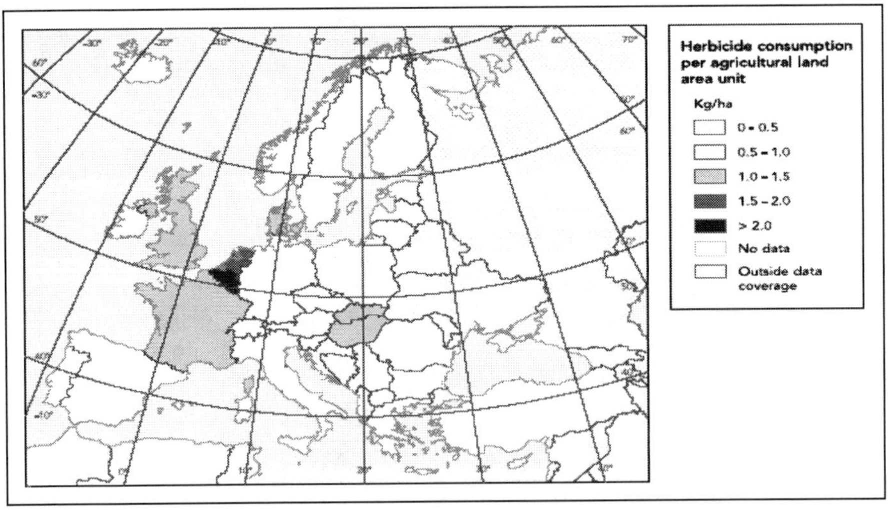

Figure 2. Use of herbicides in Europe (years: 1996 for EU and 1998 for accession countries).

Table 1. Sales of pesticides in 1998 (tons of active ingredients per year)

Country	Fungicides	Herbicides	Insecticides	Other	Total
EU-15	138.784	117.006	37.738	28.787	321.965
EURO Zone	131.354	88.067	35.873	25.489	280.783
Germany	10.530	17.269	6.276	4.809	38.884
Spain	11.984	9.413	10.173	3.500	35.070
France	58.807	36.439	4.672	7.835	107.753
Italy	24.761	9.555	8.390	4.092	46.798
Netherlands	5.127	2.921	1.577	1.097	10.722
Portugal	10.475	1.914	1.079	914.000	14.382
UK	6.612	24.063	1.607	3.073	35.354

Total sales of herbicides (Table 1) in EU-15 were estimated in 117.006 tons for 1998 while the total sales in EURO zone were calculated in 88.067 tons.[9] Herbicide sales increased in both areas since 1993 values which corresponded in 99.368 tons and 74.756 tons.

France is the largest EU pesticide market in terms of sold volume, accounting for 31% of the pesticides sales in the EU in 1996. Italy, the second most important market, is a far behind France with 16% of the European market, followed by UK (12%), Germany (12%) and Spain (11%). The top five markets, i.e., France, Germany, Italy, UK and Spain, account for over 80% of the EU sales.[10]

In considering this record it must be paid attention that the values of pesticide sales are not greatly harmonized for all the countries. For example in Belgium, France and Portugal total sales include pesticide usage for non agricultural use as biocides.

Another consideration on pesticide use that must be pointed out regards the different classes of pesticides: herbicides are applied at different rates in different countries, i.e., about 6 kg/ha in Belgium, about 4 kg/ha in UK, 3 kg/ha in Netherlands and 1 kg/ha in Italy.

European Community legislation regulates the use of plant protection products, of biocides and of some pesticides considered dangerous chemicals. European Union Directive 79/117/EEC forbids to place on the market and the use of plant protection products containing some active substances. European Union Directive 91/414/EEC (15 July 1991) regulates the authorization, the placing on the market, the use and the monitoring in the European Community of plant protection products. The first stage of the working program established a priority list of 90 active substances to be reviewed, among them are 33 herbicides. The total number of active substances authorized in EU countries in 1993 was 808. Directive 98/8/EC of the European Union Parliament regulates the placing of biocidal products on the market.

The new Water Framework Directive (WFD, Directive 2000/60/EC) marks a change in Community water policy towards a coherent and integrated framework for assessment, monitoring, and management of all surface waters and groundwater based on their ecological and chemical status. The targets and principles set out in Directive 91/414/EEC for pesticides were translated into objectives for all waters and will be implemented on a river basin scale. With the adoption of the WFD, Community water policy is based firmly on the precautionary principle and the sustainable use of water.

Updated environmental requirements of the existing surface water Directive (75/440/EEC), the Directive on discharges of dangerous substances (76/464/EEC) and the groundwater Directive (80/68/EEC) have been incorporated into the WFD. It is planned that, once the WFD will be fully operational, these Directives will be repealed in 2013.

For the protection of surface waters, the Directive introduces criteria for establishing a list of priority substances and priority hazardous substances, for which specific measures such as quality standards and emission controls must be taken in order to reduce or eliminate emissions, discharges and losses. A list of 33 priority substances was adopted in 2001 (Decision No. 2455/2001/EC); 13 were PPPs. Whilst the Community will propose the measures for priority substances by the end of 2003, Member States must prepare comprehensive programs of measures within river basin management plans by 2009, which include measures against pesticide's pollution.

In order to achieve a good groundwater status, the Commission proposed in 2002 criteria for assessing the chemical status of all pollutants and the reversal of upward trends in their concentration. As regards active substances contained in pesticides (and their relevant metabolites) the present limit value (0.1 µg/l), which is an exclusion criteria for authorization purposes, is considered as the maximum permissible concentration for defining good groundwater chemical status.

There are several risks for the environment related to pesticides. The most consistent are:

- Consumption of pesticides: the risk varies considerably, depending on specific characteristics (i.e., toxicity, persistence) of the active ingredients and pattern usages (i.e., volumes applied, application period, application method, type of crop treated, type of soil);
- Soil contamination: accumulation of residual concentrations in soil;
- Water contamination: diffuse contamination of ground and surface water.

Contamination of European Freshwater by Herbicides

Freshwater is an essential and highly vulnerable resource of human society. The irrigation practise applied in the production of agricultural crops, which currently regards the 17% of the whole agriculture, greatly contributes to increase productivity. Actually 75% of the irrigated lands are located in the developing countries.

Agriculture is the largest user of freshwater resources, using a global average of 70% of all the surface water supplies. Except for the water lost through evaporation and transpiration, the water used in agricultural practices is recycled back to the surface water or groundwater. Agriculture is therefore the main cause and the victim of water pollution. Discharge of pollutants to surface and/or groundwater caused water contamination events. The impact of pesticide contamination on the water quality is due to different factors: pollution due to the active ingredients contained in the pesticide formulation; to the impurities contained in the commercial formulates; to the additives that are mixed with the active ingredients and to the transformation products that are formed during chemical, microbial or photochemical degradation of the active ingredients.

In an agricultural ecosystem, the risk of groundwater contamination from herbicide use is ultimately determined by the relative rates of percolation and degradation within the soil profile, as well as by factors controlling these processes, such as climate, soil properties, microbial activity and chemical properties of the herbicides. Groundwater contamination is less likely to occur if the degradation rates of parent compounds and their metabolites exceed their percolation rates through the soil profile. Pesticide leaching may contaminate groundwater resources directly through the water percolation when the applied pesticides are particularly mobile.

Besides, the pesticide runoff is the main process that can contaminate the surface water. It occurs if the rainfall events prevail on the soil absorption. The active surface soil depth that interacts with runoff is about 2-10 mm.[11] Surface water also can be contaminated directly by pesticides by the drift caused by pesticide application.

A nonpoint source of water pollution is generally considered those human activities, which does not create a point pollution source into the receiving waterbodies. In contrast, a point source of water pollution is considered the human activity, which has a point input of pollution directly into receiving waterbodies. Obviously, a nonpoint pollution source is more difficult to identify, measure and monitor than a point source.

Conventionally, in the most countries, all agricultural practices and land uses are treated as a nonpoint pollution sources. The main characteristic of a nonpoint source is that it varies in function of hydrological conditions and therefore it is not easily to measure or control.

The distribution of monitoring data on pesticide contamination is generally scarce in the world and particularly in the developing countries. The pesticide analysis is included in the monitoring program of the most of western countries. However the relative high cost of analyses and the necessity to sample during definite period of the year linked to pesticide application often preclude the possibility of obtaining extensive data set. Many of the developing countries have great difficulties to carry out the adequate chemical analyses on the water samples due to problems such as inadequate instruments, impurity of reagents and high analysis costs. Pesticide monitoring requires highly flexible sampling programs that can be undertaken at the same time or after the periods of pesticide application. For example, the pesticides that are highly soluble in water must be monitored close to the periods of their use. In the United States triazines (atrazine and cyanazine) and alachlor are amongst the most widely used herbicides. They are mainly applied in May. Studies conducted by Schottler and his colleagues indicated that 55-80% of the pesticide runoff occurred in the month of June.[12] It means that many pesticides can only be detected short time after application; therefore monitoring programs that operate on a monthly basis can not be useful to quantify the presence of pesticides in the surface waters.

Surface waters, including rivers and lakes, are usually contaminated by the most used pesticides, especially herbicides. In spring and early summer, after herbicide applications, pesticide concentrations in surface waters can be much higher that in groundwater. This seasonal peak concentration of pesticides in fresh waters in the period following their use on fields indicates that runoff, the drainage of excess rainfall from the surface of fields, is the main factor responsible for water pollution.

The Environmental Protection Agency of the United States (US-EPA)[13] reported in its Report to Congress that 65% of the considered rivers in the USA were affected by nonpoint source of pollution. The US EPA identified agriculture as the main cause of water quality deterioration in rivers and lakes. In considering wetlands, the US EPA reported that "*the agriculture is the most important land use causing wetland degradation*".

Pesticides are considered as one of the first four categories of pollutants significantly associated with the agricultural practices. The US-EPA concluded that "*...more than 75% of the States reported that agriculture activities posed a significant threat to groundwater quality*". The US-EPA National Pesticide Survey pointed out the 10.4% of the public wells and 4.2% of the rural wells contained detectible levels of one or more pesticides.[14]

Since the 1970s there was a growing concern in Europe about the increase of nitrogen, phosphorous and pesticides concentrations in surface and groundwater. Intense cultivation pointed at the conclusion, already drawn by the France in 1980, that agriculture is a significant nonpoint source of surface and groundwater pollution.[15] The National Institute of Public Health and Environmental Protection in the Netherlands (RIVM) concluded that "*groundwater is threatened by pesticides in all European states. This is obvious both from the available monitoring data and calculations concerning pesticide load, soil sensitivity and leaching.... It has been calculated that on 65% of all agricultural land the EC standard for the sum of pesticides (0.5 μg/l) will be exceeded. In approximately 25% of the area this standard will be exceeded by more than 10 times...*"[16] The extent of pesticide usage in Europe suggests that environmental contamination could be widespread; however, the magnitude and distribution of this pollution is only started to be more accurately characterized. Previously, poor evidence on pesticide contamination of waters was documented but with the adoption of the Drinking Water Directive (Council Directive 80/778) the drinking water suppliers were obliged to monitor systematically drinking water supplied for human consumption for a large range of pollutants. For example, between 1985 and 1987 in the UK the analyses of pesticides showed that the Maximum Admissible Concentration (MAC) for the single pesticides was exceeded in 298 water supplies and that for the total amount of pesticides in 70 supplies.[17] It was calculated that in 1992 approximately 14.5 million of English and Welsh people lived in areas supplied by drinking water resources in which pesticide level breached the MAC.

At European scale the data on the quality of drinking water became available by 1995 when a study founded by the EU showed that approximately 30% of drinking water supplies exceeded the standard of the EU Drinking Water Directive with a large variability among the Member States (Table 2).[18] In many regions the authorities tolerated (at least for a transitional period) the exceeding of the 0.1 μg/l limit for each single pesticide.

Table 2. Drinking water contamination due to pesticides in the EU in 1995

Country	Drinking Water Production Million m^3	Percent of Groundwater Resource	Pesticide Level > 0.1 μg/l (%)	Pesticide Level > 0.1 μg/l in Groundwater (%)	Pesticide Level > 0.1 μg/l in Surface Water (%)
Germany	6052	64	15	15	15
Austria	450	49	7	15	0
France	6080	62	48	40	60
Italy	8465	48	31	50	50
The Netherlands	1227	69	48	25	100
Denmark	348	99	5	5	–
UK	7620	28	26	15	30
Greece	950	68	12	–	50

It must be noted that the MAC of 0.1 μg/l for each individual pesticide was frequently criticized, because that it does not take into account the toxicological significance of the different pesticide presence. The United States, for example, applies differentiated values for each pesticides[19] as it is suggested by the WTO Guidelines.[20] However, many toxicologists argued that the MCL value used in the EU should be retained, since the knowledge of the environmental risks on pesticide use is mainly incomplete, particularly concerning synergistic or interaction effects.

In fact, it must be considered that there is a complete lack of systematic monitoring data of pesticide levels in European waters and where the data exist, usually it does not belong to a systematic monitoring network. In many European countries monitoring programs are still under development and in many others available information often can not allow to assess and predict a general trend. Some basic monitoring networks have been developed for rivers and lakes focussing on nutrients and organic pollution following the water questionnaires reported by member states to the EU under the standard reporting directive 91/662/EC. The results related to some herbicides are showed in Table 3.

Table 3. Number of countries where some herbicides are allowed (A) and monitored in the EU-15

Herbicides	Number of Countries	
	A	P/N
Atrazine	10	7/7
Simazine	13	7/7
Diuron	13	5/5
Isoproturon	14	4/4
Linuron	14	4/4
Bentazone	15	4/4
Propazine	10	4/5
Alachlor	5	4/5
Metolachlor	10	4/5
Atrazine-Desethyl	3/3	
Metobromuron	11	3/3
Bromoxynil	14	2/2
Ioxynil	15	2/2
Terbuthylazine	14	2/3

P = presence of herbicide in waters; N = number of countries in which herbicide was investigated.

Pesticide presence in groundwater is of particular concern. On average, 65% of European drinking water is supplied from aquifers and they often take a long time to recover an acceptable quality levels. According to the report Groundwater quality and quantity in Europe out of approximately 600 pesticides usually applied, only about 30 have been monitored.[3]

Pesticide Contamination of Water Resources in the United Kingdom

All the pesticides used in the UK must be approved by the official regulatory authority and extensive data on fate and effect of pesticides in the environment are required.

The data concerning the use of PPPs are widely available. UK long-term trends in pesticide use showed that the weight of the applied active substances decreased for 19% from year 1986 to year 1996. These data also showed that a significant change in the type of active substances was observed. For example, organochlorine pesticide usage decreased because of the revocation or the restriction in use. Isoproturon, a cereal herbicide, showed an increase in its use of 22% from 1986 to 1996.

The recent published reports on drinking water monitoring for pesticides in England and Wales published by the Drinking Water Inspectorate (DWI) showed that the drinking water standards was exceeded by 23 individual pesticides in 1996, 15 in 1997 and 12 in 1998.[21] Among herbicides, the maximum detected levels was 0.21μg/l for atrazine, 0.2 μg/l for diuron, 0.13 μg/l for isoproturon and 0.14 μg/l for simazine in 1997.

A great effort of monitoring programs is focused in the surveillance of pesticide levels in surface waters. During 1997 a total number of 1419 sites were monitored for pesticides by the Environment Agency in England and Wales. Considering the 163 individual pesticides analysed in surface water, 58% were detected above 0.1 μg/l on at least one occasion in 1997, while 22% were detected below the 0.1 μg/l level and 20% were never detected. Isoproturon, mecoprop, diuron and MCPA exceeded the MAC level most frequently (Table 4).

Figure 3, shows some pesticides exceeding 0.1 μg/l in surface freshwaters between 1993 and 2002.[23] The great majority of the substances were used for agricultural purposes suggesting that agriculture continued to be the main source of freshwater pesticide pollution.

In 2002, for the first time since 1999, the cereal herbicide isoproturon is not the most frequently occurring pesticide. The nonagricultural herbicide diuron occurs most frequently, with herbicide mecoprop second and isoproturon third. Diuron is a persistent herbicide used on noncrop areas, particularly railways lines, recreational areas and other hard surface. Use of diuron as an antifoulant has been ceased in November 2002. Simazine shows an increasing level that can be explained as much targeted monitoring as by variation in agricultural practice.

Groundwater monitoring is increasing in importance as part of the Water Framework Directive (WFD). In 2001, over 600 groundwater sites were monitored. Figure 4 shows some pesticides between the most frequently detected in groundwater at concentrations greater than 0.1 μg/l.[24]

Table 4. Herbicides which most frequently exceede 0.1 μg/l surface waters[22]

Herbicide	Total Number of Samples	% of Samples >0.1 μg/l
Isoproturon	3571	17.4
Mecoprop	3526	12.6
Diuron	3759	11.9
MCPA	2120	5.7
Simazine	6284	5.3
Atrazine	6409	4.6
Bentazone	1638	1.5
Chlorotoluron	3619	1.4

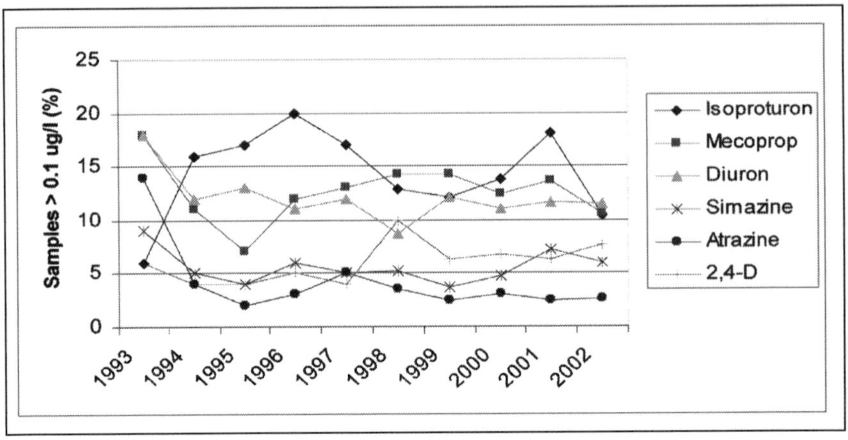

Figure 3. Trends in the pesticides occurring in surface waters, 1993-2002.

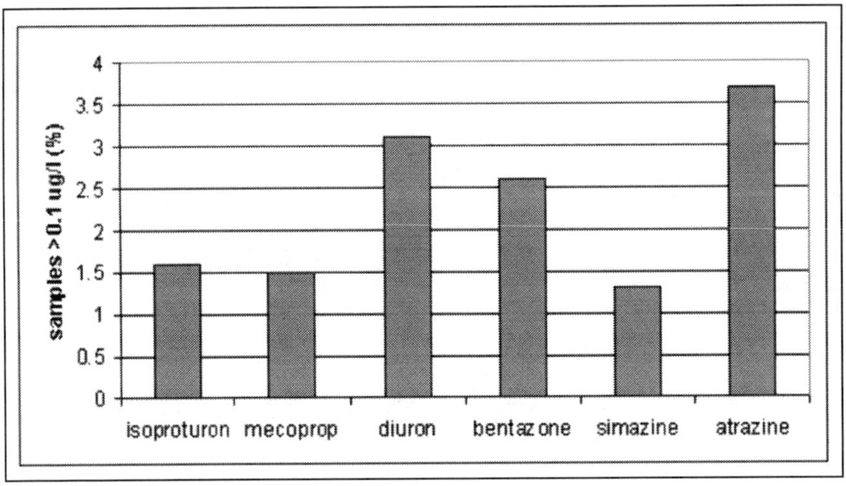

Figure 4. Percentage of groundwater analyses >0.1 μg/l in 2001.

The most frequently found pesticide was atrazine, mainly because of its historical use as herbicide in noncrop areas. Atrazine has been the most frequently found pesticide in groundwater for many years. While the ban on use of atrazine in noncrop areas may have reduced herbicide level, its increasing use on maize fields may reverse the trend.

Pesticide Contamination of Water Resources in Denmark

Denmark has an extensive legislation in the environmental field and it is now redoubling its efforts against the spreading of dangerous chemicals. The Danish approach lies in reduction of the consumption of hazardous chemicals, so that future generations can inherit a best environment for living. The problems caused by dangerous chemical substances extend across national boundaries, which is why Denmark gives such high priority to the European and international efforts. Danish policies and legislation in the environment are based on a holistic point of view, which embraces the human health and environment. Danish legislation covers the industrial use of chemicals, as well as the chemicals used in domestic products.

Danish regulation is particularly severe in two areas: agricultural pesticides and biocides, because these substances are xenobiotics, deliberately spread in the environment, e.g., for killing weeds, fungi and insects. The greatest Danish efforts in safeguarding the environment are directed towards restricting the use of harmful substances, as those that are persistent, accumulate in living organisms and suspected of being carcinogens or hormone disrupters.

In Denmark, the special action plans have been drafted for limiting the consumption of a dangerous substance, e.g., for PVC, phthalates and agricultural pesticides. Danish regulation of chemical use is tightly bound to international agreements and, not only, to EU legislation. The overall responsibility in this field is kept by the Danish Environmental Protection Agency (Danish EPA), which is subordinate to the Ministry of Environment and Energy.

In Denmark, the environmental impact of chemicals is regulated by two main acts, i.e., the Environmental Protection Act, published in 1974 and the Chemical Substances and Products Act, published in 1980.

The Environmental Protection Act deals with releases into air, water, soil and subsoil some chemical substances, which can be dangerous to the human health or environment.

The Chemical Substances and Products Act gives the precise regulations on the distribution, consumption and disposal of chemical substances. The purpose of this act is to prevent damages to the human health and environment and to promote the use of cleaner technologies.

Agricultural pesticides can be used in Denmark only after the approval of the Danish EPA. Nowadays, about 190 pesticides are approved and used as ingredients in several hundred products, while a large number of pesticides have been banned.

Biocides are substances manufactured for the purpose of killing living organisms. They can be found, for instance, in certain preservatives, disinfectants and pesticides. The main problem of these poisons is that they can be dangerous to untargeted organisms into the environment. The use of biocides must be approved by the Danish EPA before they can be sold and used. In recent years, Denmark has forbidden the use of arsenic in impregnated wood and banned the use of irgarol and diuron, which are harmful to the aquatic environment, in paintings for boats.

Besides, Denmark has pursued action plans in the field of pesticide use since 1986. The goal was that of reducing the consumption of agricultural pesticides through the application of a number of restriction measures.

Agriculture is the largest consumer of pesticides in Denmark, therefore this area represents one of the country's major challenges. About two-thirds of the pesticides used in Denmark are used in agriculture, where they are of considerable significance to productivity and therefore to the economy. Farming employs about 84.000 operators, which corresponds to 3.5% of the total employment. In 1997 the number of farms was 60.900 with an average size of 44 ha.

Pesticides are first and foremost a serious threat to the quality of drinking water resources. The Pesticides Action Plan II is operating now in Denmark. This action plan was drafted on the basis of a report published by The Bichel Committee, which thoroughly investigated all of the aspects involved in the use of pesticides in agriculture.[25] The committee concluded that the consumption of pesticides could be halved in Denmark without causing any reduction in crop production. Besides a special tax was introduced in Denmark to reduce the pesticide consumption. This tax is actually used to support research studies for developing alternatives to pesticides.

As regard pesticide contamination, residual concentrations (Table 5) were found in one third of the samples taken from surface groundwater. By considering the positive samples, the MAC limit for drinking water was exceeded in 13 of the samples.[25]

A large number of other chemicals, which was used in past years in agriculture, were also found in a relatively large number of wells. Triazines, including atrazine, and their metabolites was included in this monitoring program, because they are equally used in agriculture and urban areas. The presence of positive samples is particularly high if we also include the triazine metabolites: DEA and DIA are in fact present in more than 20% of the considered samples.

These data (Table 5) are available from the LOOP program, a Danish acronym for "Land Monitoring Catchment Areas" in which the monitoring is carried out in agricultural areas with known farming practices.

Table 5. Pesticides and relative metabolites found in groundwater in the five land monitoring catchment's areas during the period 1990-1997

Groundwater Monitoring	N° of Samples	Samples Containing Herbicides	Samples > 0.1 µg/l	Concentration Max.
Pesticide	**Number**	**%**	**%**	**µg/l**
Atrazine	471	6.7	1.0	0.12
Desethylatrazine	173	22.0	1.7	0.22
Desisopropylatrazine	150	28.3	9.4	0.24
Hydroxyatrazine	46	5.4		0.02
2,4-D	386	4.3	1.1	0.12
Bentazone	223	22.6		0.05
Cyanazine	173	3.4		0.02
Dichloroprop	466	6.7		0.04
Dinoseb	467	3.8	1.0	0.12
DNOC	467	4.8	1.0	0.10
Isoproturon	236	8.1		0.05
MCPA	467	9.5		0.07
Mecoprop	463	11.4		0.08
Metamitron	143	5.3		0.01
Pirimicarb	23	18.2		0.01
Propyzamide	18	11.1	11.1	0.11
Simazine	461	2.9		0.05

Pesticide Contamination of Water Resources in Italy

Agriculture is by far the largest water user in Italy; it is estimated that nearly 2/3 of the available water resources are used for irrigation. Nonetheless, groundwater resource is exceptionally used for irrigation.

A first notable feature of Italian agriculture is the *fragmentation*. There are more than 3 millions of farms with an average surface of 7.5 ha. Italian agriculture is typically concentrated at the territorial level, even as a consequence of orography. The agriculture is thus concentrated in few areas with a diffuse segmentation.

Italian agriculture is among the largest users of fertilizers and pesticides in the OECD.[26] Total consumption of mineral fertilizers has reached a peak in 1992; in the following years the total quantities of sold fertilizers have somewhat decreased, even if the annual mean quantity was substantially stable in 1991-1994, estimated in about 65 kg/ha of nitrogen, 46 kg/ha of phosphorus and 27 kg/ha of potassium.

A slightly better trend was estimated for the pesticide use (Table 6). After a dramatic increase in the period 1985-1988 (reaching a mean annual pesticide use of 16,5 kg/ha), pesticide consumption decreased on an average of 12 kg/ha arable land. Nonetheless a huge regional variability was observed: regions like Emilia Romagna and Trentino-Alto Adige, where fruit production is concentrated, reached sold volumes 2-3 times higher than the mean average volume. In any case, Italy must be considered one of the highest pesticide consumers in the EU: 766 kg/km^2 against 502 kg/km^{2}of France, 442 kg/km^2 of Germany and 417 kg/km^2 of the UK. Other Mediterranean countries as Spain and Portugal have higher unitary consumption than Italy. It can be argued that these differences maybe due to the different productive specializations and to the highest importance of crops like fruit, grapes etc.

The number of active ingredients sold in 1998 that overall the volume of one ton per year were 309. Total volumes of sold pesticides reduced from 84.130 tons in 1996 to 79.026 in 1997 and to 56.199 tons in 1998 (Fig. 5).[27] Fungicides were the pesticides more sold in Italy: 39.5% of the total volume followed by insecticides with 29.7% and herbicides 16.8% (Fig. 6).[24]

Table 6. Pesticide's consumption in Italy (thousands of tons)[26]

	1990	1994	1995	1996	1997
Herbicides	*27.8*	*25.9*	*25.9*	*25.0*	*24.9*
Insecticides	36.5	33.4	33.4	31.4	30.5
Fumigants	6.7	4.1	4.7	4.9	5.1
Fungicides	65.7	46.8	49.4	48.3	45.8
Other	4.5	4.1	4.3	4.5	4.4
Total	**141.2**	**114.2**	**117.7**	**114.1**	**110.7**

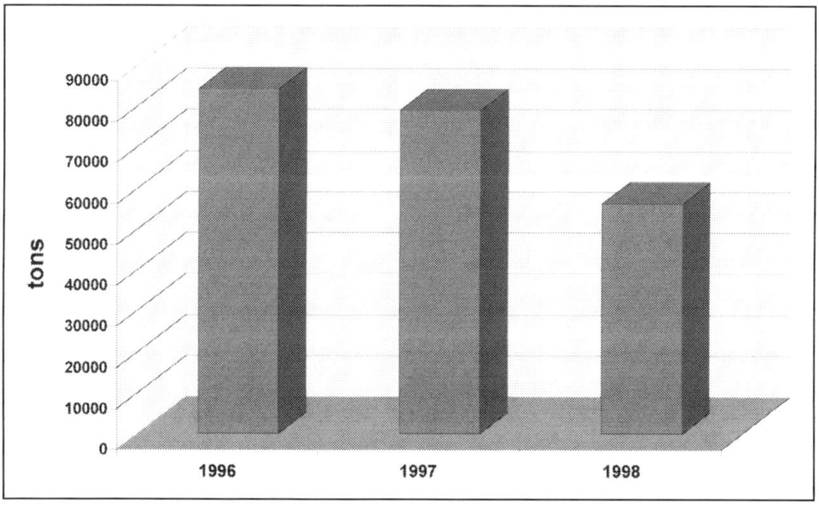

Figure 5. Active ingredients (tons) of pesticide sold in Italy in 1998.

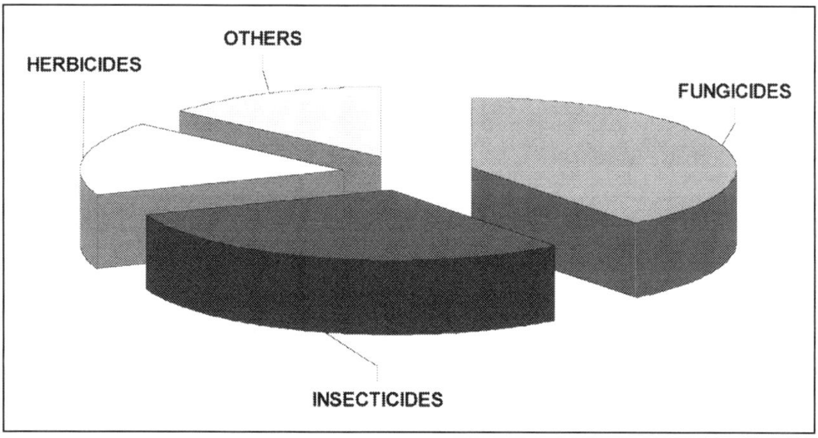

Figure 6. Percent of pesticide sales in Italy in 1998.

Table 7. Priority index (P.I.) and monitoring data on water samples

Active Ingredient	P.I.	%R	%P	P/R
Dalapon	10	1	0	0
Metamitron	10	3	0	0
Cloridazon	10	3	0	0
Dazomet	10	1	0	0
Dimetoate	9.5	49	3	0.06
Metolachlor	9	63	32	0.51
Alachlor	9	75	24	0.32
MCPA	9	13	1	0.08
Bentazone	9	19	3	0.16
Metribuzin	9	25	3	0.12
Propanil	8	26	1	0.04
Terbuthylazine	8	68	46	0.68
Molinate	8	50	15	0.30
Linuron	8	31	3	0.10
Simazine	8	85	29	0.34

% R—% of laboratories that search for pesticide; % P—% of positive samples.

Italian monitoring data of freshwater samples related to pesticide contamination in 1997 are reported in Table 7.[28] The pesticides are listed in a decreasing order in respect to Priority Index (P.I.), an index calculated on the basis of pesticide sold quantity, of the field application rates and of the water repartition evaluated with the Mackay model. In Table 7, the percent of laboratories that searched pesticides (%R) and the percent of positive samples (%P) are also showed. It is evident that the monitoring of some pesticides is very poor, even if they have an elevate P.I. Some active ingredients, like metolachlor and terbuthylazine, are contaminants widespread in freshwaters.

Conclusion

Review of pesticide contamination in the EU and in particular in UK, Denmark and Italy showed a general concern about pesticide contamination of fresh and groundwater resources. It is evident that a diffuse contamination of herbicides used in the past (i.e., atrazine) is still persistent and that the metabolites of triazine herbicides greatly contribute to the pollution of water resources.

The UK and Italian data showed that metolachlor and isoproturon are also widespread herbicides in aquatic ecosystems.

Intensive and specific monitoring programs carried out in agricultural areas, as done by Denmark with the LOOP program on Land Monitoring Catchment Areas, can greatly contribute to reduction and prevention of the water contamination in EU countries.

References

1. Haskoning N. Danube integrated environmental study. Final report of the EU-Phare environmental programme for the Danube Basin. Haskoning Royal Dutch Consulting Engineers and Architects, Nijmegen: 1994.
2. Stephenson GA, Solomon KR. Pesticides and the environment. Guelph, Ontario: Department of Environmental Biology, University of Guelph, 1993.
3. EEA. Groundwater quality and quantity in Europe. Technical Report No. 22. Denmark: European Environmental Agency, 1999.
4. Eurostat. Environmental Pressure Indicators for the EU 2001.
5. Eurostat and European Crop Protection Association, 1999.

6. Commission of the European Communities. Towards a thematic strategy on the sustainable use of pesticide. Communication from the Commission to the Council, the European Parliament and the Economic and Social Committee, Brussels 1.7.2002.
7. Nixon S, Trent Z, Marcuello C et al. Europe's water: An indicator-based assessment. EEA Topic Report 1/2003.
8. Wossink GA, Feitshans TA. Pesticides policies in the European Union. Paper presented at the 20th Annual American Agricultural Law Association (AALA), Symposium New Orleans, 1999.
9. Eurostat. Environment & Energy, Agriculture, 2003.
10. Redbond M. Agrow's Complete Guide to Agrochemical Marketing Strategies in the EU. Richmond: PJB Publications, 2000.
11. Triegel E, Guo L. Overview of the fate of pesticides in the environment, water balance; runoff vs. leaching. In: Honeycutt R, Shabacker eds. Mechanisms of Pesticide Movement into Groundwater. Lewis, 1994:1-13.
12. Schottler SP, Elsenreich SJ, Capel PD. Atrazine, alachlor and cyanazine in a large agricultural river system. Environ Sci Tech 1994; 28:1079-1089.
13. US-EPA. National Water quality Inventory. Washington DC: 1992 Report to Congress. EPA-841-R-94-001. Office of Water, 1994.
14. US-EPA. National Pesticide Survey: Update and summary of Phase II results. Washington DC: Office of Water & Office of Pesticides and Toxic Substances, EPA570/9-91-021, 1992.
15. Ignazi JC. Improving nitrogen management in irrigated, intensely cultivated areas: The approach in France. In: Prevention of Water Pollution by Agriculture and Related Activities. Santiago: Proceedings of the FAO Expert Consultation, 1992. Rome: Water Report 1, FAO, 1993:247-261.
16. RIVM. The environment in Europe: A global perspective. Netherlands: National Institute of Public Health and Environmental Protection (RIVM), 1992.
17. Neil W. An evolutionary perspective on pesticide use and water pollution in Europe. Wageningen: Policy Measures to Control Environmental Impacts from Agriculture, 519 Workshop on Pesticides, EU Concert Action AIR3-CT93-1164, 1995.
18. Heinz I. Cost and benefits of pesticide reduction in agriculture: Best solutions. In: Grada A, Wossink, et al, eds. Economics of Agro-Chemicals. 1998:333.
19. EPA. Current Drinking Water Standards, 1999.
20. WHO. Guidelines for drinking water. Recommendations. 1993:1.
21. DWI. Drinking Waters 1997. London: The Stationery Office, 1998.
22. Environment Agency. The state of the environment of England and Wales freshwaters. London: The Stationery Office, 1998.
23. Environment Agency. Url: http://www.environment-agency.gov.uk/industry/agri/pests/.
24. Environment Agency. Url: http://www.environment-agency.gov.uk/commondata/ 105385/ pests_report_554125.
25. Bichel Committee. Danish Environmental Protection Agency, 1999.
26. Inea. L'agricoltura italiana conta 1998.
27. Sesia E. Dati di vendita dei prodotti fitosanitari: Elaborazioni per sostanze attive - anno 1998. Gruppo di lavoro ANPA-ARPA-APPA Fitofarmaci. 3 Seminario Fitofarmaci e Ambiente, Napoli 2001.
28. Franchi A. Controllo dei residui di fitofarmaci nelle acque: Schema di rilevazione dell'attività per la verifica degli obiettivi. Fitofarmaci e ambiente. ARPAT Firenze 2000.

Chapter 12

Application of Chloroplast D1 Protein in Biosensors for Monitoring Photosystem II-Inhibiting Herbicides

Elena V. Piletska,* Sergey A. Piletsky and Regis Rouillon

Abstract

Environmental pollution by toxic chemicals has become one of the world's most serious problems. Among the most widespread pesticides is photosynthesis inhibiting herbicides, such as atrazine, metribuzin, diuron, bromacil, ioxynil and dinoseb. They all belong to different families but have a common mode of action: binding specifically to the chloroplast D1 protein with subsequent interruption of the electron and proton flow through Photosystem II. The goal of this chapter is to evaluate the possibility of application of the natural receptor properties of D1 protein in various biosensor systems for herbicide detection.

Introduction

Environmental pollution by pesticides is a very serious problem. The appearance of herbicides in ground water and in agricultural products poses considerable problems in the control of drinking water and product quality. Regulations governing herbicide concentration in drinking water are very strictly enforced by the European Community with a maximum allowable at 0.1 µg/L level for each individual substances and 0.5 µg/L for sum of photosynthesis inhibiting herbicides.[1]

These herbicides can be monitored by means of high performance liquid chromatography (HPLC) and gas chromatography (GC) with mass-spectrometry (GC/MS).[2-5] Although the different types of chromatography permit the determination of several kinds of herbicides at one time with high sensitivity, these methods are time consuming and expensive. Taking into account large amount of samples, which have to be measured, the development of fast and inexpensive tests is very important. Recently the immunoassays based on antibodies have been developed for herbicide detection.[6,7] They permit the detection of a broad range of herbicides, but due to antibody's individual specificity, it is difficult to use the immunosensor to monitor herbicide in samples containing unknown herbicides or several different herbicides.

Great attention has been paid to the application of thylakoid membranes and photosynthetic microorganisms in environmental pollution control.[8] The biorecognition system based on the binding of certain herbicides to the photosynthetic reaction center of plants and microorganisms seems to be the most direct and simple method for herbicide detection. These systems used as sensor's recognition elements allow the detection of a broad range of herbicides. Unfortunately, their stability and sensitivity are insufficient in the most cases. From this point of view, the D1 protein, which binds specifically

*Corresponding Author: Elena V. Piletska—Cranfield University, Institute of Bioscience and Technology, Silsoe, Bedfordshire, MK45 4DT, U.K. Email: e.piletska@cranfield.ac.uk

Biotechnological Applications of Photosynthetic Proteins: Biochips, Biosensors and Biodevices, edited by Maria Teresa Giardi and Elena V. Piletska.

many herbicides inhibiting photosynthesis and subsequently electron and proton flow are interrupted through the chloroplast membrane, in combination with high stability of D1 protein preparations, opens a unique possibility for using this protein in sensor systems for herbicide detection.

D1 Protein Properties

The 32-kDa protein (D1 protein) has been a major focus of research in plant molecular biology for many years.[9] It results from interesting behavior and undefined completely functions. As for now, researches have found that:

- D1 protein is a core part of the reaction center of Photosystem II (Fig. 1).[10]
- D1 protein is trans-membrane protein of thylakoid membranes, which has five inter-membrane loops and two short "parallel" helices which do not span the membrane.
- This protein is product of chloroplast *psbA* gene, so called "photogene". It is the main product of chloroplast protein synthesis in light and its mRNA is the most abundant in the chloroplasts.
- It is synthesized and degraded in the light with rates exceeding those for other known chloroplast proteins and has some regulatory functions.
- It was observed that D1 protein is a primary target for photo-inhibition.
- It was found that even few damaged and degraded copies of D1 protein could lead to extensive proton leakage and cause very rapid loss of trans-thylakoid proton gradient.[11]
- The secondary plastoquinone acceptor Q_B is reversibly associated with D1 protein to perform important role in Photosystem II electron chain. This secondary acceptor acts to connect the single electron transfer events of the reaction center with the pool of free plastoquinone in the membrane by operating as a two-electron gate. Several classes of photosynthesis-inhibiting herbicides, such as triazines, ureas, uraciles, anilides are able to compete with Q_B for binding site within the D1 protein thereby inhibit the electron transfer through PSII and lead to plant damage and death (Fig. 1).[12]

The natural functions of D1 protein such as mediator properties and herbicide binding can be basis for application of D1 protein in biosensors for environmental monitoring.

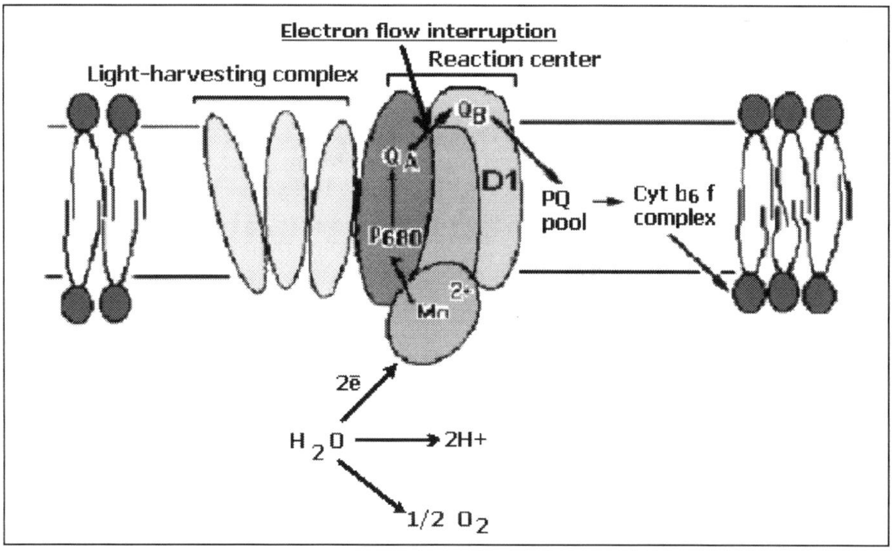

Figure 1. Reaction center of Photosystem II with electron transport pathway. Arrow shows the site of interruption the electron flow by herbicides.

D1 Protein Isolation and Purification

Isolation of the D1 protein from pea (*Pisum sativum L*) was performed by method described by Piletskaya et al.[13] Mature green leaves of pea were collected and washed in distilled water, dried with filter paper and frozen in liquid nitrogen. The frozen leaves were homogenized with isolation buffer containing 0.4 M sucrose, 10 mM NaCl and 50 mM Tris-HCl, pH 8.0. The homogenate was filtered through two sieves (Mesh No. 32 and 20) and centrifuged at 1000 *g* for 15 min at + 4°C.

The pellet was washed twice with 50 mM Tris-HCl buffer, pH 8.0, containing 10 mM NaCl, to purify the membranes from stroma proteins. The thylakoid membranes were then homogenized with a small amount of washing buffer and sonicated in a cold ultrasonic bath for 30 min. For future solubilization of the membrane proteins an equal volume of cold n-butanol (-20°C) was used. The phases were separated by centrifugation for 5 min at 1000 *g* and water phase contained the proteins was collected. The butanol extraction was repeated twice to obtain a lipid-free protein preparation.

The crude D1 protein extract was further purified by reverse-phase chromatography. The chromatography was performed on a C_{18} column with a linear gradient of iso-propanol (0-30%). The D1 protein was lyophilized and stored at + 5°C. It was found that the preparations of D1 protein were stable and suitable for the sensor design at least for 2 years.

The D1 protein without plastoquinone was also studied. Plastoquinone extraction was performed with hexane. 1 ml of hexane was added to 3 mg of the D1 protein and the mixture was shaken for 5 min, after which the pellet was centrifuged (5 min at 1000 *g*), dried and resuspended in 10 ml of 25 mM sodium phosphate buffer, pH 7.5.

Several approaches, such as electrochemistry, optical methods and assays, were tested to evaluate the possibility of D1 herbicide-binding protein application in biosensors as a recognition element.

Assays

Application of assay format is attractive because it allows a mass-screening of the samples for rapid and inexpensive monitoring of the important classes of herbicides. Three assay systems based on specific properties of D1 protein have been tested, such as ELISA-type assay, DELFIA fluorometric assay and assay based on the liposomes incorporated D1 protein.

D1 Protein as a Substitute of Antibodies in ELISA Assay

The idea of this system was to substitute the antibodies in ELISA protocol by herbicide-binding D1 protein.[13] The mechanism of action was based on the competition between free- and HPR-conjugated metribuzin for binding to the Q_B site of the D1 protein, which was immobilized by physical adsorption onto the microtitre plate surface (see Fig. 2). The increase of free herbicide concentration in the sample caused the decrease in quantity of the conjugate bound to D1 protein.

Several herbicides, representing different classes: metribuzin (triazinone herbicide), atrazine, simazine (triazine herbicides) and diuron (arylurea) were tested for their ability to generate an assay response. It was shown that all these herbicides were able to compete with the metribuzin-HRP conjugate in binding to the D1 protein (see Fig. 3). The detection limits (herbicide concentration which gives the response with a value three times higher than the STD for a water blank) for atrazine, diuron and metribuzin were estimated as $5 \cdot 10^{-8}$ M. It was found that simazine had a lower affinity to D1 protein than other herbicides and the detection limit for simazine was $5 \cdot 10^{-7}$ M. All herbicides could be measured at concentrations ranging from $5 \cdot 10^{-7}$ M to $5 \cdot 10^{-5}$ M. The method proposed was less sensitive than the ELISA procedure based on antibodies. However, the advantages of the proposed method are the group specificity for all kinds of Photosystem II inhibiting herbicides, simplicity of the D1 protein isolation and its high stability.

Liposomes

Second tested assay system was based on liposomes containing D1 protein incorporated into lipid wall and pH-sensitive dye used as a marker encapsulated inside of liposomes (Fig. 4).

The basic idea was to monitor the loss of trans-membrane proton gradient, which happens due to D1 protein-herbicide interaction,[11] and to use it for herbicide detection.

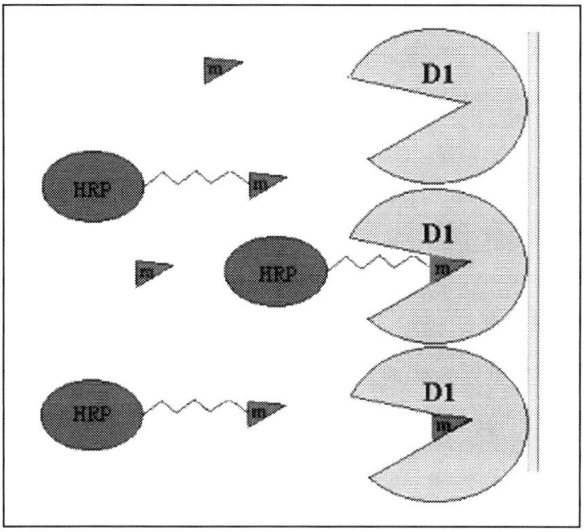

Figure 2. A scheme of the competition reaction between metribuzin-HRP conjugate and free metribuzin for the binding to the D1 protein.

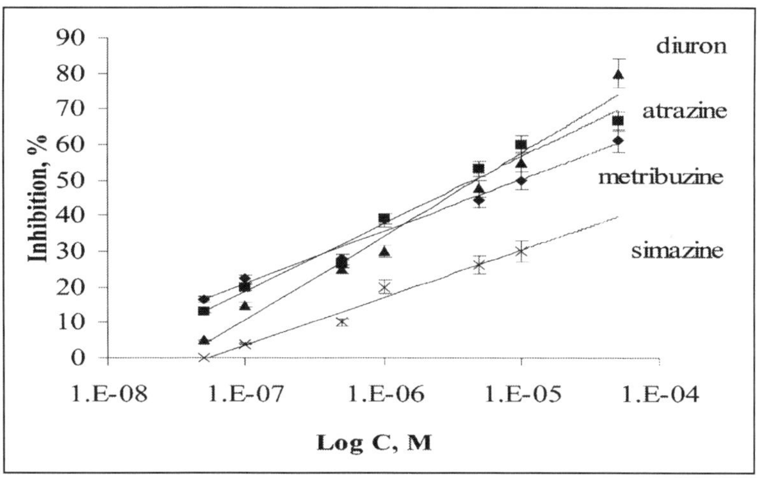

Figure 3. Inhibition of the metribuzin-HRP conjugate binding to the D1 protein by different herbicides.

2 mg of methyl orange and 2 mg of lyophilized D1 protein were dissolved in 1 ml of 0.1 M Tris-HCl buffer, pH 8.0 and added to vial contained 3 mg of egg lecithin. The mixture was ultra-sonicated for 30 min in precooled ultrasonic bath at full output. Subsequently liposomes were separated from external methyl orange on a 15-cm Sephadex G-75 column, equilibrated with 0.1 M Tris-HCl, pH 7.5. Buffer contained purified liposomes (100 μl) and 3 μl of sample contained herbicide solution were mixed in microtitre plate. This mixture was titrated by 0.1 M HCl (final volume 25-30 μl). Changes of optical absorption were measured using the microtitre plate reader (Dynex, USA) at 560 nm. This method demonstrated extremely high sensitivity (up to $1 \cdot 10^{-16}$ M) (see Fig. 5) but reproducibility between different liposome batches was low. This approach is very promising but requires further development.

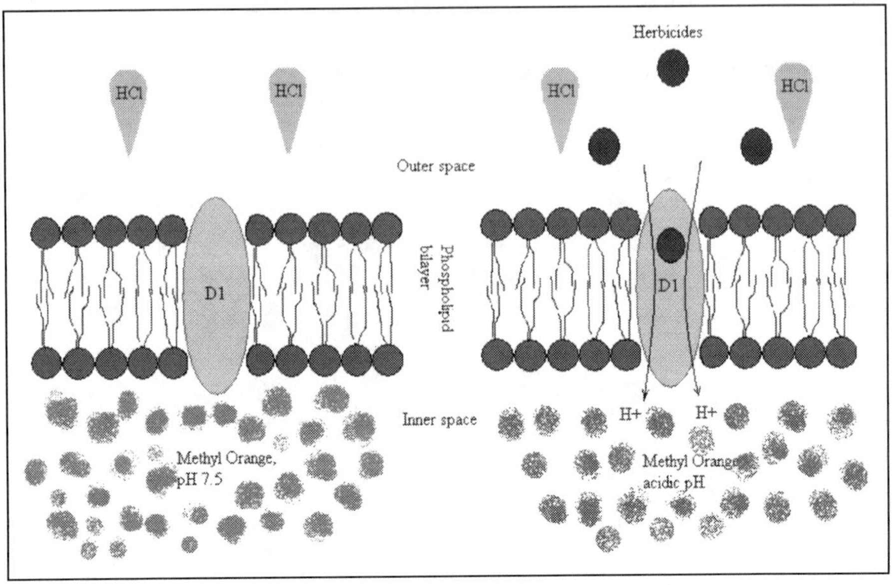

Figure 4. Scheme of D1 protein-liposome assay in action.

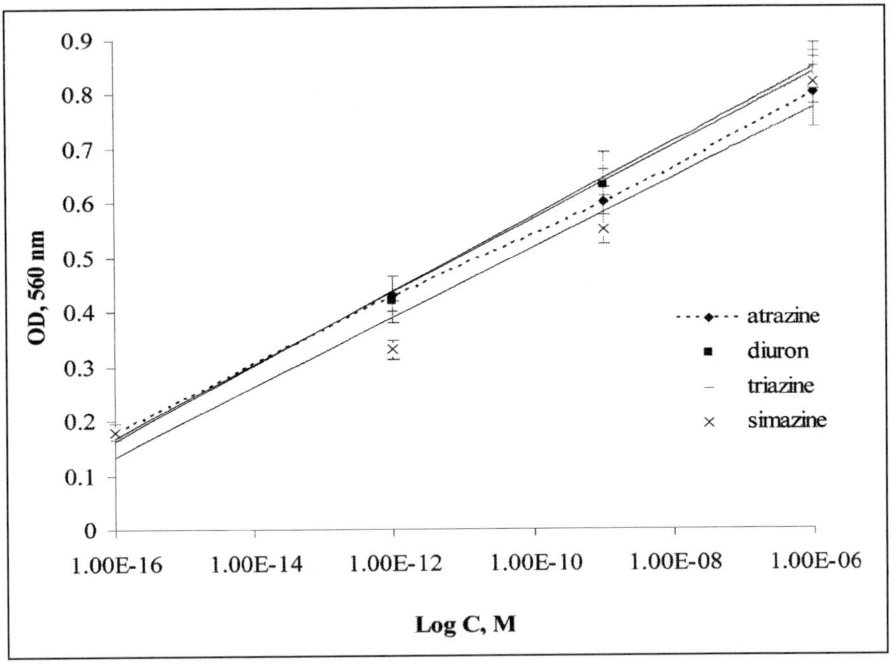

Figure 5. Dependence of optical absorbance of D1 protein- and methyl orange-contained liposomes on herbicide concentration during titration.

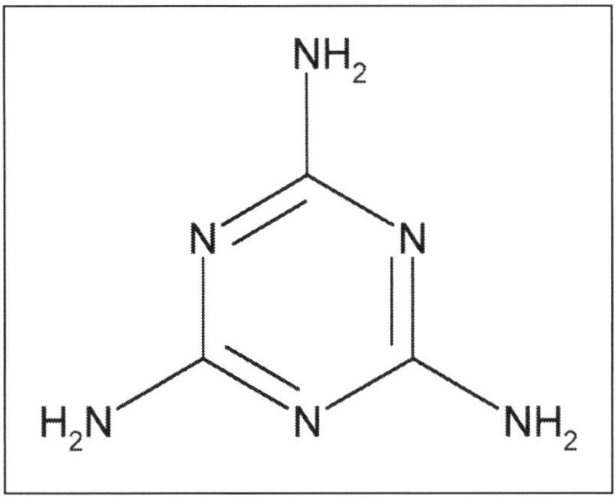

Figure 6. Molecular structure of melamine.

DELFIA Fluorometric System in Herbicide Monitoring

DELFIA or time-resolved fluorometry (TRF) is a well-established technology that exploits the unique fluorescence properties of lanthanide chelates and provides a powerful alternative to radioisotopic assays in many high throughput-screening applications.

The proposed TRF assay for herbicides detection was based on binding of D1 protein labeled with europium (Eu) to herbicide- mimicking polymer surface in presence of different concentrations of free herbicide in solution. D1 protein was labeled with Wallac- Eu chelate using protocol recommended by producer (Instruction Manual for DELFIA® Eu-N1-ITC Labeling Kit, product number 1244-302). The degree of D1 protein labeling was 1:1 (one Eu atom per one D1 protein molecule) accordingly to ratio between D1 protein concentration and concentration of europium bound to protein. Fluorescent measurements were carried out using a 1232 DELFIA Research Fluorimeter (Wallac, USA). The bottom of microtitre plate was covered with specific (Nylon 11 conjugated with melamine) or nonspecific (Nylon 11) polymer layer. Melamine can be considered as a close analogue of triazine herbicides due to triazine ring, which is the core part of all triazine herbicides (Fig. 6). Although melamine is not used as photosynthesis inhibiting herbicide in agricultural practice, it appears to demonstrate a weak herbicide action. Due to presence of three amino groups, melamine is a very convenient molecule for conjugation and covalent immobilization.

In order to reduce the nonspecific binding of D1 protein to hydrophobic surface of microtiter plate, the polystyrene surface was blocked by bovine serum albumin (BSA). Eu- labelled D1 protein (1 mg/ml) was incubated with different concentrations of atrazine ($1 \cdot 10^{-16}$ -$1 \cdot 10^{-6}$ M) for 1 h. The plate was washed 8 times by phosphate buffered saline (PBS), pH 7.5, followed by incubation with enhancing solution and measurement of fluorescence. Despite the high hydrophobic properties of D1 protein and Nylon, the adsorption of D1 protein on the Nylon 11 surface was 3-5 times smaller than on Nylon modified with melamine, which suggests that specific matrix contained herbicide molecules was created. The binding reaction of the Eu- labelled D1 protein to specific polymer demonstrated some competitive effect in presence of atrazine but it was not sufficiently sensitive (Fig. 7). The maximum inhibition of binding reaction was in the range of 20%.

In order to verify the specificity of this reaction, different concentrations of unlabelled D1 protein have been used. It was found that inhibition of the binding reaction does not depend strongly on the amount of unlabelled protein (Table 1).

It can be explained by high nonspecific binding of the Eu-contained reagent to polystyrene surface. Nevertheless, it is possible to conclude that Nylon-melamine polymer can be the useful material as triazine-mimicking coating of piezo-crystals and electrodes.

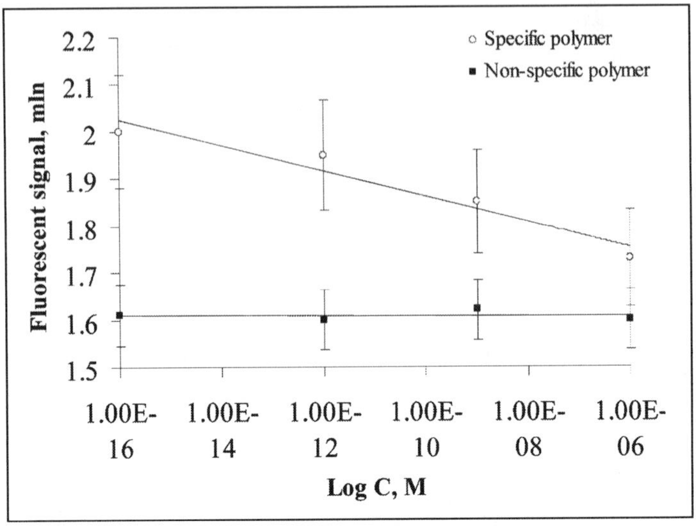

Figure 7. Dependence of D1 protein binding to specific and nonspecific surface on different concentration of atrazine presented in sample.

Table 1. The binding of the Eu-labelled D1 protein to melamine surface in absence and presence of unlabelled D1 protein (fluorescent signal, in mln)

Labelled Protein	Unlabelled Protein/ Labeled Protein (ratio 1:1)	Unlabelled Protein/ Labeled Protein (ratio 5:1)
2.7	2.5	2.4

Optical Methods

Surface Plasmon Resonance (SPR)

SPR is a phenomenon, which occurs when light is reflected by thin metal films. This optical method now is one of the basic methods for the investigation of biomolecules interactions. A fraction of the light energy incident at a sharply defined angle can interact with delocalized electrons in the metal film (plasmon) thus reducing the reflected light intensity. The adsorption of biomolecules on the metal surface results in increase of a surface plasmon wave vector and in displacement of a minimum position in reflective curve. It is possible to estimate the optical thickness of a dielectric layer by calculation of displacement values. Among the advantages of SPR technology are high sensitivity, the possibility to observe the binding kinetics in real time and possibility to work without labeling of analyte. However, there are some limitations of this method due to small size of molecules. Detection of molecules with mass smaller than 200 Da is almost impossible in direct registration scheme. Two approaches were used in order to facilitate the herbicide detection using SPR: first- the detection of plastoquinone displacement from D1 protein by herbicide and second- the competitive approach.[14]

The computer equipped SPR-set-up with Kretchmann-configuration of sensor block (Fig. 8) was used. Set-up consisted of a He-Ne (2mw) laser, 68^0 glass microprizm (K8 glass, refractive index 1.51), a flow cell (silicone rubber, 8 mm in diameter with thickness 1 mm) and a register photodiode.

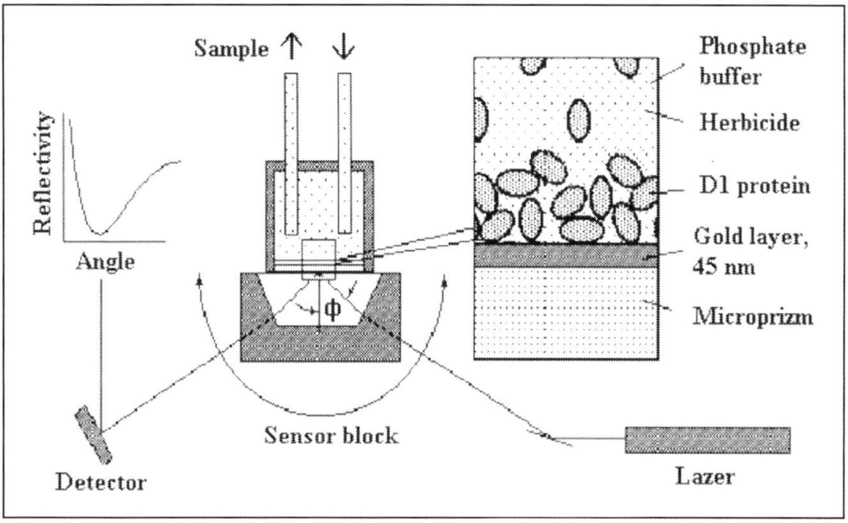

Figure 8. Scheme of the SPR set-up designed for the kinetics studies and model of the layer structure used in SPR-experiments.

Disposable microprizm and photodiode were placed on a rotary table (10 arc sec resolution). Before depositing metal layers, the glass surface of the microprizm was treated with a chromium mixture, rinsed in bi-distilled water and then dried in a dust-free airflow. Au layer (thickness 45 nm) was deposited onto the quartz surface of microprizm in special vacuum sets equipped with oil pumps and nitrogen traps (pressure 10^{-3} Pa). Before deposition, the mikroprizm surface was ion-bombarded in a vacuum (pressure 1 Pa). In order to increase the chip stability in the aqueous environment, Au film was modified by covalent binding of the thiol contained 12-carbohydrate spacer.

Detection of Plastoquinone Displacement

D1 protein containing plastoquinone was dissolved in PBS, pH 7.4 (protein concentration 5 mg/ml) and 50 μl was added to microprizm surface. The kinetic of protein adsorption was observed until it reached the saturation (approximately 15 min). The sensor surface with immobilized D1 protein was washed with PBS until the system reached the equilibrium. Solution of atrazine was injected and kinetic of the plastoquinone replacement reaction and SPR -curves were monitored (Figs. 9,10). It was found that the direct replacement of plastoquinone (molecular weight is 750 Da) from layer of D1 protein adsorbed on the gold surface by smaller molecule of atrazine (216 Da), changes the dielectric characteristics of molecular complex on the sensitive gold surface and can be registered by the device. The experiments performed with D1 protein with extracted plastoquinone shown no sensor signal.

The SPR response was proportional to the concentration of herbicide in solution (Fig. 11). The sensitivity increased (with shift 110-170 s for concentration 0.1 μg/ml) if the protein was adsorbed from the solution with pH 2.2-5.2 probably due to partial denaturation of the protein-plastoquinone complex which facilitates the access of the atrazine molecules to binding sites.

Second approach was defined as a competition assay, where small herbicide molecules (in our case, metribuzin) were immobilized on the gold surface and competed with free herbicide for the binding to the D1 protein in solution (Fig. 12).

In order to avoid the steric hydrance, herbicide was immobilized covalently on the gold surface through 17-carbohydrate spacer contained thiol group. In order to accelerate the binding, the plastoquinone was extracted by hexane from D1 protein before the experiment. For herbicide detection the solution of D1 protein was incubated with different concentrations of the metribuzin. The SPR

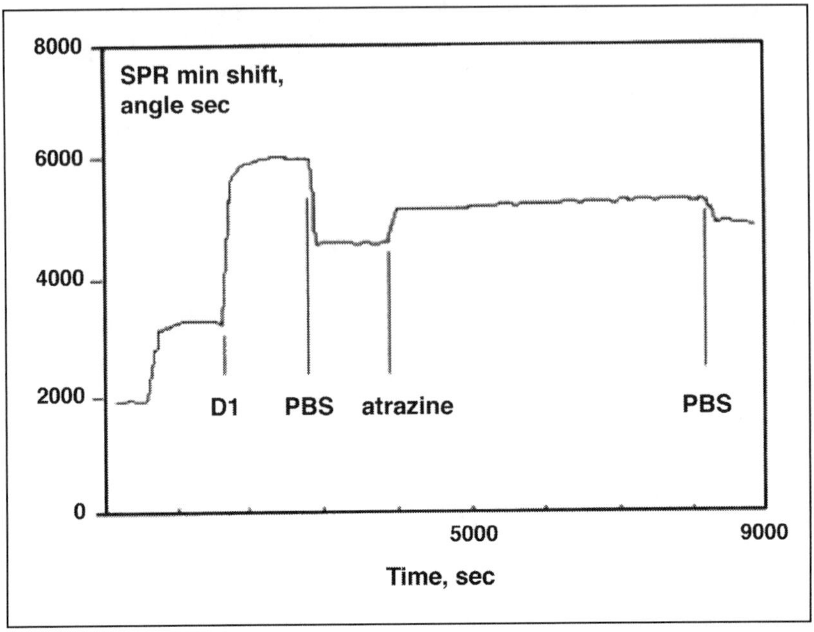

Figure 9. The sensogram shows the binding reaction between D1 protein and atrazine. D1 protein does not contain plastoquinone which is extracted by hexane. PBS- 10 mM phosphate buffer, pH 7.4 containing 2.7 mM KCL, 0.137 M NaCl.

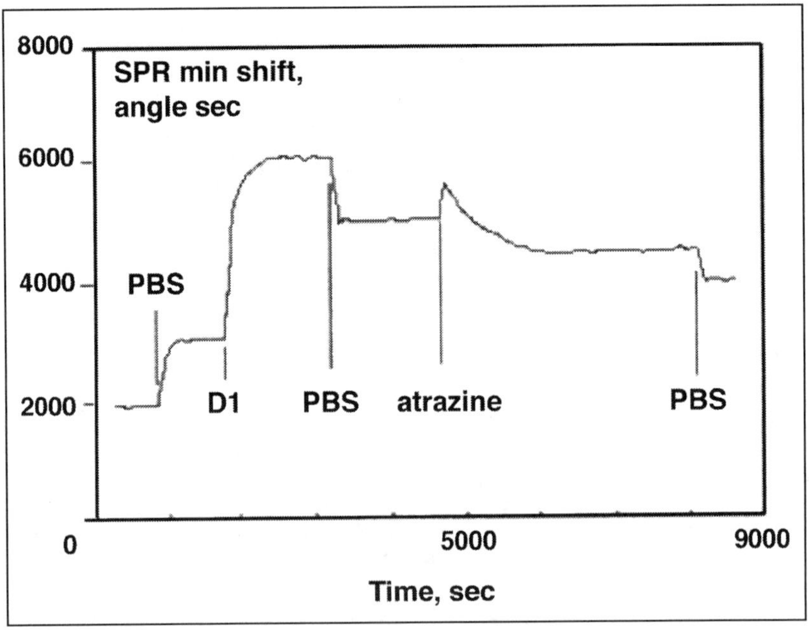

Figure 10. The sensogram shows the plastoquinone replacement from D1 protein by atrazine.

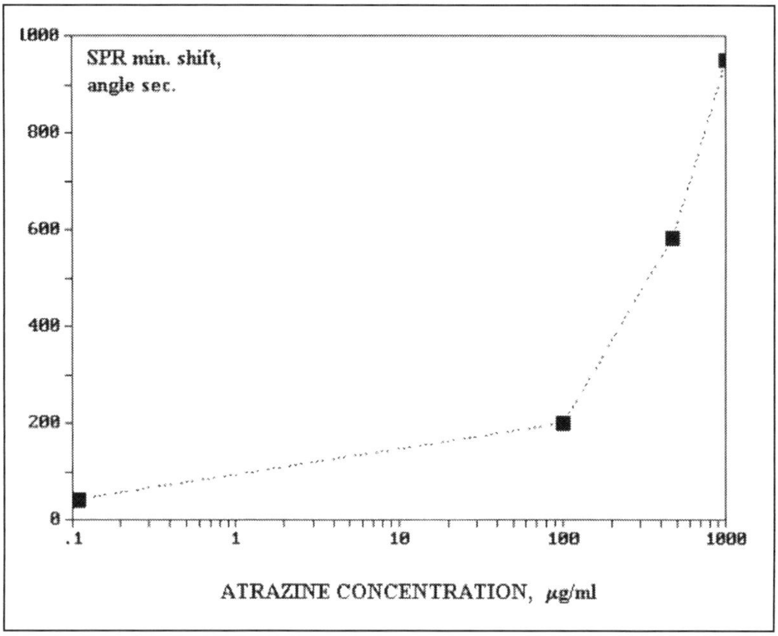

Figure 11. The calibration curve for atrazine concentrations measured by surface plasmon resonance sensor with D1 protein as natural herbicide receptor.

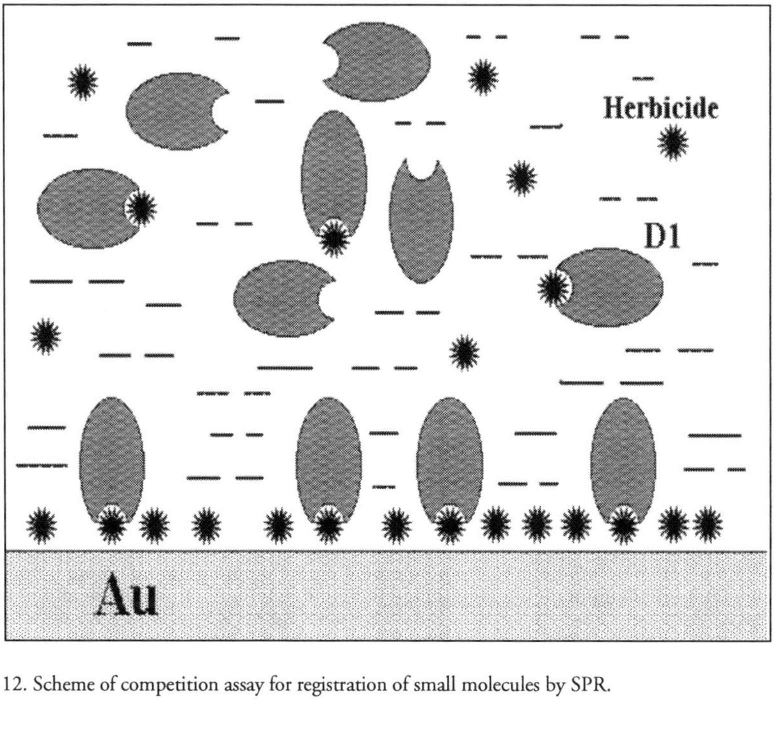

Figure 12. Scheme of competition assay for registration of small molecules by SPR.

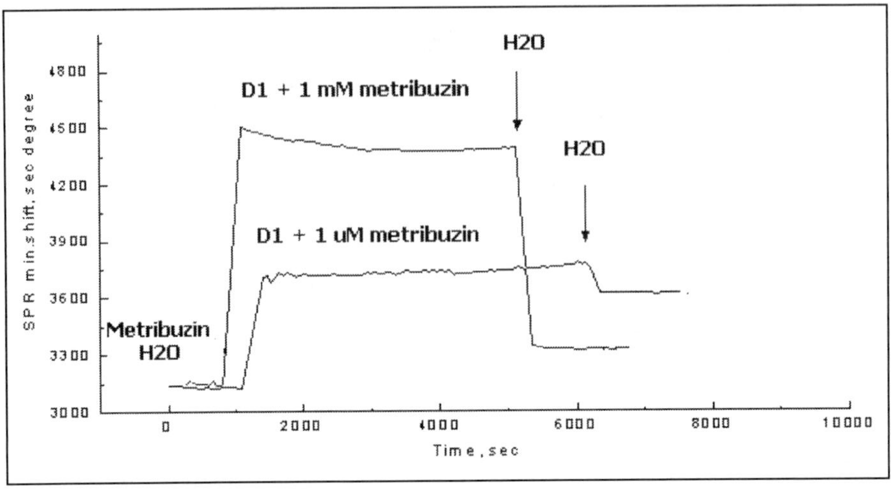

Figure 13. The inhibition of D1 protein binding to metribuzin, immobilized on a gold surface by free metribuzin in solution.

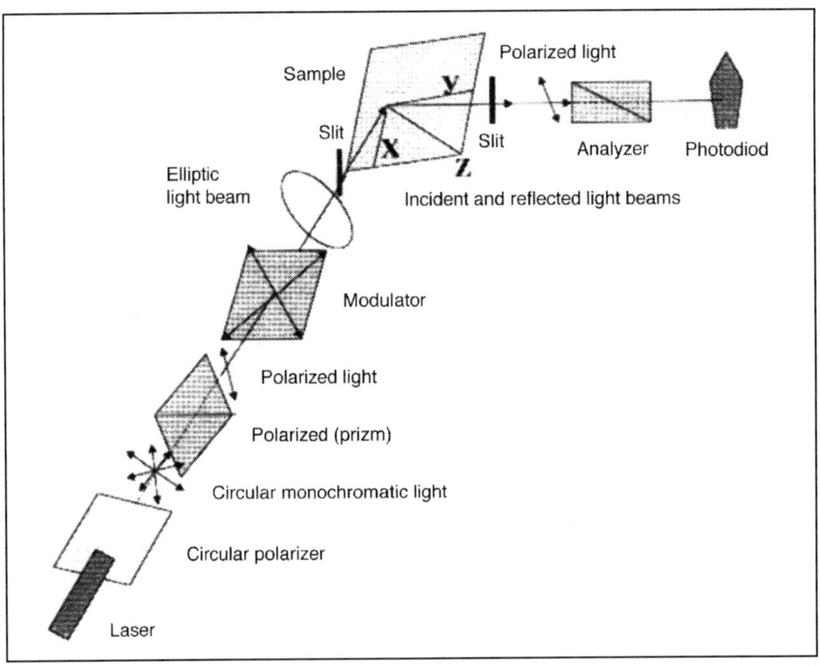

Figure 14. Ellipsometry set-up.

response was dependent on concentration of free herbicide in solution (Fig. 13). Free metribuzin suppressed the ability of the D1 protein to interact with immobilized metribuzin on the gold surface. The maximal response of SPR-sensor was observed for the minimal concentration of the herbicide in sample. In comparison with direct measurements, competitive assay allowed to increase the sensitivity of a method from 0.5 mM (direct measurement) to 0.5 μM (competitive assay).

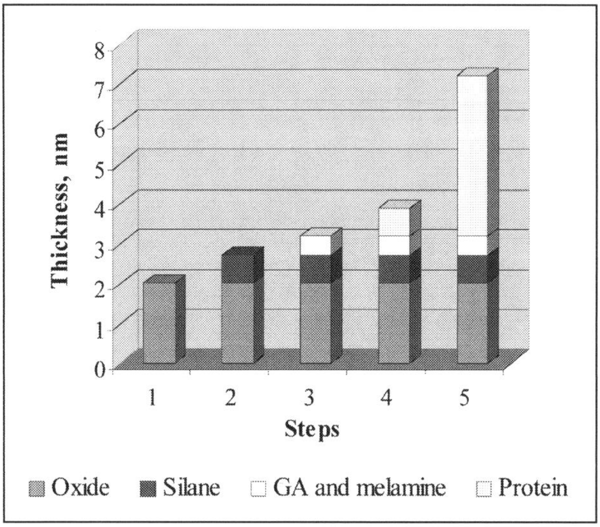

Figure 15. The optical thickness of various layers deposited on the silica after surface modification and consequent binding of D1 protein.

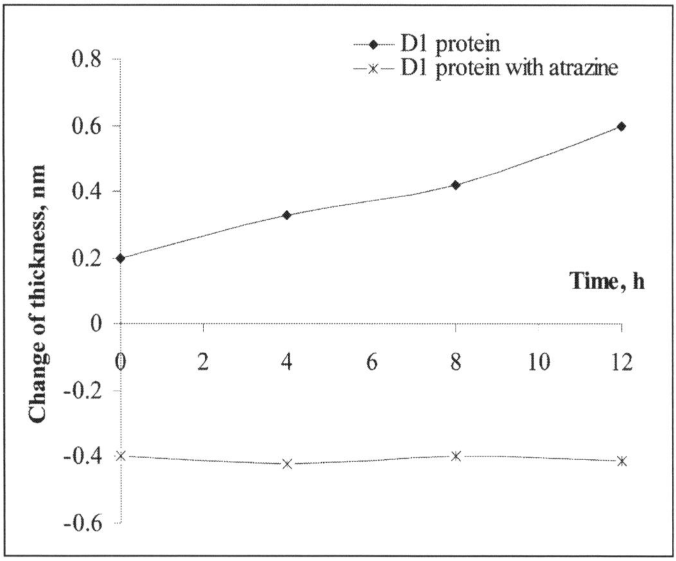

Figure 16. Inhibition of D1 protein binding to melamine-modified surface by atrazine ($1 \cdot 10^{-7}$ M) in solution.

Ellipsometry

Ellipsometry is an optical reflection measurement using polarized light (Fig. 14). Ellipsometry can provide the information about thickness, refractive index, and overall morphology of thin films, surfaces, and multi-layers. The technique has been known for almost a century and has many standard applications. It is mainly used in semiconductor research and fabrication to determine properties of layer stacks of thin films and the interfaces between the layers. However, ellipsometry is also becoming an important tool for research in other disciplines such as biology and medicine.

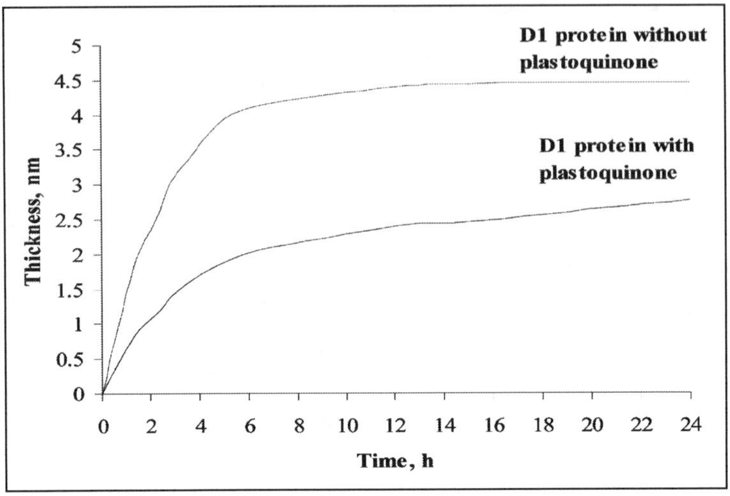

Figure 17. Kinetics of D1 protein binding to melamine surface.

The ellipsometry experiments with D1 protein were based on competition principle.[15] The melamine molecules were immobilized on the silica surface (see Fig. 15).

It was demonstrated that the presence of triazine herbicides can inhibit the binding of D1 protein to melamine layer immobilized on the silica surface (Fig. 16). The sensitivity of ellipsometry detection for atrazine was in the range 10^{-6}-10^{-8} M.

The slow kinetics of D1 protein binding to melamine surface was improved by extraction of plastoquinone (Fig. 17).

Electrochemical Methods

Among the electrochemical methods potentiometry and cyclic voltammetry have been used to monitor the specific interaction between the D1 protein and herbicide.[16]

Development of the Sensor Unit, Based on Potentiometric Detection of Protein-Herbicide Interaction

The isolation of D1 protein was conducted as mild as possible in order to isolate the D1 protein without loss of secondary acceptor of electrons- plastoquinone Q_B. The D1 protein and BSA were dissolved in 100 mM sodium phosphate saline buffer (PBS) at the concentrations 5 mg/ml to prepare relatively thick (10-50 μm) gel-like films containing the protein. A drop of resulting mixture was cast onto Si-SiO_2-Si_3N_4 (silica nitride) surface and then they were exposed to glutaraldehyde vapors in a closed vessel for 30 minutes at + 10°C. After that, samples were rinsed in PBS for 30 min.

Silica nitride structures are known as highly sensitive to surface charges. Voltage-dependent impedance of such structures in contact with an electrolyte reflects the possible changes in both density and mobility of charges within the membrane. Thus, a change in the charge density should lead to a shift of the capacitance-voltage dependency while the mobility of ions is in relation with the impedance maximum. Impedance of the electrolyte/cross-linked protein membrane onto Si-SiO_2-Si_3N_4 surface was studied in the frequency range 30-3000 Hz, under the bias -3-+ 1 V. The measuring set-up was shown schematically in Figure 18.

The frequency of 80 Hz was found to be the most suitable for characterization of the membrane itself. Thus, the capacitance of Si-SiO_2-Si_3N_4 insulating layer had significant contribution to the total impedance at the lower frequencies while the electrolyte resistance predominated at the higher frequencies.

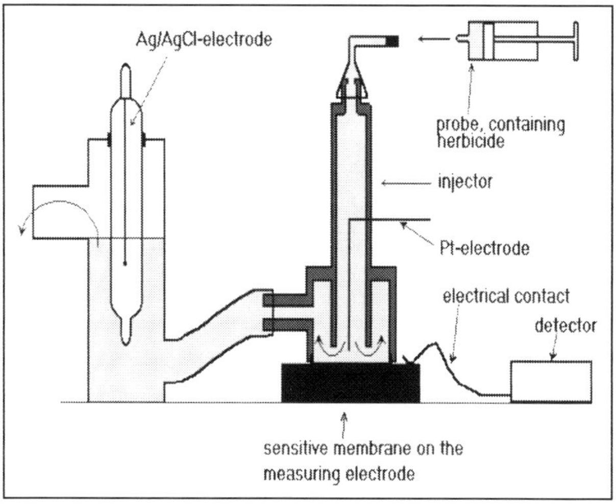

Figure 18. Scheme of amperometric measuring arrangement for the investigation of the electrochemical process during the binding reactions.

The measurements were performed in 100 mM sodium phosphate buffer solutions and pH varying from 4.0 to 7.0. The structures with protein membranes exhibited pH sensitivity similar to those of bare $Si\text{-}SiO_2\text{-}Si_3N_4$ structures: capacitance-voltage dependencies shifted along the voltage axis on pH-changes. No additional shift of the capacitance-voltage dependencies was observed for $1{\cdot}10^{-7}$ M of atrazine. This means that atrazine binding does not alter the net charge in the membrane.

At the same time, some increase of the maximum impedance was observed when atrazine was added to the membrane. The membrane impedance was higher in the acidic solution and the presence of atrazine further increased it (Fig. 19).

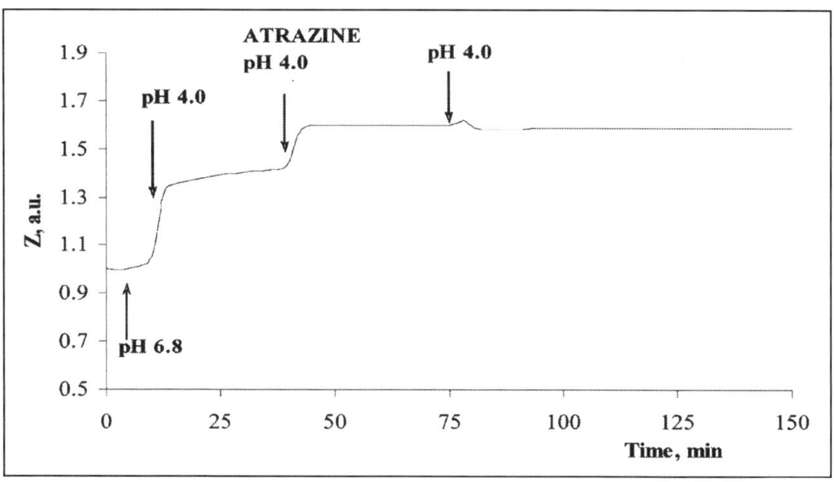

Figure 19. Response of the total membrane impedance on buffer changing and atrazine adding (atrazine concentration is $1{\cdot}10^{-7}$ M).

These data indicate that the membrane became more compact and less ion-permeable at lower pH and this permeability decreased even more as a result of atrazine binding. These results can be explained by protein conformation changes upon interaction with herbicide. It is noteworthy that the responses to atrazine were noticeable in the acid media only and that these responses were irreversible. Because of poorly controllable properties and disordered structure of the cross-linked BSA matrix, reproducibility of sensor performance was insufficient.

Development of the Sensor Unit, Based on Cyclic Voltammetry-Measurements (CV) of Protein-based Membrane

The electrochemical sensor was based on changes in the mediator properties of D1 protein in presence and in absence of herbicides detected by cyclic voltammetry. CV-measurements has been carried out in a supporting electrolyte which contained electroactive marker-ions, to monitor charge transport through membranes embedding the protein and associated redox processes. In these experiments, a working electrode coated with a protein-embedded multi-bilayer was used. Peaks in voltammograms should be in relation with charges and mediator properties of the protein.

The Langmuir-Blodgett (LB) technique has been employed to prepare membranes with well-defined molecular packing.[17] Octadecylamine was selected as an amphiphile which forms charged mono-layers on the water surface and has good insulating properties when deposited onto electrically conductive substrates, indium-tin oxide (ITO) layer.

The D1 protein (0.1 mM) was embedded in LB films spreading onto the aqueous sub-phase containing $CaHCO_3$ (0.2 mM) with dodecylamine solution in chloroform (0.5 mM). After evaporation of the solvent for 10 min, surface pressure and surface area isotherms were recorded three times with intervals of 20 minutes between measurements. The observed shift of the isotherms to the larger areas on time indicated that substantial amount of the protein had been embedded into the monolayer at the third recompression. The monolayers have been slowly recomposed up to 15 mNm^{-1} twice, during 1 h-exposure on the protein-containing sub-phase, before they were transferred onto the substrates at 25 mNm^{-1}. For comparative experiments, samples with dodecylamine monolayers without the protein were also prepared. In this case pure water was used as a sub-phase.

Electrochemical measurements were performed with 3-electrode system, described earlier.[18] Ag/AgCl and platinum were used as reference electrode and counter-electrode respectively. The characteristics of the D1 protein/LB electrodes as herbicide sensors were measured by applying a constant potential.

Ion permeability of thin multi-bilayer membranes (10-20 nm) with and without protein was monitored in PBS, pH 6.9, in presence of 10 mM potassium ferricyanide ($K_3[Fe(CN)_6]$). Working electrodes coated with ten monolayers of dodecylamine with and without protein, exhibited almost similar voltammograms; the peaks were reduced approximately by factor 100 in comparison to the case of the bare ITO. After addition of atrazine, the current increased for the multi-layer embedding the protein around 5 times (from 0.2 to 1 μA). The effect was found to be fast and non reversible (Fig. 20).

The calibration curves of the amperometric sensor for atrazine, simazine and metribuzin (measured at potential -400 mV) in Figure 21 demonstrate the possibility of the herbicide detection in the range concentration 10^{-7}-10^{-9} M (standard deviation is about 5 %).

The data suggested that the toxic action of herbicides on chloroplasts traditionally interpreted by inhibition of electron flow along the chloroplast membrane also may be the result of the thylakoid membrane depolarization.

Conclusion

Presented experimental results suggest that application of herbicide-binding protein in sensor technology has a high potential. Several detection systems were tested in combination with D1 protein: electrochemical (amperometry and cyclic voltammetry), optical (surface plasmon resonance and ellipsometry) and assays (ELISA and D1 protein- containing liposomes and DELFIA fluorimetric assay). The main mechanisms of D1 action are either on the ability of herbicides to replace the plastoquinone molecule in D1 protein and in this way change the electrochemical and optical

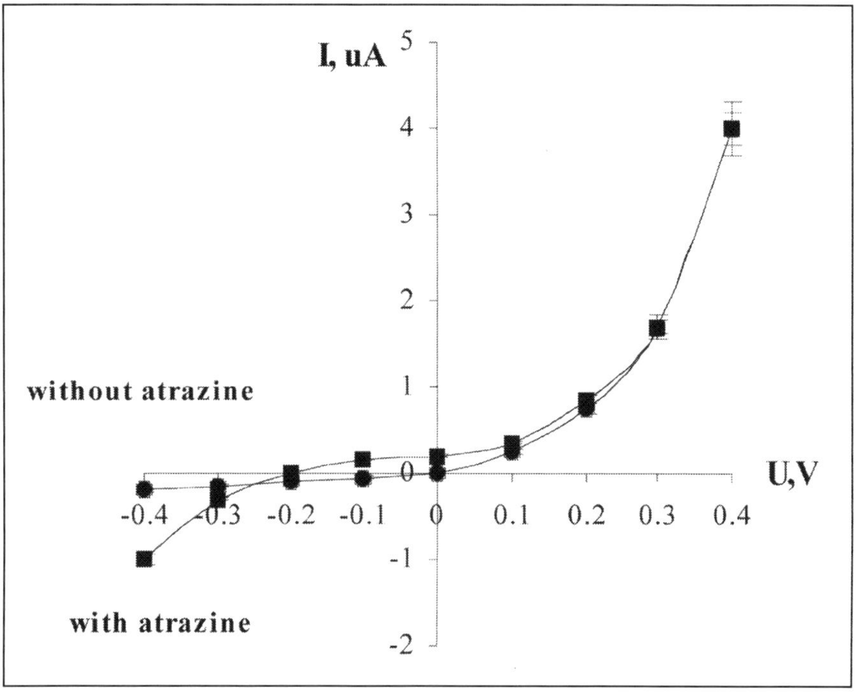

Figure 20. Influence of atrazine on volt-amperometric characteristics of SnO_2-electrode covered with a LB-membrane containing D1 protein.

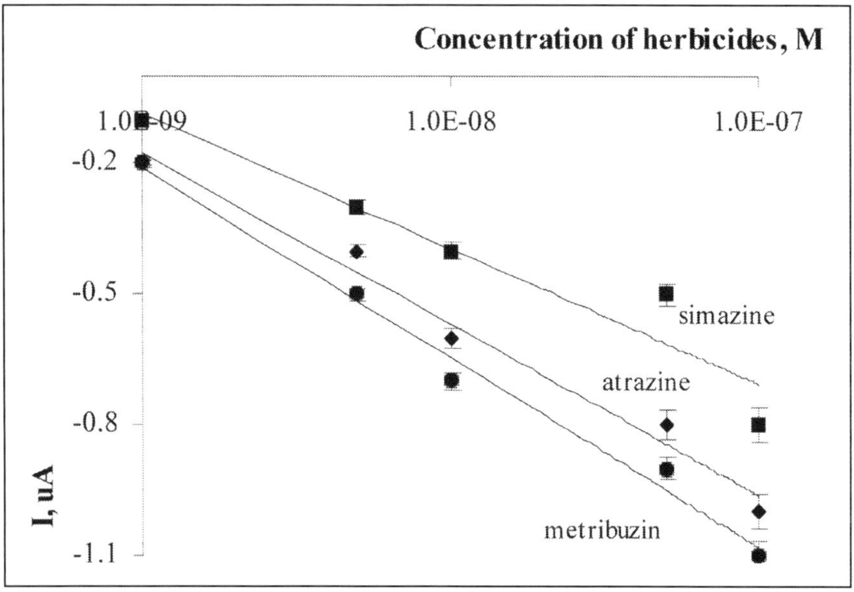

Figure 21. Calibration curves for atrazine, metribuzin and simazine measured at potential -400 mV by cyclic voltammetry on D1 protein-embedded LB-membranes in presence of herbicides.

properties of the system or secondly, on the antibodies-like specific binding properties. The experimental results indicate a high potential of D1 protein application in sensor technology. Among the advantages are high stability, simplicity of D1 protein preparation and its broad specificity towards several different families of photosynthesis inhibiting herbicides. The sensitivity for atrazine with the optical end electrochemical detection was in the range of $1\cdot10^{-6}$-$1\cdot10^{-8}$ M. It can be further increased by using preconcentration of samples by solid-phase extraction technique (SPE) which allows the application of reported sensors for herbicide control in drinking water.[20]

References

1. European Communities, Drinking Water Directive L330 1998, 32.
2. Carabias-Martínez R, Rodríguez-Gonzalo E, Herrero-Hernández E et al. Determination of herbicides and metabolites by solid-phase extraction and liquid chromatography: Evaluation of pollution due to herbicides in surface and groundwaters. J Chromatogr A 2002; 950:157-166.
3. Balinova A. Solid-phase extraction followed by high-performance liquid chromatographic analysis for monitoring herbicides in drinking water. J Chromatogr A 1993; 643:203-207.
4. Halko R, Hutta M. Study of high-performance liquid chromatographic separation of selected herbicides by hydro-methanolic and micellar liquid chromatography using Genapol X-080 nonionic surfactant as mobile phase constituent. Anal Chim Acta 2002; 466:325-333.
5. Monson SJ, Ma L, Cassada DA et al. Confirmation and method development for dechlorinated atrazine from reductive dehalogenation of atrazine with Fe^0. Anal Chim Acta 1998; 373:153-160.
6. Ballesteros B, Barceló D, Dankwardt A et al. Evaluation of a field-test kit for triazine herbicides (SensioScreen® TR500) as a fast assay to detect pesticide contamination in water samples. Anal Chim Acta 2003; 475:105-115.
7. Stöcklein WFM, Rohde M, Scharte G et al. Sensitive detection of triazine and phenylurea pesticides in pure organic solvent by enzyme linked immunosorbent assay (ELISA): Stabilities, solubilities and sensitivities. Anal Chim Acta 2000; 405:255-265.
8. Giardi MT, Koblízek M, Masojídek J. Photosystem II-based biosensors for the detection of pollutants. Biosens Bioelectr 2001; 16:1027-1033.
9. Marder JB, Mattoo AK, Edelman M. Identification and characterization of the psbA gene product: The 32-kDa chloroplast membrane protein. In: Weissbach A, Weissbach H, eds. Methods Enzymol, Vol. 118. New York: Academic Press Inc., 1986:384-396.
10. Barber J. Photosystem two. Biochim Biophys Acta 1998; 1365:269-277.
11. Tjus S, Andersson B. Loss of the trans-thylakoid proton gradient is an early event during photoinhibitory illumination of chloroplast preparations. Biochim Biophys Acta 1993; 1183(2):315-322.
12. Ort DR. Energy transduction in oxigenic photosynthesis. In: Staehelin LA, Arntzen CJ, eds. Encyclopedia of Plant Physiology, Vol 19. New York: Springer-Verlag, 1986:145-165, (178).
13. Piletskaya EV, Piletsky SA, El'skaya AV et al. D1 protein- an effective substitute for immunoglobulins in ELISA for the detection of photosynthesis inhibiting herbicides. Anal Chim Acta 1999; 398:49-56.
14. Chegel VI, Shirshov YuM, Piletskaya EV et al. Surface plasmon resonance for pesticide detection. Sensor Actuat B 1998; 48:456-460.
15. Piletska E, Piletsky S, Lavrik M et al. Development optical system for herbicide detection based on D1 protein. Agroecol Biotechnol 1998; 74-79.
16. Piletskaya E, Piletsky S, Lavrik N et al. Towards the D1 protein application for the development of sensors specific for herbicides. Anal Lett 1998; 31:2577-2589.
17. Tatsuma T, Tsuzuki H, Okawa Y et al. Bifunctional Langmuir-Blodgett film for enzyme immobilization and amperometric biosensor sensitization. Thin Solid Films 1991; 202:145-150.
18. Starodub NF, Piletsky SA, Lavryk NV et al. Template sensors for low weight organic molecules based on SiO_2 surfaces. Sensor Actuators B 1993; 13-14:708-710.
19. Mouvet C, Harris RD, Maciag C et al. Determination of simazine in water samples by waveguide surface plasmon resonance. Anal Chim Acta 1997; 338:109-117.
20. Piletskaya EV, Piletsky SA, Sergeyeva TA et al. Thylakoid membranes -based test-system for detecting of trace quantities of the photosynthesis inhibiting herbicides in drinking water. Anal Chem Acta 1999; 391:1-7.

Chapter 13

Photosystem II-Based Biosensors for the Detection of Photosynthetic Herbicides

Maria Teresa Giardi and Emanuela Pace*

Abstract

Photosystem II (PSII) is the supramolecular pigment-protein complex in the chloroplast, which catalyses the light-induced transfer of electrons from water to plastoquinone in a process that evolves oxygen. The PSII complex is also known to bind several groups of herbicides. In the world, pesticide pollution of soil and ground water is a widespread problem. The objective of our work is the development of biosensors using PSII isolated from photosynthetic organisms to monitor polluting chemicals. This should lead to the set-up of a low cost, easy-to-use apparatus, able to reveal specific herbicides, and eventually, a wide range of organic compounds present in industrial and urban effluents, sewage sludge, landfill leak-water, ground water, and irrigation water. Within the framework of sustainable development, a number of developed countries will have to make strong efforts during the coming years to meet the directives and standards in the areas of environmental monitoring, pollution control, waste management, water and soil quality.

In this study we provide an overview of the systems available for the bioassay of herbicide pollutants using biosensors that are based on the photochemical activity.

Introduction

Toxic chemicals originating from sources such as industrial waste effluents and agricultural run-off can contaminate soils, and surface and ground waters. Some toxic compounds with slow degradation rates accumulate in the soil and water, and subsequently in plant tissues. These chemical residues (i.e., pesticides, herbicides, heavy metals) are then stored in animal and human tissues where their toxicity represents a serious health risk. Environmental monitoring is important to guarantee the health of ecosystems. Practical monitoring programs require rapid, simple, and low-cost screening procedures for the detection of harmful chemicals in aquatic and soil environments.

Traditional chemical methods of pollution monitoring such as gas and high-performance liquid chromatography, atomic absorption and mass spectrometry are sensitive and effective. However, biosensors better reflect the real physiological impact of active compounds present in the sample because even low concentrations of pollutants affect living organisms by disturbing physiological processes. A biological component is therefore essential for biosensors that aim to ascertain the general state of an ecosystem.[1]

Photosystem II is the multi-enzymatic chlorophyll-protein complex (water-plastoquinone oxido-reductase) located in the thylakoid membrane of algae, cyanobacteria and higher plants. It is an integral part of the electron transport chain that catalyses primary charge separation. This protein complex consists of over 25 polypeptides, which make up a light-harvesting chlorophyll protein

*Corresponding Author: Emanuela Pace—Institute of Cristallography, CNR, National Council of Research-IC, Via Salaria km 29.300 - 00016, Monterotondo Scalo, Rome, Italy.
Email: pace@mlib.cnr.it

Biotechnological Applications of Photosynthetic Proteins: Biochips, Biosensors and Biodevices, edited by Maria Teresa Giardi and Elena V. Piletska. ©2006 Landes Bioscience.

complex (LHCII), a reaction centre and the water-splitting system, also called the oxygen evolving complex (OEC). The PSII complex also contains the target site of the most widely used photosynthetic herbicides.[2] The preparation of a biosensor for the detection of polluting compounds is based on the specific characteristics of PSII. Under illumination PSII drives electron transfer which is inhibited by specific chemical compounds (e.g., herbicides). Thus, the photosynthetic membrane isolated from higher plants and photosynthetic micro-organisms, immobilised and stabilised, will serve as the biosensor biomediator. The effect of compounds that alter or inhibit photosynthetic activity, measured as oxygen evolution, electron transport or fluorescence can then be translated and monitored by amperometric or optical systems. Recent experiments that were performed by our this team resulted in amperometric, potentiometric, optical biosensors for herbicides based on isolated and immobilised PSII particles exhibiting stable biological material and a highly sensitive monitoring response (limit of detection in the nanomolar range and for diuron herbicide in the picomolar range).[3-5] However, the system is specific to photosynthetic herbicides but not very selective since several classes of compounds (herbicides represented by triazines, ureas, diazines, phenols) can bind to proteins of the PSII complex, most of them to the D1 protein of the reaction centre.[6,7]

The objective of this study is to provide an overview of the systems available for the bioassay of herbicide pollutants affecting directly photosynthetic activities. Since biological detection reflects the true physiological consequence of pollutants, such a system makes it possible to monitor even trace amounts of chemicals. It is foreseeable that advanced biosensing systems based on PSII may eventually substitute for the current costly chemical techniques.

This paper will contribute to this objective by presenting data of the biosensors already built and that constitute a versatile, low-cost technology with a presumably lower environmental impact than many traditional techniques for herbicide detection. Photosystem II based biosensors will avoid costly and often environmentally unsound analyses in two ways: (i) biosensors that respond to a range of pollutants can be used for rapid, low cost prescreening of large numbers of samples to determine samples that will then undergo more detailed analysis. This should avoid the use of large quantities of organic solvents and the need for expensive apparatus to extract herbicide compounds; (ii) the use of an array of PSII based biosensors that are each highly specific to a given class of pollutant in conjunction with sophisticated data elaboration will enable the user to generate an identikit of the pollutants present in samples.

PSII-based biosensors should eventually serve as sensitive, low cost detection systems that are capable of detecting specific pollutants. The arising need to comply with EU and EPA directives[8-10] and standards for environmental monitoring makes such a system increasingly attractive.

Herbicides

Phenylurea, triazine and diazine, heavy metals and phenolic compounds represent economically important compounds and are widely used in chemical, pharmaceutical and agricultural industries. These compounds have been synthesised since the second World War and have been applied in great quantities, particularly as herbicides. The application of herbicides in agriculture has increased appreciably during the past few decades, resulting in the massive pollution of water and soils. Although new herbicides are now available, the above compounds are still the most widely used products for weed control. About 30% of herbicides (derivatives of phenylurea, triazine, diazine and phenolic types) that are currently in use inhibit the light reactions of photosynthesis, usually by targeting PSII-dependent electron flow.[7,11] All the compounds of these classes have a world-wide usage of the million thousands of tons. We should consider that the surface utilised for maize production is about 150 millions of hectares in world (FAO 1986) and that the surface is spread of photosynthetic herbicides for 2-5 kg per hectare twice a year! These compounds act by inhibiting Photosystem II-dependent electron transfer and therefore block production of ATP and NADPH. This inhibition is specifically due to the binding of herbicides to the Q_B binding domain of D1.[7,12] Some crop species have developed mechanisms of resistance to the attack of herbicides on the D1 domain of Q_B. This trait can be advantageous in that it allows the application of pre and post-emergence spraying without harming the crop. A widespread use of these substances, especially atrazine, may drive natural selection of resistant weed species. The subsequent development of the resistant weed species may in turn require higher doses of herbicides. Although the use of atrazine is banned by law in a number of European countries, similar compounds, such as simazine (which differs from atrazine by the addition of a methyl group in the lateral chain) are widely used. Thus, it is important to have systems available for rapid analysis of soils, solid substrates and aqueous systems in order to monitor the presence of increased amounts of herbicides.

Worldwide, thousands of tons of photosynthetic herbicides are applied annually. These compounds are usually absorbed through the roots and then translocated via the xylem to the leaves, although some are directly absorbed by the leaves.[7] These chemicals are quite stable in the soil and are adsorbed by colloids and organic substances on the basis of their ability to exchange cations. Although most of these herbicides remain on the surface of the soil, in sandy soils, or in the absence of organic substances, the water soluble compounds can undergo percolation. Since these herbicides can be highly toxic to humans and animals, their indiscriminate use has serious environmental implications. Consequently, the use of dinoseb was prohibited in the US and in most other countries because of its high toxicity.[9] Atrazine, a possible human carcinogen, has also been banned.[10] However, in USA atrazine is still in use and it contaminates countries where it is not even used.

In 1980, the EU introduced a directive that does not allow concentrations of pesticides in drinking water to exceed 0.1 ppb for any individual pesticide, and 0.5 ppb for total pesticides.[8] Thus, in order meet drinking water standards, concentrations around 0.1 ppb must be readily detectable. Sensitive and reliable detection methods are needed to monitor such low residual herbicide levels. Three methods are currently used in the testing of most herbicides: HPLC, GC-MS and ELISA. Although HPLC and GC-MS are both standard and reliable, they require expensive equipment, very large amounts of organic solvents and the purification of samples prior to assay, thus limiting the number of samples that can be analysed.[13,14] The immunochemical technique ELISA has a high sensitivity (46 ppt for diuron; 22 ppt for atrazine).[15,16] The preparation of monoclonal and polyclonal antibodies has been optimised and ELISA tests can be specific to either one compound or several structural analogues (see www.biosensor.it). The comparison of both chemical and biological methods can provide valuable results.[17]

Biosensors

The most frequently used biosensing systems for monitoring photosynthetic herbicides utilise intact cells of algae, cyanobacteria and diatoms to measure either changes in photocurrent,[18-23] inhibition of electron transport with artificial mediators,[24,25] or changes in chlorophyll fluorescence.[26-31] As early as the 1950s, selection for the most effective herbicides took advantage of the fact that herbicides can inhibit the Hill reaction in isolated chloroplasts.[32,33] Mutants and adaptable strains of the cyanobacterium *Synechocystis* sp. PCC6803 have been used in phytotoxicity tests.[22] Optical whole-cell biosensor using *Chlorella vulgaris* designed for monitoring herbicides at sub-ppb concentrations was developed by Vedrine et al[34] and by Rodriguez et al.[35]

This approach has been used primarily to study photosynthesis, for example to introduce tolerance to a variety of stress conditions.[36] Many studies have been successful in verifying the effect of single amino acid modifications on herbicide binding affinity. High affinity binding to the D1 protein is a useful property for the detection of herbicides. A fluorescence biosensor based on mutants resistant to various herbicide subclasses was developed, it makes possible to distinguish between subclasses of herbicides (e.g., triazines from urea and phenolic type herbicides).[37]

Recent approach utilise isolated PSII complexes as biomediators. The PSII reaction center was isolated in 1987 and the isolated complex was found to maintain its herbicide-binding ability.[38,39] The binding affinity of the herbicide depends on the amino acid composition of the hydrophilic loop in the D1 protein. It is recognized that these herbicides act by binding D1 protein of reaction center. Two levels of action occur: displacement of the secondary quinone electron transfer PQ and inhibition of D1 protein turnover.[40]

Photosynthetic herbicides fall into three main groups: phenylureas, triazines, diazines and phenols, depending on their chemical structure and binding properties.[41] Although both classes of herbicides replace the Q_B acceptor on the D1 protein,[42,43] they interact with different amino acid residues on D1.[7]

One of the advantages in using PSII-based biosensors is the simplicity of the biological transduction, which can be monitored directly without requiring additional markers or transducer molecules. Another advantage is its extreme susceptibility and selectivity towards the binding of herbicides.[44] Two types of PSII-based biosensor, based on fluorescence (A) and amperometry-printed electrodes (B), respectively, are schematically represented in Figure 1.

The sensitivity of any PSII-based biosensor is determined by the binding constant of the herbicide. During the 1990s there was an increased interest in developing biological sensors to detect low levels of herbicides in water and soil using an isolated PSII complex or reaction center. Isolated chloroplasts and thylakoids have been widely used in biosensors for herbicide detection. Such biosensors test either the inhibition of the Hill reaction,[45-47] the inhibition of DCPIP photoreduction,[48,49] or changes in chlorophyll fluorescence.[27,28,30] Hall and coworkers developed an elegant technique to generate photocurrent using isolated PSII particles immobilised on TiO_2 electrodes in the presence of a quinone as an electron acceptor.[50] They found that this reaction is inhibited by the addition of diuron. In principle, they utilise an electrochemical biosensor based on PSII particles for herbicide detection.

The practical applications of herbicide biosensors based on PSII preparations were earlier limited by their instability, particularly upon illumination. Significant stabilisation of photochemical activity was achieved by the entrapment of cells or PSII preparations, e.g., chloroplasts and thylakoids, on either an albumin-glutaraldehyde crosslinked matrix,[51-53] polyurethane-albumin polymers or polyvinylalcohol bearing styrylpyridinium groups.[54-56] Techniques for immobilisation range from the entrapment of whole cells or isolated membranes or pigment-protein complexes in an agarose, alginate or gelatine matrix, to crosslinking in a glutaraldehyde-serum-albumin matrix.

Another approach to the construction of a stable and sensitive biosensor for herbicide detection is to use PSII particles from thermophilic species, e.g., the thermophilic cyanobacterium *Synechococcus elongatus*.[57] Two types of biosensors for the detection of photosynthetic herbicides have been constructed using these particles: a Clark's oxygen electrode and a screen-printed carbon-silver electrode.[47,58,59] The Clark electrode traces oxygen produced by PSII particles, while the positively polarised carbon-silver electrode registers the reduction of an artificial electron acceptor as electrical current. In the latter, the presence of herbicides is detected as a decrease in the current produced by the PSII particles compared with a control. The following detection limits were measured with the Clark-electrode biosensor for classical herbicides: 116 ppb for diuron (DCMU), 430 ppb for atrazine and 4 ppm for simazine.[47,58] The value for diuron was comparable with the detection limit of ELISA.[60] In the Clark-electrode biosensor, the signal to noise ratio was maximised by mounting onto a flow system, which resulted in a high concentration of PSII oxygen evolution in the microenvironment of the measuring probe.[47,58] Moreover, since the herbicides could be washed out, the biomediator was reused for several samples.

The reaction centers of photosynthetic microorganisms have also been shown to bind herbicide and could potentially be used for the detection of herbicides as part of sensing devices based on Langmuir-Blodgett monolayer films[1,61] or artificial membrane forming liposomes.[63] Even isolated D1 protein has been embedded on a working electrode for the potentiometric monitoring of the specific interaction between a protein and an herbicide.[63]

Recent application of this technology provides collateral applications. For instance the *Synechocystis* sp. strain PCC6803 was chromosomally marked with the luciferase gene *luc* from the firefly *Photinus pyralis* to create a novel bioluminescent cyanobacterial strain. Successful expression of the *luc* gene during growth of the cyanobacterium was characterized by measuring optical density and bioluminescence. Bioluminescence was optimized with regard to uptake of the luciferase substrate, luciferin, and the physiology of the cyanobacterium. The novel luminescent cyanobactyerial biosensor has been developed which responded to a range of compounds including different herbicides type and other toxins.[64]

Interesting it is the use of the technology based on photosynthetic proteins for detecting microbiological foodborne hazards reported by Hall.[65] Also interesting it is the development of biosensors based on photosynthesis to detect TNT (a triazine trinitro-compound or similar explosives).

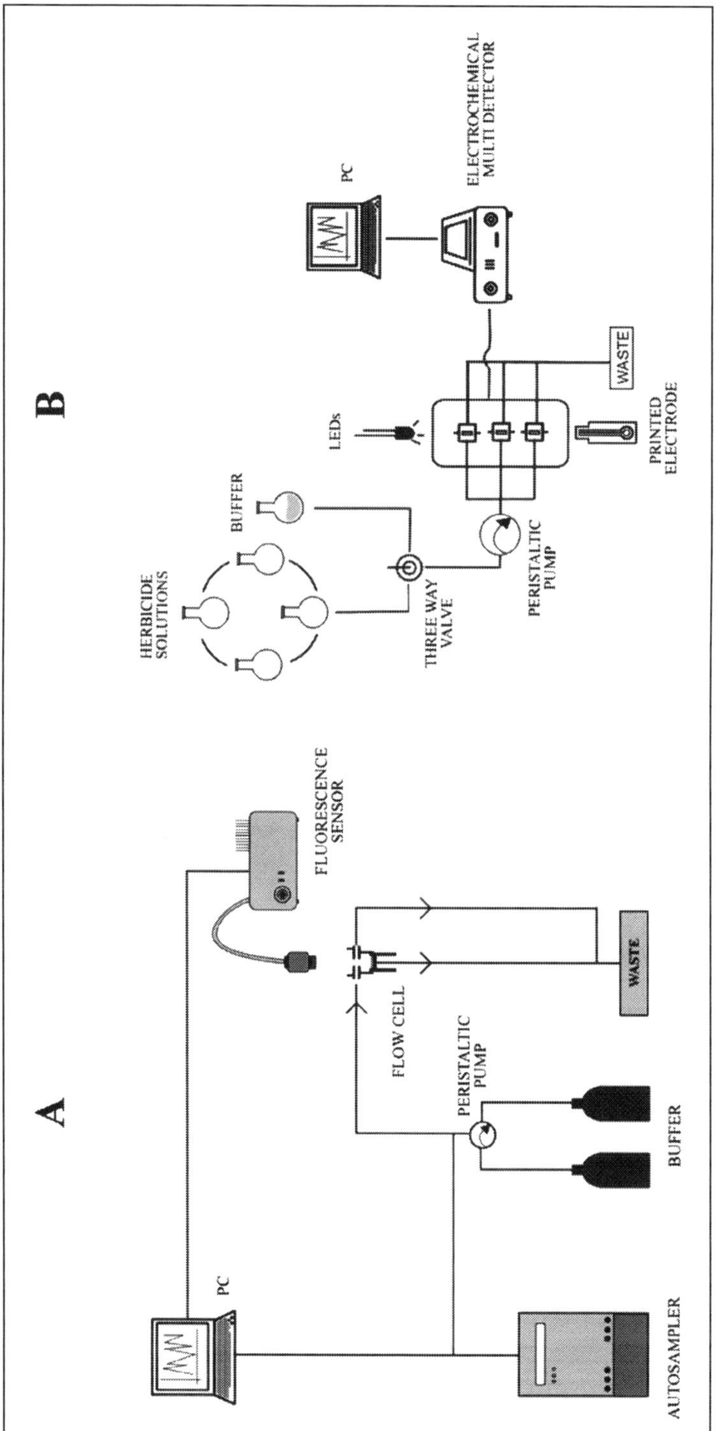

Figure 1. Two types of PSII-based biosensor, based on fluorescence (A) and amperometry-printed electrodes (B).

The main purpose of the recent work on integrated biosensors is to render the highly sensitive, relatively stable Photosystem II from algae and cyanobacteria, highly stable and specific to different subclasses of pollutants (herbicides and heavy metals). This is achieved in various ways: (i) use of the alga *Chlamydomonas reinhardtii*, mutated at the Q_B herbicide binding site by side-directed mutagenesis; (ii) modification of organisms by site-directed mutagenesis to produce a highly stable specific biomediator (iii) overexpression of *psbA* to produce the D1/Q_B site in *Escherichia coli* from selected mutated organisms (Johanningmeier U, Torzillo G, Esposito D, Pace E and Giardi MT, in preparation).

The specificity of the whole cells and isolated photosynthetic materials was obtained by applying the knowledge available on the relationships between herbicide binding activity and the structure of the D1 protein. For example, distinctions among classes of chemicals could be achieved through mutations in amino acid residues of D1 which can impart resistance to individual triazine herbicides. Realisation of new, sophisticated transduction systems based on printed electrodes, fluorescence and chemiluminescence as well as alternative systems such as the reconstitution of Q_B site in overexpressed D1 protein utilizing chromophore quinones to enhance sensitivity and specificity for detected signals.

It was possible to obtain real information on the type and concentration of subclasses of herbicides through sophisticated signal elaboration as several mutated stable biomediators with different sensitivity to pollutant subclasses were positioned in series (See www.biosensor.it).[37]

Conclusions

Compared with classical analytical methods such as gas chromatography, HPLC, atomic absorption or mass spectrometry, the detection of pollutants by biosensors is generally less specific. These devices provide valuable information on a class of pollutants rather than a mere information about a specific compound, although the use of mutants that are resistant to specific pollutants can render them more selective. Biosensors provide valuable information about the real biological effects of the pollutants in a sample since phytotoxicity is determined from the measurement of electron transport activity, photocurrent or photosynthetic oxygen evolution. It is important to note that although the PSII complex is sensitive to various pollutants (herbicides, heavy metals, sulphites, nitrates, carbonates), its susceptibility to these compounds is highly variable, ranging from nanomolar to millimolar concentrations.

PSII-based biosensors have potential for use in the evaluation of overall general toxicity and the prescreening of large numbers of samples in order to focus more complex laboratory analyses.

Acknowledgements

This research was supported by the EU grant QLRT-2000-01629.

References

1. Yatsenko V. Determining the characteristics of water pollutants by neural sensors and pattern-recognition methods. J Chromatogr 1996; A722(1-2):233-243.
2. Mattoo A, Giardi MT, Raskind A et al. Dynamic metabolism of photosystem II reaction center proteins and pigments. A review. Physiol Plant 1999; 107:454-461.
3. Giardi MT, Rigoni F, Barbato R. Photosystem II core phosphorylation heterogeneity, differential herbicide binding, and regulation of electron transfer in photosystem II preparations from spinach. Plant Physiol 1992; 100:1948-1054.
4. Oettmeier W. Herbicide resistance and supersensitivity in photosystem II. A review. CMLS 1999; 55:1255-1277.
5. Giardi MT. Phosphorylation and disassembly of photosystem II as an early stage of photoinhibition. Planta 1993; 190:107-113.
6. Pesticide Chemistry. In: Frehse H, ed. Weinheim, New York, Basel, Cambridge: VCH, Weinheim, 1991.
7. Draber W, Tietjen K, Kluth JF et al. Herbicides in photosynthesis research. Angew Chem Int Ed Engl 1991; 3:1621-1633.
8. European Communities, Drinking Water Directive L229, 1980:11.
9. US Federal Register, 1986, 51 FR 36634.

10. US EPA. Atrazine: Health Advisory. Washington DC: Office of Drinking Water, US EPA, 1988, (http://www.epa.gov/oppbead1/pestsales/97pestsales/table8.htm).
11. Moreland DE. Mechanisms of action of herbicides. Annu Rev Plant Physiol 1980; 31:597-638.
12. Geiken B, Masojídek J, Rizzuto M et al. Incorporation of [^{35}S]methionine in higher plants reveals that stimulation of the D1 reaction centre II protein turnover accompanies tolerance to heavy metal stress. Plant Cell Environ 1998; 21:1265-1273.
13. Pacakova′ V, Stulik K, Jiskra J. High-performance separation in the determination of triazine herbicides and their residues. J Chromatogr 1996; 754:17-35.
14. Bushway RJ, Perkins LB, Fukal L et al. Comparison of enzyme-linked immunosorbent assay and high-performance liquid chromatography for the analysis of Atrazine in water from Czechoslovakia. Arch Environ Contam Toxicol 1991; 21:365-370.
15. Schneider P, Hammock BD. Influence of the ELISA format and the hapten-enzyme conjugate on the sensitivity of an immunoassay for s-triazine herbicides using monoclonal antibodies. J Agric Food Chem 1992; 40:525-530.
16. Giersch T. A new monoclonal antibody for the sensitive detection of Atrazine with immunoassay in microtiter plate and dipstick format. J Agric Food Chem 1993; 41:1006-1011.
17. Iwanzik W, Egli H. Comparison of bioassay and chemical analysis for triasulfuron quantification in soil samples: Proceedings. Brighton Crop Protection Conference 1989; 3:1145-1150.
18. Rawson DM, Willmer AJ, Cardosi MF. The development of whole cell biosensors for on- line screening of herbidice pollution of surface waters. Toxicity Assess 1987; 2:325-340.
19. Pandard P, Rawson DM. An amperometric algal biosensor for herbicide detection employing a carbon cathode oxygen-electrode. Environ Toxicol Water Qual 1993; 8(3):323-333.
20. Pandard P, Vasseur P, Rawson DM. Comparison of 2 types of sensors using eukaryotic algae to monitor pollution of aquatic systems. Water Res 1993; 27(3):427-431.
21. Preuss M, Hall EAH. Mediated herbicide inhibition in a PET biosensor. Anal Chem 1995; 67(13):1940-1949.
22. Avramescu A, Rouillon R, Carpentier R. Potential for use of a cyanobacterium Synechocyatis sp. Immobilized in poly(vinylalcohol): Application to the detection of pollutants. Biotechnol Tech 1999; 13:559-562.
23. Rouillon R, Tocabens M, Carpentier R. A photoelectrochemical cell for detecting pollutant-induced effects on the activity of immobilized cyanobacterium Synechoccuc sp PCC7942. Enz Microb Technol 1999; 25:230-235.
24. Rawson DM, Willmer AJ, Turner APF. Whole-cell biosensors for environmental monitoring. Biosensors 1989; 4:299-311.
25. Haggett BGD. Mathematical-model of toxicity monitoring sensors incorporating microbial whole cells. Analyst 1994; 119(2):197-201.
26. Arsalane W, Parésys G, Duval JC et al. A new fluorometric device to measure the in vivo chlorophyll a fluorescence yield in microalgae and its use as a herbicide monitor. Eur J Phycol 1993; 28:247-252.
27. Conrad R, Buchel C, Wilhelm C et al. Changes in yield in in-vivo fluorescence of chlorophyll a as a tool for selective herbicide monitoring. J Appl Phycol 1993; 5:505-516.
28. Yoneyama K, Nakajima Y, Maejima N et al. Simple and rapid screening method for Photosystem II inhibitory herbicides using photoautotrophically cultured plant cells with chlorophyll fluorescence monitoring. Biosci Biotech Biochem 1993; 57(8):1389-1390.
29. Merz D, Geyer M, Moss DA et al. Chlorophyll fluorescence biosensor for the detection of herbicides. Fresenius J Anal Chem 1996; 354:299-305.
30. El Jay A, Ducruet JM, Duval JC et al. A high-sensitivity chlorophyll fluorescence assay for monitoring herbicide inhibition of photosystem II in the chlorophyte Selenastrum capricornutum: Comparison with effect on cell growth. Arch Hydrobiol 1997; 140:273-286.
31. Van der Heever JA, Grobbelaar JU. In vivo chlorophyll a fluorescence of Selenastrum capricornutum as a screening bioassay in toxicity studies. Arch Environ Contam Toxicol 1998; 35:281-286.
32. Wessels JSC, Van der Veen R. The action of some derivatives of phenylurea and of 3-phenyl-1,1-dimethylurea on the Hill reaction. Biochim Biophys Acta 1956; 19:548-549.
33. Good NE. Inhibitors of the Hill reaction. Plant Physiol 1961; 36:788-803.
34. Védrine C, Leclerc JC, Durrieu C et al. Optical whole-cell biosensor using Chlorella vulgaris designed for monitoring herbicides. Biosens Bioelectron 2003; 18:457-463.
35. Rodriguez M, Sanders CA, Greenbaum E. Biosensors for rapid monitoring of primary-source drinking water using naturally occurring photosynthesis. Biosens Bioelectron 2002; 17:843-849.
36. Vermaas W. Molecular genetics of the cyanobacterium Synechocystis sp. PCC 6803: Principles and possible biotechnology applications. J Appl Phycol 1996; 8:263-273.

37. Giardi MT, Esposito D, Leonardi C et al. Portable system for selective monitoring of herbicides based on fluorescence and resistant mutants to herbicides. Patent, Upica, Italy: 2000:112.
38. Nanba O, Satoh K. Isolation of a photosystem II reaction center consisting of D-1 and D-2 polypeptides and cytochrome b559. Proc Natl Acad Sci USA 1987; 84:109-112.
39. Giardi MT, Marder JB, Barber J. Herbicide binding to the isolated Photosystem II reaction centre. Biochim Biophys Acta 1989; 934:64-71.
40. Franco E, Alessandrelli S, Masojidek J et al. Modulation of D1 protein turnover under cadmium and heat stresses monitored by [^{35}S] methionine incorporation. Plant Science 1999; 144:53-61.
41. Trebst A, Draber W. In: Greissbuehler H, ed. Advances in Pesticide Science. Oxford: Pergamon Press, 1979:223.
42. Pfister K, Steinback KE, Gardner G et al. Photoaffinity labeling of an herbicide receptor protein in chloroplast membranes. Proc Natl Acad Sci USA 1981; 78:981-985.
43. Mattoo AK, Pick U, Hoffman-Falk H et al. The rapidly metabolized 32kDa polypeptide of the chloroplast is the "proteinaceous shield" regulating photosystem II electron transport and mediating diuron herbicide sensitivity. Proc Natl Acad Sci USA 1981; 78:1572-1576.
44. Giardi MT, Masojidek J, Godde D. Effects of abiotic stresses on the turnover of the D1 reaction centre II protein. Physiol Plant 1997; 101:635-642.
45. Loranger C, Carpentier R. A fast bioassay for phytotoxicity measurements using immobilized photosynthetic membranes. Biotechnol Bioeng 1994; 44:178-183.
46. Rouillon R, Tocabens M, Marty JL. Stabilization of chloroplasts by entrapment in polyvinylalcohol bearing styrylpyridinium groups. Anal Lett 1994; 27:2239-2248.
47. Koblizek M, Masojidek J, Komenda J et al. A sensitive photosystemII-based biosensor for detection of a class of herbicides. Biotechnol Bioengg 1998; 60:664-669.
48. Brewster JD, Lightfield AR. Rapid biorecognition assay for herbicides in biological matrices. Anal Chem 1993; 65:2415-2419.
49. Brewster JD, Lightfield AR, Bermel PL. Storage and immobilization of photosystem II reaction centers used in an assay for herbicides. Anal Chem 1995; 67:1296-1299.
50. Rao KK, Hall DO, Vlachopoulos N et al. Photoelectrochemical response of photosystem II particles immobilized on Dye-derivatized TiO2 films. J Photochem Photobiol B 1990; 5:379-389.
51. Thomasset B, Thomasset T, Vejux A et al. Immobilized thylakoids in a cross-linked albumin matrix. Plant Physiol 1982; 70:714-722.
52. Carpentier R, Loranger C, Chartrand J et al. Photoelectrochemical cell containing chloroplast membranes as a biosensor for phytotoxicity measurement. Anal Chim Acta 1989; 249:55-60.
53. Carpentier R, Lemieux S, Mimeault M et al. A photoelectrochemical cell using immobilized photosynthetic membranes. Bioelectrochem Bioenerg 1991; 22:391-401.
54. Rouillon R, Mestres JJ, Marty JL. Entrapment of chloroplasts and thylakoids in polyvinylalcohol-SbQ. Optimalization of membrane preparation and storage conditions. Anal Chim Acta 1995; 311:437-442.
55. Rouillon R, Sole M, Carpentier R et al. Immobilization of thylakoids in plyvinylalcohol for the detection of herbicides. Sensors and Actuators B 1995; 26-27:477-479.
56. Rouillon R, Gingras Y, Carpentier R et al. In: Mathis P, ed. Photosynthesis: From Light to Biosphere, Vol. 5. The Netherlands: Kluwer, 1995:933–936.
57. Komenda J, Masojidek J, Setlikova E. An indication of different functions of CP47 and CP43. In: Argyroudi-Akoyunoglou JH, ed. Regulation of Chloroplast Biogenesis. New York: Plenum Press, 1999:411-415, (NATO ASI series).
58. Koblizek M, Mela J, Giardi MT et al. Book of abstracts, 2nd workshop on biosensors and biological techniques in environmental analysis. Lund, Sweden: 1998.
59. Clark LC. Monitor and control of blood and oxygen in tissue. Trans Am Soc Artif Intern Organs 1956; 2:41-48.
60. Schneider P, Goodrow MH, Gee SJ et al. A highly sensitive and rapid ELISA for the urea herbicides diuron, monuronand linuron. J Agric Food Chem 1994; 42:413-422.
61. Alegria G, Dutton PL. Langmuir-Blodgett monolayer films of bacterial photosynthetic membranes and isolated reaction centers: Preparation, spectrophotometric and electrochemical characterization. Biochim Biophys Acta 1991; 1057:239-257.
62. Jockers R, Bier FF, Schmid RD. Specific binding of photosynthetic reaction centres to herbicide-modified grating couplers. Anal Chim Acta 1993; 280:53-59.
63. Piletskaya E, Piletsky S, Lavrik N et al. Toward the D1 protein application for the development of sensors specific for herbicides. Anal Lett 1998; 31:2577-2589.
64. Shao CY, Howe CJ, Porter AJR et al. Novel cyanobacterial biosensor for detection of herbicide. Appl Environ Microbiol 2002; 68:5026-5033.
65. Hall HR. Biosensor technologies for detecting microbiological foodborne hazards. Microbes Infect 2002; 4:425-432.

Chapter 14

Mimicking the Plastoquinone-Binding Pocket of Photosystem II Using Molecularly Imprinted Polymers

Florent Breton, Elena V. Piletska, Khalku Karim, Régis Rouillon* and Sergey A. Piletsky

Abstract

The photosystem II (PSII) is a complex system consisting of at least 10 proteins. The electron-flow events in PSII are mediated via prosthetic groups (plastoquinones Q_A and Q_B) bound to two proteins called D1 and D2. A large group of photosynthesis-inhibiting herbicides consisting of arylureas, triazines, triazinones and phenylic herbicides has an ability to replace the plastoquinone Q_B from its binding pocket located in the D1 protein, interrupting the electron flow between the photosystems and causing the plant's death. The plastoquinone-binding pocket continues to be a subject of intense research by specialists working on the design and testing of new herbicides, as well as scientists and engineers developing new sensors for herbicide detection. The goal of this review is to analyze the structure of the herbicide-binding pocket of the D1 protein in comparison to known natural and synthetic receptors of herbicides with the aim of developing efficient synthetic mimics useful in the generation of new stable environmental sensor devices.

Introduction

Photosystem II (PSII) is a large, heteromeric enzyme complex with more than twenty different protein subunits and an array of cofactors, that participate in the photosynthetic electron transport in the thylakoid membranes of chloroplasts and cyanobacteria.[1] PSII demonstrates the oxido-reductase activity and couples the oxidation of H_2O with the reduction of plastoquinones through a series of intermediate redox reactions. The central core of the PSII reaction center is composed of D1 and D2 proteins associated with redox active components. These compounds include a tetra-manganese cluster, two redox-active residues, four to six chlorophyll *a* molecules, two pheophytins, and plastoquinones Q_A and Q_B.[2,3] The secondary plastoquinone Q_B is a two-electron carrier which during the operation cycle is reduced by Q_A^-. The semiquinone form of Q_B is bound at the Q_B site of the D1 protein but the oxidized or reduced forms are exchanged easily with a pool of free plastoquinones providing the electron transport between photosystems. The same time Q_A is tightly associated with the D2 protein and it is not exchanged with the plastoquinone PQ pool.[4]

A large number of commercial herbicides such as arylureas, triazines, triazinones and phenolic compounds act as competitors to plastoquinones (Fig. 1). They occupy the Q_B-binding site of the D1 protein, thereby displacing Q_B from its binding niche[5] and prevent the oxidation of reduced Q_A.[6] The displacement of electron mediator Q_B from the D1 protein leads to interruption of the electron flow and, consequently, results in plant's death.

*Corresponding Author: Régis Rouillon—Université de Perpignan, Centre de Phytopharmacie, 52 Avenue Paul Alduy, 66860 Perpignan, France. Email: rouillon@univ-perp.fr

Biotechnological Applications of Photosynthetic Proteins: Biochips, Biosensors and Biodevices, edited by Maria Teresa Giardi and Elena V. Piletska. ©2006 Landes Bioscience.

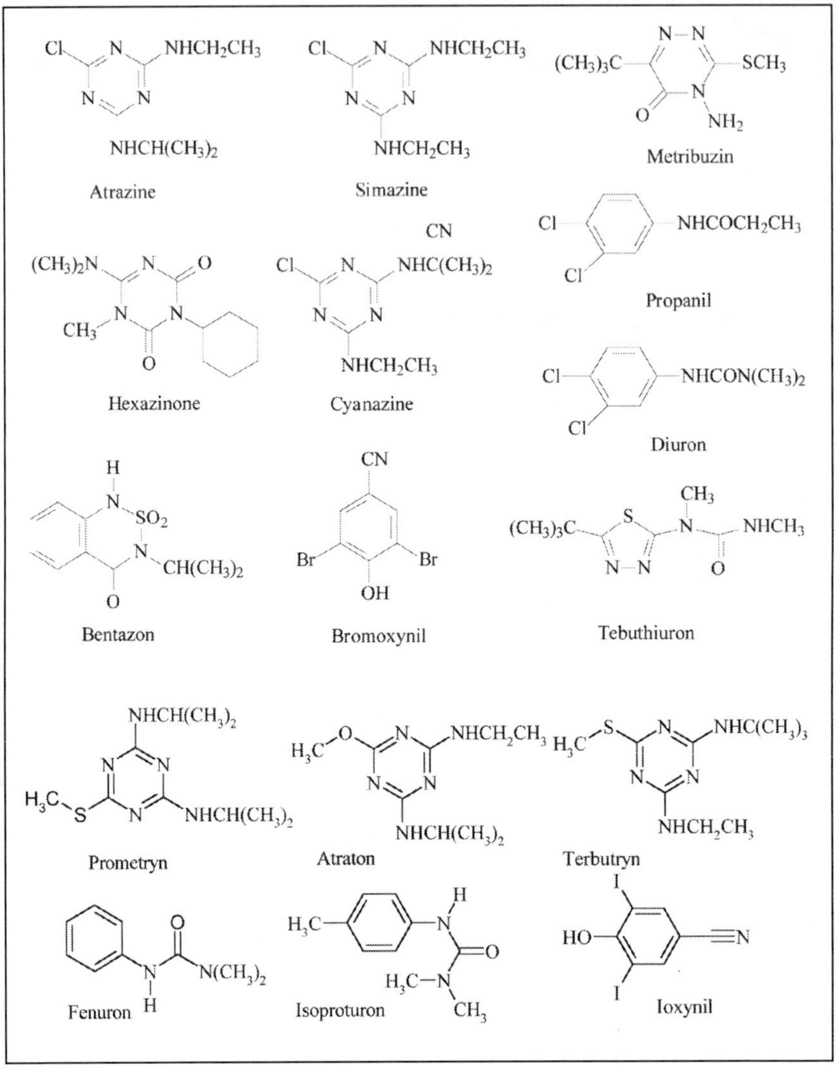

Figure 1. Structures of photosynthesis-inhibiting herbicides.

The first photosynthesis-inhibiting herbicides such as arylurea (e.g., diuron) and triazines derivatives (e.g., atrazine) were identified in 1956 even before the photosynthetic reactions and two photosystems were known and before plastoquinone had been discovered.[7,8] Surprisingly, this group of herbicides still dominates the field. The second group, which includes phenolic compounds such as bromoxynil and ioxynil, were recognized later. Although phenolic herbicides inhibit the PSII reaction centre differently from triazine herbicides, they also interfere with the Q_B function and bind the D1 protein.

The combination of the mediator properties of the D1 protein and the natural affinity towards herbicides make D1 proteins an interesting object for incorporation into optical and electrochemical sensors for herbicide monitoring.[9,10] Despite the advantages which the D1 protein as the natural herbicide receptor possess, the sensor application requires higher specificity and stability. From this point of view, it seems necessary to develop new synthetic materials, which would mimic the herbicide-binding pocket of the D1 protein. In this chapter we would like to study the interactions between photosynthesis-inhibiting herbicides and natural receptors such as the D1 protein and atrazine-specific antibodies on molecular level.

Natural Receptors for Photosynthesis-Inhibiting Herbicides

D1 Protein

The arrangement of the two reaction center proteins D1 and D2,[11] was based on studies of its structural and functional analogues, L and M subunits of the reaction centers of *Rhodobacter sphaeroides and Rhodopseudomonas viridis.*[12,13] The D1 protein is membrane protein which contains five trans-membranes helices (A-E) and two short "parallel" helices that do not span the membrane. One of them, the DE helix is located between the fourth and fifth (*D* and *E*) transmembrane regions. It was exposed to the stromal side (the acceptor side) of the photosynthetic membrane.[14,15] It was found that this region of the D1 protein formed the binding niche of the secondary quinone (Q_B). There were a number of D2 protein residues that were also identified as part of the Q_B-niche.[16]

Plastoquinone (Q_B) binding to the D1 protein was characterized by a hydrogen bond between each quinone carbonyl group and delta-nitrogen of His 215 and hydroxyl group of Ser 264 residues. The binding with the D1 protein was also stabilized by ring stacking of the quinone head group and the phenyl side chain of Phe 255. Residues though to be in close proximity to bound Q_B are Phe 211, Met 214, His 215, Leu 218, Val 219, Ala 251, His 252, Phe 255, DE loop (residues 259-267), Ser 268 and Leu 271 (Table 1).[16]

Table 1. Modeling results of the PSII-inhibiting herbicides binding to different natural receptors

	D1 Protein	*Rps. viridis*	Antibodies
Plastoquinone	Pro 196, Phe 197, Leu 200, Ala 203, Gly 207, Phe 211, Met 214, His 215, Leu 218, Val 219, Phe 246, Ala 251, His 252, Phe 255, Ile 259, Ala 263, Ser 264, Phe 265, Asn 267, Ser 268, Leu 271, Trp 278, Ile 281, Phe 285[16]		
Terbutryn	Phe 211, Met 214, Tyr 262, Ser 264, Phe 265, Phe 274[14]	Ser L223, Ile L224, Glu L212, Ile K229, Val L220, Phe L216[23]	
Atrazine	Phe 211, Met 214, Phe 255, Tyr 262, Ser 264, Phe 265 and Phe 274[15]	Leu L189, His L190, Glu L212, Asn L213, Phe L216, Arg L217, Tyr L222, Ser L223, Ile L224[48]	Tyr L96, His H95, Gly H100a, Gln L98, Val H37, Tyr L36, Phe L98, Trp H47, Phe H100b, Trp H33, Glu H50, His H95, Tyr L96[32]
Diuron	His 215, Phe 255, Ser 264, Phe 265 or Ser 268[27]	Tyr L222[26]	
Ioxynil	Val 249[49], Asn 266[50], His 215[14]		
Metribuzin	Val 219, Asn 247, Ala 251, Phe 255, Ser 264, Ser 268, Leu 275[8]		

The D1 protein was found to be encoded by the *psbA* gene of the chloroplast genome. The psbA gene sequences from several organisms were determined and its in vitro expression in *Escherichia coli* was successfully performed.[17-19] A number of mutations in the gene *psbA* that led to herbicide resistance was mapped in a region of the polypeptide between amino acid residues 211 and 275. Molecular modeling studies, combining molecular graphics and computational chemistry were used to investigate the intermolecular interactions governing the binding of diuron[20] and triazines[15] to the D1 protein of photosystem II. Based on the calculated intermolecular energy of interaction between the herbicide and its binding site and on the correlation with the inhibition behavior of mutant species, the favored orientations of the various types of herbicides within the binding site were proposed. The results supported the concept of different though overlapping binding sites for various PSII inhibitors near the Q_B binding niche.[21] The inhibitor-binding domain was thought to reside between helixes IV and V (D and E) of the D1 protein, which corresponded the Q_B-binding niche.[22]

The mechanism by which triazines bind to the bacterial reaction centre of *Rps. viridis* has been studied by X-ray crystallography. It was found that the herbicide terbutryn binds to the L subunit of the reaction centre in the Q_B binding pocket through two hydrogen bonds. One is between the N3 triazine ring nitrogen and the backbone NH group of the Ile L224 residue. The other is between the side chain hydroxyl of Ser L223 and the aminoethyl NH group of terbutryn.[23] In addition, Phe L216 was involved in stacking interactions with the triazine ring system and close contacts also exist between the herbicide and Glu L212, Ile K229 and Val L220.

In the absence of high-resolution crystal structures for PSII, molecular models have been constructed based upon functional and partial sequence homologies found between the L and M subunits of the photosynthetic bacteria and the D1 and D2 proteins of higher plants. It was suggested that a hydrogen bond existed between the ethylamino group of atrazine and the side chain of Ser 264, the equivalent residue to Ser L233 in the bacterial L subunit. It was also proposed that second hydrogen bond involved the isopropylamino NH of atrazine and the carbonyl group of Phe 265. The triazine ring system formed energetically favorable interaction with both the aromatic Phe 255 (equivalent to Phe L216 in bacterial reaction centers) and Tyr 262 residues. The narrowness of the binding region around the ethylamino functional group was evident from the repulsive energy between the ethyl group and the Phe 255 and Ser 264 residue which was counterbalanced by its favorable interaction with Ala 251, Tyr 262 and Asn 266, and between the amino group and Phe 255 and Ser 264.[15] The labeling of the D1 protein with azido [^{14}C]atrazine showed that the chloro substituent of the triazine ring of atrazine was located in proximity to Met 214, similar to methylthio substituent of terbutryn. The t-butyl group of terbutryn was found to be enclosed in a hydrophobic pocket consisting of the Phe 211, Tyr 262, Phe 265 and Phe 274 residues.[14]

The differences in size and hydrophobicity of the substituents on the basic triazine ring system determined the different interactions with the protein[24] but the binding site of atrazine on the Q_B site of PSII was found very similar to that of terbutryn (Table 1).[14] The studies of *Synechococcus PCC7942* showed that replacement of Phe 255 by Tyr led to atrazine-resistance but not diuron-resistance, Ser 264 was essential for binding of atrazine, diuron and quinone, whereas Phe 255 was involved in atrazine binding but not Q_B or diuron.[25] In both cases, the same hydrogen bonds with the protein matrix were formed and the chloro substituent was close to Met 214. It was also proposed that the hydrophobic pocket consisting of the residues Phe 211, Tyr 262, Phe 265 and Phe 274 could be the stereoselective region of the binding site.[15]

Due to the fact that the arylurea type herbicides, such as diuron and monuron, did not inhibit the wild type bacterial reaction centers, the predictions have been based mainly upon mutations of the Q_B-binding domain, which was affected by interaction with diuron. For example, the characterization of the herbicide-resistant mutants from *Rps. viridis* has revealed that one of the mutants, T4 (Tyr L222 to Phe) was sensitive to the urea type inhibitors similar to the D1 protein of PSII reaction centre.[26] The semiquinone-iron electron paramagnetic resonance (EPR) signal of Q_B in *Rps. viridis* T4 mutants was also similar to that reported for photosystem II.[13]

The molecular modeling was made in order to study the interactions determining the diuron's binding to the D1 protein. Previously, modeling discussions concerning this herbicide have been focused on the formation of hydrogen bonds with the top of the Q_B binding pocket of PSII, involving the Ser 264, Phe 265 or Ser 268 residues.[27] Although the trans-amide isomer of diuron was considered to be the more stable conformation, this modeling study required the adoption of the cis-amide conformation by the herbicide for the formation of a hydrogen bond between the phenylamino NH of diuron and the side chain oxygen of Ser 264 in order to maximize van der Waals dispersion forces between the phenyl ring and the side chain of Phe 255 and to minimize the repulsive forces between the dimethylamino group and the protein matrix. In another model, the phenyl ring of diuron was mapped in proximity to Phe 255 and a hydrogen bond is formed between the carbonyl group of the urea and His 215.[15] In this case, the herbicide was able to form a more stable planar trans-amide form. It was suggested that water molecules acted as bridges connecting the diuron hydrogen functions and the D1 protein's His 215 and Ser 264 together.[28]

Due to their difference in chemistry, all PSII-inhibiting herbicides demonstrate different binding properties. For example, urea/triazine type inhibitors were proposed to be oriented towards Ser 264, triazinones towards Ala 251 and phenolic herbicides were oriented towards His 215 (Table 1).[14]

Antibodies Specific for Photosynthesis-Inhibiting Herbicides

One of the established methods of herbicides monitoring and quantification is enzyme-linked immunosorbent assay (ELISA) which is based on specific antigen-antibodies interactions. There is a big choice of existing antibodies specific for triazines and other photosynthesis-inhibiting herbicides which can bind effectively these herbicides.[29-31] Similar to the D1 protein mapping, the mutations of antibodies were introduced in order to better understand the molecular interactions between antibodies and antigens and to design the antibodies with increased affinity and specificity. The mapping and modeling of the atrazine-binding site of antibodies showed that the triazine ring was sandwiched between Tyr L96, His H95 and Gly H100a residues. The hydrophobic isopropyl-binding pocket was formed by Gln L89, Val H37 and aromatic residues Tyr L36, Phe L98, Trp H47 and Phe H100b, and the chlorine-specific binding pocket was formed by Trp H33, Glu H50, His H95 and Tyr L96.[32] The chlorine-binding site of antibodies was very chlorine-specific. The chlorine substitution by hydroxyl, methoxyl or thiomethyl groups led to two orders of magnitude lower cross-reactivity between antibodies and the herbicides. The results of the ELISA suggested that better recognition of isopropyl residues resulted in higher antibody affinity towards the atrazine molecule and consequently higher sensitivity of the assay.

Synthetic Receptors

At present, molecularly imprinted polymers are the only existing examples of synthetic receptors for PSII-inhibiting herbicides. Molecularly imprinted polymers (MIPs) were invented by Günter Wulff with later contribution from Klaus Mosbach more than 30 years ago.[33,34] The basic principle of MIP design included the formation of specific functional sites by complexation of the target molecule (template) with functional monomers (Fig. 2). The presence of the template in solution helped to coordinate mutual positioning of the selected monomers in the synthesized

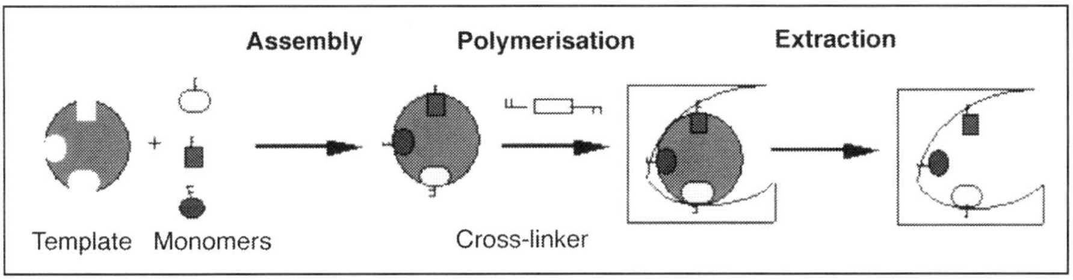

Figure 2. Schematic representation of the molecular imprinting protocol.

polymer leading to the formation of selective binding sites. The molecular complex between template and functional monomers was preserved using an excess of polymerizable cross-linker. Thermal or photochemical initiated polymerization gave a highly cross-linked insoluble polymer. The high degree of cross-linking allowed the cavities to maintain their shape after removal of the template, and gave the polymer a high mechanical stability. The extraction of the template from the MIP created the cavities in the matrix, which were complementary in both shape and chemical functionality to those of the template (Fig. 2).

Based on their chemical and mechanical properties, imprinted materials were applied in a variety of techniques. Solid phase extraction (SPE) has become a common method used to concentrate analytes in order to improve their detection. Among the advantages of SPE material based on MIPs were specificity and compatibility with both aqueous and organic solvents. MIPs as SPE matrixes were useful for sample preconcentration and purification. The superior performance of MIPs' SPE for routine analysis was demonstrated on a variety of compounds such as drug and pesticides.[35-37]

Chemical sensors are another interesting area of MIP application. Due to the increasing demand of rapid and sensitive detection in food and environment analysis, drug detection, clinical diagnosis, chemical- and biosensors MIPs have generated a great interest. In particular, biosensors which incorporate the biological macromolecules such as enzymes or antibodies, have already shown their ability to detect many pollutants and drugs at low concentrations.[38,39] Together with high affinity and specificity, the recognition materials based on MIPs offer high stability even under particularly harsh conditions such as high temperature and/or pressure with the added advantage of low cost of preparation.

Another interesting example of MIP application was the development of a MIP catalyst for the decomposition of a soil and water pollutant atrazine to atraton, from a biologically active molecule to a biologically inactive one. Five libraries were constructed with varying concentrations of MAA and different functional monomers. MAA was expected to provide the high affinity of binding sites, whereas the second monomer (2-sulfoethyl methacrylate) was responsible for catalytic activity.[40] Although the results were too modest to be considered for practical applications, nevertheless they have indicated a potentially new application for MIPs in environmental engineering.

MIPs Specific for Photosynthesis-Inhibiting Herbicides

The triazine herbicides, particularly, atrazine, were common templates for MIP preparation. One of the examples was a MIP made with methacrylic acid (MAA) as a functional monomer. In order to evaluate the selectivity and affinity towards atrazine, a MAA-based MIP was tested in high performance liquid chromatography (HPLC) and in batch-binding assays. It was found that MAA formed a strong hydrogen bond with the amino group of atrazine.[41] The polymer did not distinguish between different chlorotriazines which were structurally very similar to atrazine (terbutylazine and propazine), but at the same time its affinity towards the methoxy- or thiomethoxytriazine (atraton, prometryn and hydroxyatrazine) was much lower. During batch-binding studies, the atrazine-specific MIP was incubated with solutions of different herbicides. It was shown that the level of cross-reactivity was roughly proportional to the structural resemblance to atrazine. The binding affinity was best for atrazine, lower for the other chlorotriazines, very lower for nonchloronated triazines and no binding was observed for nontriazine herbicides like isoproturon. A similar conclusion was received during studies of atrazine-specific antibodies where the substituent of the Cl carbon atom was the most important feature determining the level of cross-reactivity of triazines, and the binding affinity decreased in the order Cl > OCH_3 > OH > SCH_3. In comparison with antibodies, the molecularly imprinted synthetic receptors prepared against atrazine, appeared more specific than antibodies. The atrazine-specific MIP recognized atrazine better than other similar herbicides whereas polyclonal and monoclonal antibodies, even though they were more specific towards some chlorotriazines, generally did not demonstrate any selectivity between propazine and atrazine. Similar results were also obtained by Muldoon and Stanker.[42]

Atrazine-specific stand-alone MIP membranes based on MAA as functional monomer were tested using conductometry.[43] The electrical conductivity of the membranes was analyzed in an electrochemical cell with two platinum electrodes, separated by the membrane under investigation. The sensor demonstrated good selectivity towards atrazine in comparison with other triazine herbicides, which generated negligible changes in the conductivity of the membrane as compared to atrazine, and a good sensitivity with a detection limit of 5 nM (about 1 $\mu g\ L^{-1}$) for atrazine was achieved.

Terbutylazine was another example of a triazine herbicide, which was used for MIP synthesis. In order to test the ability of different functional monomers to form strong interactions with terbutylazine, a combinatorial approach was used.[44] Thus, MAA, 2-(trifluoromethyl) acrylic acid (TFMAA), hydroxyethyl methacrylate (HEM), methyl methacrylate (MMA), N-vinyl-α-pyrrolidone (NVP) and 4-vinylpyridine (4VP) were used for polymer preparation. In dichloromethane, MAA and TFMAA appeared to be the best monomers able to form the terbutilazine-specific functional sites.

The arylurea herbicides were also used for MIPs synthesis. Two arylureas, fenuron and isoproturon, were polymerized with MAA as functional monomer[45] and used as adsorbents for solid phase extraction of pollutants. The isoproturon-imprinted polymer did not demonstrate the selectivity towards the template but had a high affinity towards all other arylurea herbicides. The corresponding blanks (nonimprinted polymers) did not retain any of these compounds, which suggest that the imprinting was achieved. With fenuron, very good specificity for this herbicide was observed. This result can be explained by the small size of the fenuron molecule. Indeed the other arylureas were not able to enter in the small cavities formed around fenuron due to steric interactions of the meta- or para-substituents in the aromatic ring.[46]

The polymer able to bind the pesticide bentazone was prepared for clean-up and preconcentration of aqueous samples.[47] The bentazone molecule has a rigid structure due to the presence of an aromatic ring fused with a saturated heterocyclic ring. The hydrogen bond was formed between an amide carbonyl group and sulfonamide acidic nitrogen. MAA and 4VP were used as functional monomers. MAA was used in order to form hydrogen bonds with the carbonyl group and 4VP for ionic interactions with sulphonamide group. Different polymers were synthesized by varying the quantities of each of these monomers. It was found that the imprinting effect was improved when both monomers were used simultaneously rather than separately.

In order to develop a general procedure for rational design of the imprinted polymers the new computational method was recently developed in Cranfield was used. It involved screening of a virtual library of molecular models of functional monomers, containing polymerizable residues and residues able to form e.g., electrostatic interactions with the templates (Fig. 3).

In order to simulate the interactions between the herbicide molecules and functional monomers, a Silicon Graphics Octane running the IRIX 6.5 operating system was used (SYBYL 6.9 software package (Tripos Inc., St. Louis, Missouri, USA)). The modeling results for different groups of herbicides are shown in Table 2. The PSII-inhibiting herbicides tested in this experiment varied in their properties from weak basic (atrazine, pKa = 1.7; simazine pKa = 1.62; cyanazine pKa = 1.1), to weak acidic (bromoxynil pKa = 4.06; bentazone pKa = 3.4) to neutral (metribuzin, diuron, hexazinone, propanil and tebuthiuron). The monomers giving the highest binding score were chosen for polymer preparation. The analysis of MIPs specific for PSII-herbicides showed that the best monomers for triazines-specific MIPs were MAA, IA, TFMAA and AMPSA. The monomers, which demonstrated the strong binding towards arylureas herbicides, were MBAA, Acrylamide, IA, TFMAA, and MAA. The selection of monomers clearly depended on the different chemical nature of the herbicides.

Based on the results of the molecular modeling of atrazine, acidic monomers such as MAA, itaconic acid (IA) and TFMAA demonstrated the best affinity towards atrazine (Table 3). It was possible to see that two carboxyl groups interact with the isopropylamine NH group and nitrogens

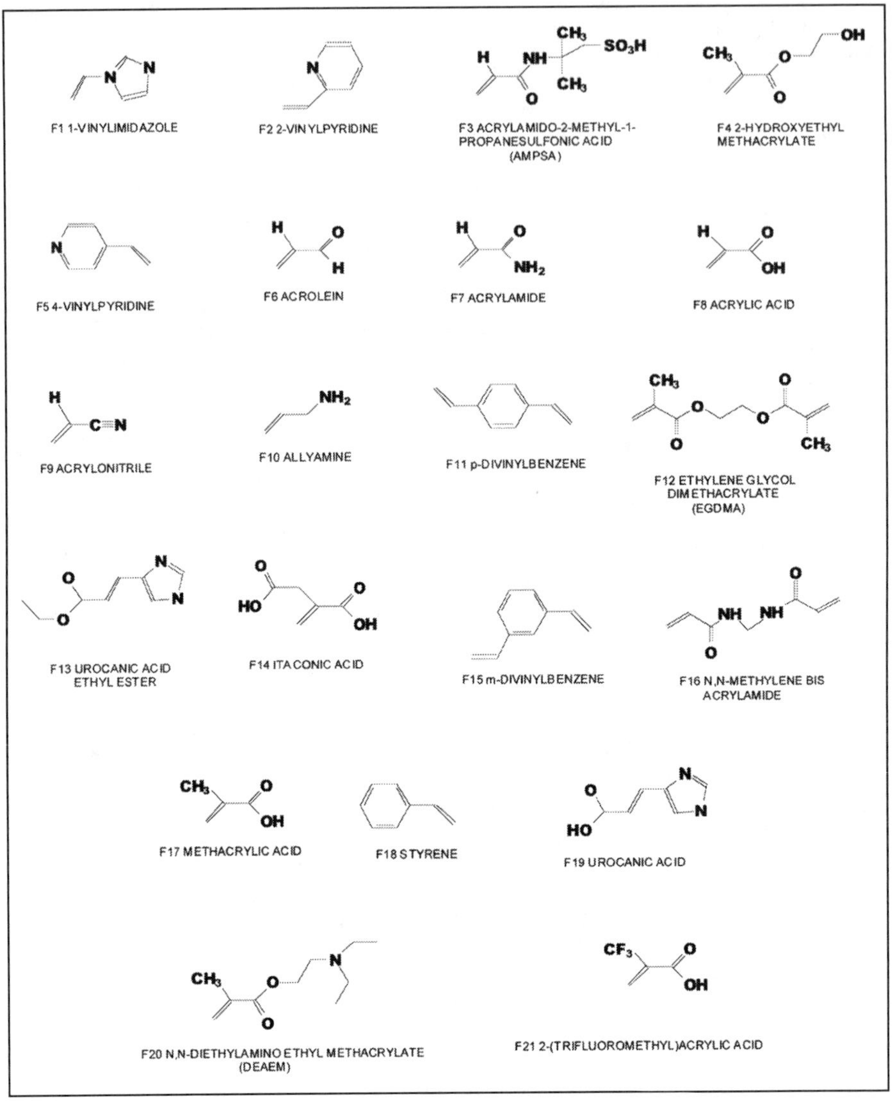

Figure 3. Structures of functional monomers used in modeling and in polymer preparation.

of the triazine ring of atrazine (Fig. 4), which could be compared to the interaction between the carbonyl group of Phe 265 and the isopropylamine NH group of atrazine proposed by Mackay.[15]

Another similarity between natural and synthetic receptors for atrazine was the high hydrophobicity of the binding pocket, which was determined by several phenylalanine residues and tyrosine (D1 protein and antibodies) and which is common for traditional MIPs with a high content of hydrophobic cross-linker.

The imprinted polymers demonstrated not only equal but better specificity than the natural receptors and antibodies and also with excellent affinity: (MIP for atrazine- Kd = 1.5 nM)[43] and stability (no changes in polymer performance were observed for a variety of polymers over 2 years of storage and use).

Table 2. Modelling results of the PSII-inhibiting herbicides binding to synthetic receptors

Herbicides		Functional Monomers
Triazines	Atrazine	Itaconic acid, Methacrylic acid, TFMAA, Acrylamide, AMPSA, DEAEM, MBAA
	Simazine	Methacrylic acid, MBAA, Itaconic acid, Acrylamide, Allylamine
	Cyanazine	Itaconic acid, Methacrylic acid, TFMAA, AMPSA, Acrylamide
Triazinones	Metribuzine	AMPSA, Methacrylic acid, Allylamine, Itaconic acid, Acrylamide
	Hexazinone	Acrylamide, Allylamine, MBAA, Methacrylic acid, Itaconic acid
Arylureas	Diuron	MBAA, Acrylamide, Itaconic acid, TFMAA, Methacrylic acid
	Propanil	TFMAA, MBAA, Acrylamide, Methacrylic acid, Allylamine
Phenolic herbicides	Bromoxynil	Itaconic acid, Methacrylic acid, TFMAA, AMPSA, Acrylamide

Table 3. Binding energy table of functional monomers with atrazine

Monomer/Polymer	Binding, kcal/mol
Itaconic acid (IA)	-94.42
Methacrylic acid (MAA)	-75.06
2-(Trifluoromethyl)acrylic acid (TFMAA)	-64.08
Acrylamide	-34.97
Acrylamido-2-methyl-1-propanesulfonic acid (AMPSA)	-26.63
N,N-diethylamino ethylmethacrylate (DEAEM)	-17.53
N,N-methylene bisacrylamide (MBAA)	-14.52

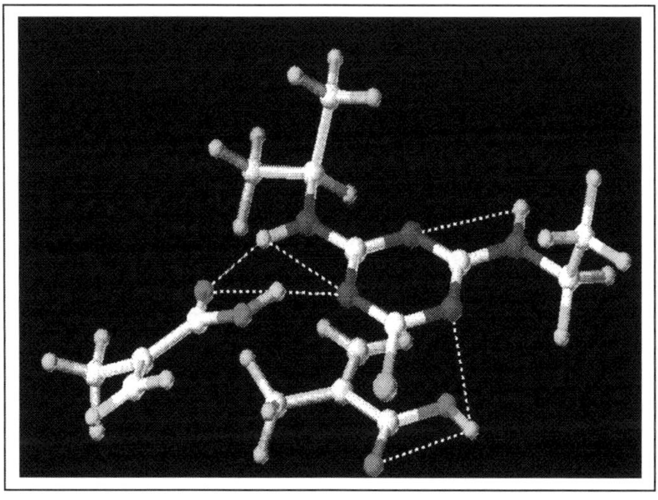

Figure 4. The computationally designed molecular complex between charged atrazine and MAA.

Conclusion

The main amino acids that participate in specific interactions between herbicide's natural receptors and photosynthesis-inhibiting herbicides were identified and analyzed. A reasonable similarity was found between the structural units and behavior of natural and synthetic receptors. Although synthetic receptors did not possessed the complexity of natural receptors, the basic mechanism governing the molecular recognition, such as complementarity for the molecule of interest, was similar to natural receptors. The development of new stable synthetic materials for PSII-herbicides could therefore be beneficial for analytical science and environmental protection.

References

1. Salih G, Wiklund R, Tyystjärvi T et al. Constructed deletions in lumen-exposed regions of the D1 protein in the cyanobacterium Synechocystis 6803: Effects on D1 insertion and accumulation in the thylakoid membranes, and on Photosystem II assembly. Photosynth Res 1996; 49:131-140.
2. Lorkovic ZJ, Schroeder WP, Pakrasi HB et al. Molecular characterization of PsbW, a nuclear-encoded component of the photosystem II reaction center complex in spinach. Proc Natl Acad Sci US 1995; 92:8930-8934.
3. Xiong J, Subramaniam S, Govindjee. Modeling of the D1/D2 proteins and cofactors of the photosystem II reaction center: Implications for herbicide and bicarbonate binding. Protein Sci 1996; 5:2054-2073.
4. Giardi MT, Cona A, Geiken B. Photosystem II core phosphorylation heterogeneity and the regulation of electron transfer in higher plants: A review. Bioelectroch Bioenerg 1995; 38:67-75.
5. Ohad N, Hirschberg J. Mutations in the D1 subunit of photosystem II distinguish between quinone and herbicides binding sites. Plant Cell 1992; 4:273-282.
6. Giardi MT. Significance of photosystem II core phosphorylation heterogeneity for the herbicide-binding domain. Z Naturforsch 1992; 48c:241-245.
7. Wessels JSC, van der Veen R. Action of some derivatives of phenylurethan and of 3-phenyl-1,1-dimethylurea on the Hill reaction. Biochim Biophys Acta 1956; 19:548-549.
8. Draber W, Tieijen K, Kluth JF et al. Herbicides in Photosynthesis Research. Angew Chem Int Ed Engl 1991; 30:1621-1633.
9. Piletskaya E, Piletsky S, Lavrik N et al. Towards the D1 protein application for the development of sensors specific for herbicides. Anal Lett 1998; 31:2577-2589.
10. Piletskaya EV, Piletsky SA, El'skaya AV et al. D1 protein- an effective substitute for immunoglobulins in ELISA for the detection of photosynthesis inhibiting herbicides. Anal Chim Acta 1999; 398:49-56.
11. Michel H, Deisenhofer J. Relevance of the photosynthetic reaction center from purple bacteria to the structure of photosystem II. Biochemistry 1988; 27:1-7.
12. Allen JP, Feher G, Yeates TO et al. Structure of the reaction center from Rhodobacter sphaeroides R-26: The protein subunits. Proc Nat Acad Sci USA 1987; 84:6162-6166.
13. Sinning I, Michel H, Mathis P et al. Characterisation of four herbicide resistant mutants of Rps. viridis by genetic analysis, electron paramagnetic resonance and optical spectroscopy. Biochemistry 1989; 28:5544-5553.
14. Trebst A. The three dimensional structure of the herbicide binding niche on the reaction center polypeptides of photosystem II. Z Naturforsch 1987; 42c:742-750.
15. Mackay SP, O'Malley PJ. Molecular modelling of the interaction between optically active triazine herbicides and photosystem II. Z Naturforsch 1993; 48c:474-481.
16. Xiong J, Subramaniam S, Govindjee. A knowledge-based three dimensional model of the Photosystem II reaction centre of Chlamydomonas reinhardtii. Photosynth Research 1998; 56:229-254.
17. Keller M, Stutz E. Structure of the Euglena gracilis chloroplast gene (psbA) coding for the 32-kDa protein of Photosysntem II. FEBS Lett 1984; 175:173-177.
18. Spielmann A, Stutz E. Nucleotide sequence of soybean chloroplast DNA regions which contain the psb A and trn H genes and cover the ends of the large single copy region and one end of the inverted repeats. NAR 1983; 11:7157-7167.
19. Efimov VA, Fradkov AF, Raskind AB et al. Expression of the barley psbA gene in Escherichia coli yields a functional in vitro photosystem II protein D1. FEBS Lett 1994; 348:153-157.
20. Mackay SP, O'Malley PJ. Molecular modelling of the interaction between DCMU and the Q_B-binding site of photosystem II. Z Naturforsch 1993; 48c:191-198.
21. Pfister K, Arntzen CJ. The mode of action of photosystem II-specific inhibitors in herbicide-resistant weed biotypes. Z Naturforsch 1979; 34c:996-1009.
22. Sobolev V, Edelman M. Modeling of the quinone-B binding site of the photosystem II reaction center using notions of complementarity and contact surface between atoms. Proteins 1995; 21:214-225.

23. Michel H, Epp O, Deisenhofer J. Pigment -protein interactions in the photosynthetic reaction centre from Rps. viridis. EMBO J 1986; 5:2445-2451.
24. Ewald G, Wiessner C, Michel H. Sequence analysis of atrazine resistant mutants from Rps. viridis. Z Naturforsch 1989; 45:459-462.
25. Gleiter HM, Ohad N, Koike H et al. Thermoluminescence and flash-induced oxygen yield in herbicide resistant mutants of the D1 protein in Synechococcus PCC7942. BBA 1992; 1140:135-143.
26. Sinning I, Michel H, Mathis P et al. Terbutryn resistance in purple bacterium can induce sensitivity toward the plant herbicide DCMU. FEBS Lett 1989; 256:192-194.
27. Bowyer J, Hilton M, Whitelegge J et al. Molecular modelling studies on the binding of phenylurea inhibitors to the D1 protein of photosystem II. Z Naturforsch 1990; 45c:379-387.
28. Egner U, Hoyer GH, Saenger W. Modelling and energy minimisation studies on the herbicide binding protein (D1) in PSII of plants. BBA 1993; 1142:106-114.
29. Winklmair M, Weller MG, Mangler J et al. Development of a highly sensitive enzyme-immunoassay for the determination of triazine herbicides. Fresenius J Anal Chem 1997; 358:614-622.
30. Goodrow MH, Hammock BD. Hapten design for compound-selective antibodies: ELISAS for environmentally deleterious small molecules. Anal Chim Acta 1998; 376:83-91.
31. Schneider P, Goodrow MH, Gee SJ et al. A highly sensitive and rapid ELISA for the arylurea herbicides diuron, monuron and linuron. J Agric Food Chem 1994; 42:413-422.
32. Kusharyoto W, Pleiss J, Bachmann TT et al. Mapping of a hapten-binding site: Molecular modelling and site-directed mutagenesis study of an anti-atrazine antibody. Protein Engineering 2002; 15:233-241.
33. Wulff G. Molecular imprinting in cross-linked materials with the aid of molecular template - a way towards artificial antibodies. Angew Chem Int Ed Engl 1995; 34:1812-1832.
34. Haupt K, Mosbach K. Plastic antibodies: Developments and applications. Trends Biotechnol 1998; 16:468-475.
35. Piletsky S, Piletska E, Karim K et al. Custom synthesis of molecular imprinted polymers for biotechnological application. Preparation of a polymer elective for tylosin. Anal Chim Acta 2004; 504:123-130.
36. Matsui J, Fujiwara K, Ugata S et al. Solid-Phase extraction with a dibutylmelamine-imprinted polymer as triazine herbicide-selective sorbent. J Chromatogr A 2000; 889:25-31.
37. Matsui J, Okada M, Tsuruoka M et al. Solid-Phase extraction of a triazine herbicide using a molecularly imprinted synthetic receptor. Anal Commun 1997; 34:85-87.
38. Mello LD, Kubota LT. Review of the use of biosensors as analytical tools in the food and drink industries. Food Chem 2002; 77:237-256.
39. Wang J. Amperometric biosensors for clinical and therapeutic drug monitoring: A review. J Pharm Biomed Anal 1999; 19:47-53.
40. Takeuchi T, Fukuma D, Matsui J et al. Combinatorial molecular imprinting for formation of atrazine decomposing polymers. Chem Lett 2001; 30:530-531.
41. Siemann M, Andersson L, Mosbach K. Selective recognition of the herbicide atrazine by noncovalent molecularly imprinted polymers. J Agric Food Chem 1996; 44:141-145.
42. Muldoon MT, Stanker LH. Molecularly imprinted solid phase extraction of atrazine from beef liver extracts. Anal Chem 1997; 69:803-808.
43. Sergeyeva T, Piletsky S, Brovko A et al. Selective recognition of atrazine by molecularly imprinted polymer membranes. Development of conductimetric sensor for herbicides detection. Anal Chim Acta 1999; 392:105-111.
44. Lanza F, Sellergren B. Method for synthesis and screening of large groups of molecularly imprinted polymers. Anal Chem 1999; 71:2092-2096.
45. Martin-Esteban A, Turiel E, Stevenson D. Effect of template size on the selectivity of molecularly imprinted polymers for phenylurea herbicides. Chromatographia 2001; 53:434-437.
46. Tamayo F, Casillas J, Martin-Esteban A. Highly selective fenuron-imprinted polymer with a homogeneous binding site distribution prepared by precipitation polymerisation and its application to the clean-up of fenuron in plant samples. Anal Chim Acta 2003; 482:165-173.
47. Baggiani C, Trotta F, Giraudi G et al. A molecularly imprinted polymer for the pesticide bentazone. Anal Commun 1999; 36:263-266.
48. Lancaster CRD, Michel H. Refined crystal structures of reaction centres from phodopseudomonas viridis in complexes with the herbicide atrazine and two chiral atrazine derivatives also lead to a new model of the bound carotenoid. J Mol Biol 1999; 286:883-898.
49. Oettmeier W, Masson K, Hohfeld J et al. $^{[125]}$I-azido-ioxynil labels Val_{249} of the photosystem II D-1 reaction center protein. Z Naturforsch 1989; 44c:444-449.
50. Creuzet S, Ajlani G, Vernotte C et al. A new ioxynil-resistant mutant in Synechocystis PCC6714: Hypothesis on the interaction of ioxynil with the D1 protein. Z Naturforsch 1990; 45:436-440.

Chapter 15

Photosystem II Biosensors for Heavy Metals Monitoring

Régis Rouillon,* Sergey A. Piletsky, Florent Breton, Elena V. Piletska and Robert Carpentier

Abstract

Biotesting based on photosynthetic material is new and a potentially commercially viable method of pollutant detection. The photosynthetic apparatus including photosystem II (PSII) is particularly sensitive to heavy metals. In this chapter, the mechanisms of heavy metals action on photosystem II are discussed. The characteristics of different biosensors based on immobilized photosynthetic material are compared and the possibility to use immobilized photosystem II sub-membranes fractions as a tool to monitor environmental sample contained the heavy metals (sewage sludge) is demonstrated.

Introduction

"Heavy metals" such as copper, mercury, lead and many others present a real danger for the environment. They are dangerous for the living organisms because of three main reasons: they are toxic at very low concentrations, they are not degraded with time, and they are tend to accumulate in the living organisms.[1]

Heavy metals are natural constituents of rocks and soils. Following the industrialization, very big quantities of metals such as copper (Cu), mercury (Hg), cadmium (Cd), lead (Pb), nickel (Ni), zinc (Zn), chromium (Cr) and arsenic (As) have been released into the environment. Some industrial sources are still responsible for the contamination of water and soil. One of the main processes, which contribute to water pollution, is industrial extraction of cadmium, arsenic, lead and zinc. Among the fields of chemical industry, which also contribute towards the release of heavy metals, are metallization processes with cadmium and chromium, production of paints (zinc, cadmium), tannery (chromium), production of pesticides (copper) and fermentation (copper, zinc).[1]

In low concentrations, some of the heavy metals are essential microelements of the plants. Other metals do not have any known biological functions. However, all heavy metals are toxic in high concentrations. The cations of toxic metals generally cause a variety of mutations when taken up by plants. The photosynthetic apparatus appears to be especially sensitive to their toxicity. The heavy metals effect the different components of photosynthetic electron transport chain. Photosystem II is affected by the heavy metals on both- oxidizing (donor) and reducing (acceptor) side.[2] Based on this fact, the use of photosynthetic material and in particular the photosystem II sub-membranes fractions as the biological receptor in a biosensor provides an excellent tool for the detection of toxic metal cations. In the following chapter, we will analyze the biological effect of individual heavy metals on the photosystem II.

*Corresponding Author: Régis Rouillon—Université de Perpignan, Centre de Phytopharmacie, 52 Avenue Paul Alduy, 66860 Perpignan, France. Email: rouillon@univ-perp.fr

Biotechnological Applications of Photosynthetic Proteins: Biochips, Biosensors and Biodevices, edited by Maria Teresa Giardi and Elena V. Piletska. ©2006 Landes Bioscience.

Effects of the Heavy Metals on Photosystem II

Copper (Cu) is an essential microelement of higher plants and algae and has a direct impact on photosynthesis. The Cu-containing protein plastocyanin is a constituent of the primary electron donor of photosystem I. However, high copper concentration inhibits the photosynthetic electron transport, especially at the level of PSII. Since copper has become a widespread pollutant due to its use as algaecide and fungicide in agriculture, the sensitivity of PSII to this metal could be exploited for the development of sensors and assays.

Several sites have been proposed as a potential target of copper action but the exact location of the binding site in PSII is still uncertain.[3] Most authors locate the target of the Cu-inhibition of PSII on its oxidizing site: at or beyond the PSII primary electron carrier donor Tyr_z (redox active Tyr of the D1 protein Fig. 1),[4,5] or very close to the oxygen evolving complex (OEC) through its binding to the residue of the membrane protein.[6] It was shown that the oxidizing side of PSII is the most sensitive site for Cu-inhibition.[7] Other studies concluded that copper ions affect the PSII electron transport on the acceptor side with targets between pheophytin and Q_A[8] or close to the Q_A and Q_B acceptors.[9] The PSII reaction center has also been considered as the Cu-inhibitory binding site.[10,11]

Zinc (Zn) is also an essential microelement, which is required as a prosthetic group for many enzymes. Zn accumulates to a toxic level in water and soil through various emission sources, such as mines and smelters. Although a direct action of Zn on photosystem II is still in question,[11] the use of electron donors shown that Zn^{2+} blocks electron transport at the oxidizing site of PSII, at a site prior to the PSII primary electron carrier donor Tyr_z.[12] The fluorescent measurements confirmed that zinc affects the electron transport flow of photosystem II.[13]

Cadmium (Cd) has received considerable attention over the past years due to increased environmental burdens from industrial, agricultural, energy and municipal sources. While Cd toxicity has been proven to be a major environmental problem, the mechanism of its action has not been fully investigated.[14] Some experiments have provided the evidence that Cd is a potent inhibitor of the photochemical activity of the chloroplasts, especially of photosystem II. A fluorescent study confirms that cadmium inhibition happens on the photo-oxidizing side and affects the electron transfer via PSII.[15] A mutation of the PSII reaction center proteins upon Cd treatment may also be observed.[16] Controversially, other studies have concluded that the light reactions of photosynthesis are not sensitive to Cd. The disagreement concerning the effect of Cd on the photosynthetic apparatus is likely to be due to the experimental conditions, which vary from one study to the other.[17] For example, it was shown that the effect of cadmium can be either a stimulation or an inhibition of PSII activity depending on the time of chemical exposure; at the first, it was observed a stimulation of D1 turnover and then an inhibition.[18]

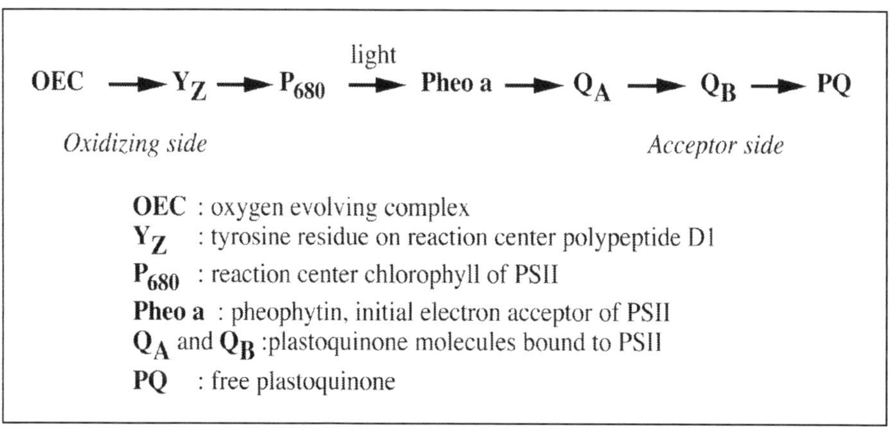

Figure 1. Electron transport components of PSII.

The effects of cadmium and lead (Pb) has been assessed in chloroplasts isolated from lucerne (Medicago sativa).[19] The presence of Cd and Pb led to the inhibition of the PSII activity. The inhibition appeared to be located on the oxidizing side of PSII because the addition of artificial donors restored the electron flow.

Mercury (Hg) is an environmental contaminant that strongly inhibits the photosynthetic electron transport. It was shown that many sites in the photosynthetic membrane especially PSII are highly sensitive towards mercury.[20] Mercury has a very high affinity towards the thyol groups of the proteins. Both, the donor[21,22] and the acceptor sides[23,24] of the photosystem II are affected. For example, it was shown that on the donor side of photosystem II, mercury acts by perturbing the chloride binding and damages the optimal functioning of the oxygen evolving complex.[25]

The toxic effect of nickel (Ni) on photosynthetic electron transport was investigated by monitoring the Hill activity, fluorescence, oxygen evolution and thermo-luminescence properties in the green algae Scenedesmus obliquus. The results obtained suggest that Ni^{2+} modify the Q_B site or interact with the nonheme iron between the Q_A and Q_B inhibiting the photosystem II functions.[26]

Similar to Hg, Pb, Cu, Cd and Cr ions, it was also shown that arsenic (As) ions decreases Hill reaction activity.[27]

Examples of Biosensors Used to Detect the Heavy Metals

Biosensors are analytical devices that consist of a biosensing element like photosynthetic materials and a transducer that transfers the chemical signal to an electrical signal. Because the isolated photosynthetic materials have a relatively short active life time, a variety of immobilization procedures have been developed to stabilize the structure and functions of this biological material. Different photosynthetic biosensors used to detect heavy metals are shown in Table 1.

Thylakoids membranes immobilized in a cross-linked albumin-glutaraldehyde matrix have been used to detect $PbCl_2$ and $CdCl_2$ in solution.[28] The photosynthetic activity was measured with an electrochemical micro-cell. The principle of measurement is based on the fact that ambient oxygen,

Table 1. Examples of photosynthetic biosensors used to detect heavy metals

Biological Material	Method of Immobilization	Method of Detection	Heavy Metals Salts Detected
Spinach thylakoid membranes[28]	Cross-linking in albumin-glutaraldehyde	Electrochemical measurement	$PbCl_2$, $CdCl_2$
Spinach thylakoid membranes[29]	Cross-linking in albumin-glutaraldehyde	Oxygen measurement	$HgCl_2$, $CuCl_2$
Whole cells of *Chlorella vulgaris*[30]	Adsorption on an alumina filter disc	Oxygen measurement	$Hg(No_3)_2$, $CuSo_4$
Spinach thylakoid membranes[31]	Entrapment in PVA-S_bQ*	Electrochemical measurement	$HgCl_2$, $CuCl_2$, $PbCl_2$, $NiCl_2$, $ZnCl_2$, $CuCl_2$
Whole cells of *Synechococcus* sp. PCC 7942[32]	Entrapment in PVA-S_bQ*	Electrochemical measurement	$HgCl_2$
Whole cells of *Synechocystis* sp. PCC 6803[33]	Entrapment in PVA-S_bQ*	Electrochemical measurement	$HgCl_2$
Spinach photosystem II submembrane fractions[34]	Entrapment in PVA-S_bQ*	Electrochemical measurement	$HgCl_2$, $Hg(No_3)_2$, $CuCl_2$, $PbCl_2$, $NiCl_2$, $ZnCl_2$, $CuCl_2$, $CrCl_3$

*PVA-S_bQ = poly(vinylalcohol) bearing styrylpyridinium groups.

which is reduced during photosynthetic process, converts into hydrogen peroxide. The reoxidation of hydrogen peroxide at the working electrode of the electrochemical cell gives a photocurrent. With this electrochemical micro-cell it is possible to detect the photocurrent generated by free or entrapped thylakoid membranes even in the absence of exogenous artificial electron acceptor usually required as an electro-active mediator. A strong inhibition of photocurrent by $PbCl_2$ and $CdCl_2$ has been monitored in the mM range using immobilized thylakoid membranes.

The same immobilized material has been used to detect the chloride salts of mercury and copper.[29] The photosynthetic activity was determined with an oxygen electrode. The PSII-specific artificial acceptor 2,5-dichlorobenzoquinone (DCBQ) was used as a final electron acceptor. After incubation for 10 min, it was shown that the concentrations of the $CuCl_2$, which inhibited 50% of oxygen production (C_{50}) were 100 μM with free thylakoids and 40 μM with immobilized photosynthetic membranes. In the case of $HgCl_2$ the corresponding figures were C_{50} = 60 μM for immobilized membrane and C_{50} = 40 μM for free thylakoids. This last result can be explained by the fact that the main inhibitory site of the mercury ions is located at the oxygen evolving complex on the lumenal side of the thylakoids.[25] This site is probably less accessible in the immobilized preparations due to the higher microviscosity of the crosslinked preparations.[25]

Whole cells of eukaryotic alga Chlorella vulgaris immobilized by filtration on an alumina filter disc allowed the detection of Hg (nitrate salt) or Cu (sulfate salt).[30] Two types of amperometric environmental sensors have been used to detect the photosynthetic activity: one by measuring the rate of reduction of a redox mediator (p-benzoquinone or 2,6-dimethylbenzoquinone) by the illuminated biocatalyst, the other by monitoring the photosynthetic oxygen production. The first system did not allow the detection of mercury and copper after 1 h of exposure. In contrast, the oxygen electrode was sensitive to these metals. A concentration of 1.5 mg/L of nitrate salt of mercury gave 80% inhibition of the oxygen production after exposure time of 180 min. The same concentration of a sulfate salt of copper gave 70% inhibition after 420 min of contact.

Different photosynthetic materials entrapped in poly(vinylalcohol) bearing styrylpyridinium polymers (PVA-S_bQ) were used to detect heavy metals.[31-34] The photosynthetic activity was evaluated using the electrochemical micro-cell mentioned before.[28] Table 2 shows the concentrations of mercuric chloride which produce a 10% inhibition of the activity (C_{10}) of different photosynthetic materials, both native or immobilized in PVA-S_bQ. The inhibition has been measured after 5 min of contact. It was possible to detect a photocurrent with free and immobilized spinach thylakoid membranes in the absence of exogenous artificial electron acceptor. The photosynthetic activity of the spinach PSII sub-membrane fractions and whole cells of cyanobacteria has been evaluated using the artificial electron acceptor DCBQ, which is reduced by the photosystem II. In all cases, immobilization in PVA-S_bQ resulted in decreased sensitivity to $HgCl_2$. Immobilization matrices provided a barrier between the thylakoid membranes working as a sensing element and the toxicants. Similar C_{10} have been obtained with both cyanobacteria and photosystem II sub-membrane fractions. The results showed that sub-membrane fraction has higher sensitivity in comparison with whole thylakoid membranes.

Table 2. Concentrations of mercuric chloride, which induced a 10% inhibition of the photosynthetic activity (C_{10}) of different photosynthetic materials native or entrapped in PVA-S_bQ

Biological Material		**Spinach Photosystem II Submembrane Functions**	**Spinach Thylakoid Membranes**	**Whole Cells Cyanobacterium *Synechococcus* PCC 7942**	**Whole Cells Cyanobacterium *Synechocystis* PCC 6803**
C_{10} mercuric (μM)	native	5	15	6	5
	entrapped	50	200	60	50

Effects of Different Parameters on the Sensitivity of Immobilized PSII Sub-Membrane Fractions towards Heavy Metals

The inhibition, which was shown in Table 2, was obtained after 5 min of exposure. It was found that the sensitivity of immobilized photosystem II sub-membrane fractions to mercury depends on contact time. In addition, the concentration of biological material can also interfere with analytical process. Figure 2 shows that 10 mg/L of mercuric chloride induced 10% inhibition after 5 min of contact with PSII (chlorophyll concentration: 3 mg/mL).[34] It is interesting that such short exposure is enough to inhibit the electron transport of isolated and immobilized PSII. It was found that chlorophyll concentration in the range between 1 and 3 mg/ml plays some role when PSII was exposed to mercury for 15 min. Thus, the chlorophyll concentration of 1 mg/mL gave low inhibition as compared with 2 and 3 mg/mL. The reason for this probably lies in the possibility that membranes made with low concentration of PSII are more dense or more accessible for nonspecific binding of heavy metal which shields PSII from the inhibition. In addition, the prolonged exposure of the PVA-S_bQ-immobilized samples to aqueous solution deteriorates the immobilization matrix. Keeping in mind the low stability of the biological material, it is important to have a control measurement of electron flow in absence of pollutant.

The choice of salt is also important (Table 3). For example, the amount of Hg nitrate $Hg(NO_3)_2$ producing 10% inhibition of immobilized photosystem II sub-membranes fraction was ten times smaller in comparison with similar concentration of Hg chloride ($HgCl_2$).[34] This difference may be attributed generally to dissociation properties of these two salts. $HgCl_2$ is present in solution mainly in the form of complex. For comparison, the Hg $(NO_3)_2$, is almost completely dissociated.[25] As a result, $Hg(NO_3)_2$ was much more powerful inhibitor than $HgCl_2$.

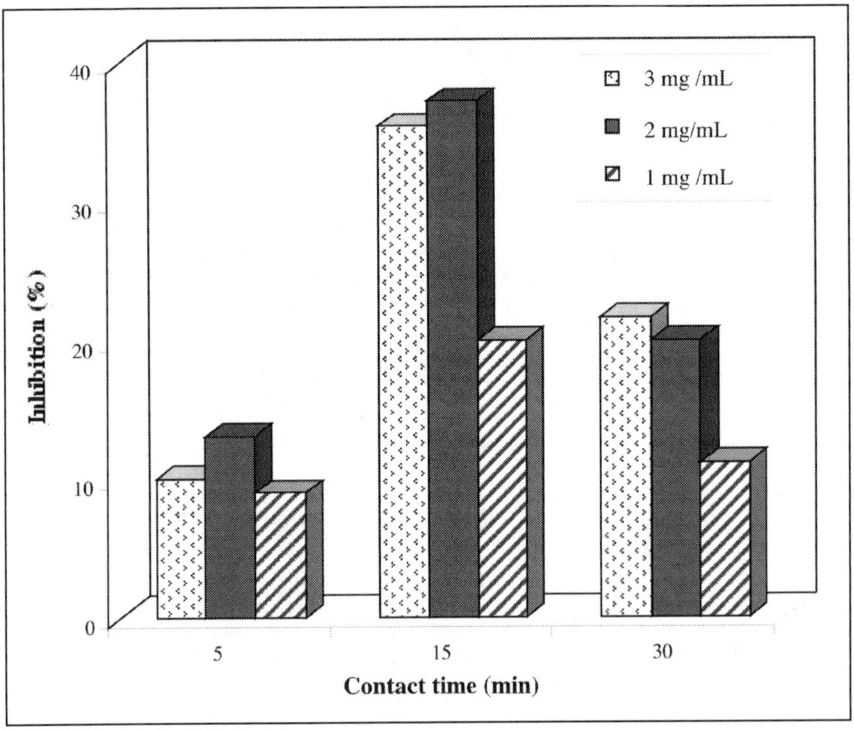

Figure 2. Effect of the chlorophyll concentration and contact time on the inhibition of photosynthetic activity of PVA-S_bQ-entrapped PSII sub-membrane fractions by 10 mg/L mercuric chloride salt.[34]

Table 3. Concentrations of different mercuric salts induced a 10% inhibition of the photocurrent generated by entrapped (in PVA-S_bQ) photosystem II sub-membrane fractions[34]

Element	$C_{10,}$ (mg/L) *Chloride Salt*	$C_{10,}$ (mg/L) *Nitrate Salt*
Hg	20	2

It is worth to mention that chloride is a cofactor required for the optimal function of the oxygen-evolving complex. Its presence may increase the activity of the photosystem II sub-membrane fractions[35] and in such way to modify the extent of inhibition. In addition, it has been shown that the inhibitory effect of mercury is reversed by chloride.[25]

The effect of the heavy metals on the photosynthetic activity also depends on which solution is used to dissolve the metal.[34] The uptake of the metallic elements is related to the concentration of free metal ions. For example, the addition of chelating agent such as ethylenediaminetetraacetic acid (EDTA) reduces the metal ion availability. EDTA decreased the concentrations of the free aqueous ionic form of the metals.[36] The influence of the EDTA complexation on mercury availability was also noticed. The concentrations of mercuric chloride leading to 10% inhibition of the activity (C_{10}) photosystems II sub-membrane fractions immobilized in PVA-S_bQ was about 10 mg/L in the absence of EDTA (Fig. 2) and 20 mg/L in the presence of the chelating agent.[34]

Analysis of the Toxicity of Environmental Samples with PSII Sub-Membrane Fractions Immobilized in PVA-S_bQ

Sewage sludge is a valuable organic manure and soil conditioner and has been used successfully as a fertilizer for many decades.[37] Sewage sludge, when applied to land, contributes the sufficient amounts of nitrogen (N), phosphorus (P) and organic matter.[38] However sewage sludge also contains some amounts of potentially toxic elements, for example heavy metals.

In order to demonstrate the application of the PSII sub-membrane fractions immobilized in PVA-S_bQ for detection of metal pollutants in environmental samples, certified residential sewage sludge have been analyzed. Most of the sewage sludge samples had the metal concentrations higher than the limits of detection obtained with the metals in solution (Table 4).

Table 4. Comparison between heavy metal concentration in a certified residential sewage sludge and metal concentrations (chloride salts) causing a 10% inhibition of the photocurrent generated by entrapped (in PVA-S_bQ) photosystem II sub-membranes fractions

Element	C_{10} (mg/L)	Metal Concentrations in the Sewage Sludge CRMPR 9472 (mg/kg)
Hg	20	3.68
Cr	26	62.0
Pb	104	134.5
Cd	90	12.4
Zn	327	812.1
Ni	441	30.8
Cu	445	589.7
As	–	12.4

Table 5. Inhibition of the photocurrent generated by entrapped photosystem II sub-membrane fractions by the sewage sludge CRMPR 9472 and a reference solution containing the same concentrations of metals

	Sewage Sludge	Reference Solution
Inhibition % of the photocurrent	37	46

Consequently, the photocurrent produced by the PSII-entrapped in PVA-SbQ film was greatly inhibited in the presence of the homogenate extracted from the sewage sludge using different extraction methods (Table 5).[34] To evaluate the extraction method used, the percentage of inhibition obtained from a reference solution (distilled water) containing the same concentrations of metals was measured.

Conclusion

It is agreed that the need in detection of toxic chemicals in aquatic and soil environments requires a development of rapid, simple, and low-cost toxicity screening procedures.[39] Due to the fact that photosynthesis is a primary target of most toxic compounds, the use of photosynthetic material in phytotoxicity biotests should present an interesting approach. This chapter showed the potential of photosynthetic material in the development of a rapid and low cost method for the detection of heavy metals. The time required for measurement of the photosynthetic activity inhibition could be very quick (5 min). The volume required for measurement can be very small. Also among the advantages of the use of photosynthetic material is its ability to react only with those metal ions, which can cause harm to living organisms. Although the limits of detection of the heavy metals with PSII-based biosensors do not appear adequate for some natural samples it can be enhanced using appropriate extraction protocols.[40] They also can be useful in some applications where the concentration of metals is potentially high (e.g., sewage sludges).

References

1. Duverneuil P, Fenouillet B, Chaffot C. In: Lavoisier Tec and Doc, ed. Récupération des métaux lourds dans les déchets et boues issues des traitements des effluents. Paris: 1997:1-134.
2. Krupa Z, Baszynski T. Some aspects of heavy metals toxicity towards photosynthetic apparatus direct and indirect effects on light and dark reactions. Acta Physiol Plant 1995; 17(2):177-190.
3. Baron M, Arellano JB, Gorge JL. Copper and photosystem II: A controversial relationship. Physiol Plant 1995; 94:147-180.
4. Samson G, Morissette JC, Popovic R. Copper quenching of the variable fluorescence in Dunaliella tertiolecta. New evidence for a copper inhibition effect on PSII photochemistry. Photochem Photobiol 1988; 48(3):329-332.
5. Shioi Y, Tamai H, Sasa T. Inhibition of photosystem II green alga Ankistrodesmus falcatus by copper. Physiol Plant 1978; 44:434-438.
6. Vierke G, Struckmeier P. Binding of copper (II) to protein of the photosynthetic membrane and its correlation with inhibition of electron transport in class II chloroplasts of spinach. Z Naturforsch C 1977; 32(7-8):605-610.
7. Samuelsson G, Oquist G. Effects of copper chloride on photosynthesis electron transport and chlorophyll-protein complexes of Spinacia oleracea. Plant Cell Physiol 1980; 21:445-454.
8. Yruela I, Alfonso M, Ortiz de Zarate I et al. Precise location of the Cu(II)-inhibitory binding site in higher plant and bacterial photosynthesis reaction centers as probed by light-induced absorption changes. J Biol Chem 1993; 268(3):1684-1689.
9. Mohanty N, Vass I, Demeter S. Copper toxicity affects photosystem II electron transport at the secondary quinone acceptor, Q_B. Plant Physiol 1989; 90:175-179.
10. Hsu BD, Lee JY. Toxic effects of copper on photosystem II of spinach chloroplasts. Plant Physiol 1988; 87:116-119.

11. Rijstenbil JW, Derksen JMW, Poortvliet TCW et al. Oxidative stress induced by copper: Defense and damage in the marine planktonic diatom Ditylum brightwellii, grown in continuous cultures with high and low zinc levels. Mar Biol 1994; 119(4):583-590.
12. Tripathy BC, Mohanty P. Zinc-inhibited electron transport of photosynthesis in isolated barley chloroplats. Plant Physiol 1980; 66:1174-1178.
13. Ivorra N, Barranguet C, Jonker M et al. Metal-induced tolerance in the freshwater microbenthic diatom Gomphonema parvulum. Environ Pollut 2002; 116:147-157.
14. Ouzounidou G, Moustakas M, Eleftheriou EP. Physiological and ultrastructural effects of cadmium on wheat (Triticum aestivum L.) leaves. Arch Environ Contam Toxicol 1997; 32(2):154-160.
15. Ouzounidou G. Changes of photosynthetic activities in leaves as a result of Cu-treatment: Dose-response relations in Silene and Thlapsi. Photosynthetica 1993; 29(3):455-462.
16. Li EH, Miles CD. Effects of cadmium on photoreaction II of chloroplasts. Plant Sci Lett 1975; 5:33-40.
17. Baryla B, Carrier P, Franck F et al. Leaf chlorosis in oilseed rape plants (Brassica napus) grown on cadmium-polluted soil: Causes and consequences for photosynthesis and growth. Planta 2001; 212:696-709.
18 Geiken B, Masojidek J, Rizzuto M et al. Incorporation of methionine in higher plants reveals that stimulation of the D1 reaction centre II protein turnover accompanies tolerance to heavy metal stress. Plant Cell Environ 1998; 21:1265-1273.
19. Becerril JM, Munoz-Rueda A, Aparicio-Tejo P et al. The effects of cadmium and lead on photosynthesis electron transport in clover and lucerne. Plant Physiol Biochem 1988; 26(3):357-363.
20. de Fillipis LF, Rudiger H, Ziegler H. The effects of sublethal concentrations of zinc, cadmium and mercury on Euglena. II. Respiration, photosynthesis and photochemical activities. Arch Microbiol 1981; 128:407-411.
21. Samson G, Morissette JC, Popovic R. Determination of four apparent mercury interaction sites in photosystem II by using a new modification of the Stern-Volmer analysis. Biochem Biophys Res Co 1990; 166(2):873-878.
22. Kimimura M, Katoh S. Studies on electron transport associated with photosystem II. Functional site of plastocyanin; inhibitory effects of $HgCl_2$ on electron transport and plastocyanin in chloroplasts. Biochim Biophys Acta 1972; 283(2):279-292.
23. Samson G, Popovic R. Inhibitory effects of mercury on photosystem II photochemistry in Dunaliella tertiolecta under in vivo conditions. J Photochem Photobio B 1990; 5(3-4):303-310.
24. Murthy SDS, Bukhov NG, Mohanty P. Mercury-induced alterations of chlorophyll a fluorescence kinetics in cyanobacteria: Multiple effects of mercury on electron transport. J Photochem Photobio B 1990; 6(4):373-380.
25. Bernier M, Popovic R, Carpentier R. Mercury inhibition at the donor side of photosystem II is reversed by chloride. FEBS 1993; 321(1):19-23.
26. El-Sheekh MM. Inhibition of photosystem II in the green alga Scenedesmus obliquus by nickel. Biochem Physiol Pflanzen 1993; 188:363-372.
27. Barua B, Jana S. Effects of heavy metals on dark-induced changes in Hill reaction activity, chorophyll and protein contents, dry matter and tissue permeability in detached Spinacea oleracea L. leaves. Photosynthetica 1986; 20(1):74-76.
28. Carpentier R, Loranger C, Chartrand J et al. Photoelectrochemical cell containing chloroplasts membranes as a biosensor for phytotoxicity measurements. Anal Chim Acta 1991; 249:55-60.
29. Loranger C, Carpentier R. A fast bioassay for phytotoxicity measurements using immobilized photosynthetic membranes. Biotechnol Bioeng 1994; 44:178-183.
30. Pandard P, Vasseur P, Rawson DM. Comparison of two types of sensors using eukaryotic algae to monitor pollution of aquatic systems. Wat Res 1993; 27(3):427-431.
31. Rouillon R, Gingras Y, Carpentier R et al. Detection of heavy metals using thylakoids entrapped in polyvinylalcohol bearing styrylpyridinium groups. In: Mathis P, ed. Photosynthesis from Light to Biosphere. Dordrecht, Boston, London: Kluwer Academic Press, 1995:933-936.
32. Rouillon R, Tocabens M, Carpentier R. A photoelectrochemical cell for detecting pollutant-induced effects on the activity of immobilized cyanobacterium Synechococcus sp. PCC 7942. Enzyme Microb Tech 1999; 25:230-235.
33. Avramescu A, Rouillon R, Carpentier R. Potential for use of a cyanobacterium Synechocystis sp. immobilized in poly(vinylalcohol): Application to the detection of pollutants. Biotechnol Tech 1999; 13:559-562.
34. Rouillon R, Boucher N, Gingras Y et al. Potential for the use of photosystem II submembrane fractions immobilised in poly(vinylalcohol) to detect heavy metals in solution or in sewage sludges. J Chem Technol Biot 2000; 75:1003-1007.

35. Homann PH. Structural effects of chloride and other anions on the water-oxidizing complex of chloroplasts photosystem II. Plant Physiol 1996; 19:124-131.
36. Laurie SH, Tancok NP, McGrath SP et al. Influence of EDTA complexation on plant uptake of manganese (II). Plant Sci 1995; 109:213-235.
37. Towers W, Paterson E. Sewage sludge application to land-a preliminary assessment of the sensitivity of Scottish soils to heavy metals inputs. Soil Use Manage 1997; 13:149-155.
38. Aitken MN. Short-term leaf surface adhesion of heavy metals following application of sewage sludge to grassland. Grass Forage Sci 1997; 52:73-85.
39. Rawson DM, Allison JW, Cardosi MF. The development of whole cell biosensors for on-line screening of herbicide pollution of surface waters. Toxic Assessment: An internationally quarterly 1987; 2:325-340.
40. Saran R, Basu Baul TS, Srinivas P et al. Simultaneous determination of trace heavy metals in waters by atomic absorption spectrometry after preconcentration by solvent extraction. Anal Lett 1992; 25(8):1545-1547.

Chapter 16

Development of Biosensors for the Detection of Hydrogen Peroxide

Louisa Giannoudi,* Elena V. Piletska and Sergey A. Piletsky

Abstract

Biosensors are a term used for a number of devices either used to monitor living systems or incorporating biotic elements. In this work, the principal applications in the history of their development are reviewed primarily for their use as sensors for detection of analytes such as hydrogen peroxide (H_2O_2). The important areas of H_2O_2 application include industrial (pharmaceutical, food, clinical) and environmental analyses. Its use as an antibacterial agent added to milk, and the environmental need to avoid halogenated substances for disinfection purposes, makes H_2O_2 an important substance in the food and beverage industry. This has raised extensive demands for establishing protocols for H_2O_2 detection depending on its application. Additionally, it is one of the most important products or substrates of enzyme-catalysed oxidation reactions. Besides being a product/substrate of enzymatic reaction, hydrogen peroxide is by itself an important analyte. It plays an important role in natural oxidation processes as it is found in air, solids and water.

For the application of PS-II, hydrogen peroxide has been reported in the past to act as an electron donor. This process is through the two-electron oxygen reduction to H_2O_2 by various synthetic quinones that can react directly to the reducing part of photosystem II. In the case of higher plants, this would require treatment of the thylakoids in the chloroplast cell in order to achieve hydrogen peroxide production and recognition.

Abbreviations

H_2O_2—hydrogen peroxide; HRP—horseradish peroxidase; PVP—poly(4-vinylpiridine); $C_pFeC_2B_9H_{11}$—π-cyclopentadiennyl-π-dicarbollyliron; $Ru(NH_3)_5py(PF_6)$—pentaammine-pyridineruthenium(II); ttb-CuPc—tetra-tert-butyl-copper phtalocyanine; GA—glutaraldehyde; BSA—bovine serum albumin; mPEG—methoxypolyethylene; $[Os(bpy)_2pyCl]^+$—Osmium dibypyridine pyridine chloride cation.

Introduction

Biosensors

Biosensors can be defined as devices that intimately associate a biological/biomimetic sensing element with a transducer.[1] These analytical instruments with exclusive capacities combine a recognition power, which naturally exists, in biological systems with sensitivity, flexibility and user-friendliness of advanced microelectronic transducer devices. The role of the latter in a biosensor is to convert an observed change, either physical or chemical into a measurable signal. The magnitude of this signal (usually electrical) is proportional to the concentration of a specific chemical or a

*Corresponding Author: Louisa Giannoudi—Institute of Bioscience and Technology, Cranfield University, Silsoe, Bedfordshire, MK45 4DT, U.K. Email: l.l.giannoudi.s01@cranfield.ac.uk

Biotechnological Applications of Photosynthetic Proteins: Biochips, Biosensors and Biodevices, edited by Maria Teresa Giardi and Elena V. Piletska.

Table 1. Comparison of different electrochemical transducers used in biosensors (from Cunningham et al)[5]

Type of Energy Transduction	Advantages	Disadvantages
Potentiometric	ISE translation is relatively easy Easily miniaturised	Requires reference electrode Limited linear range Often pH sensitive
Amperometric	Wide variety of biochemical redox mechanisms as basis for signal generation Easily miniaturised Good dynamic range, controllable by membrane thickness Relatively good sensitivity	Requires a reference electrode Multiple membranes or enzymes may be necessary for required selectivity and sensitivity
Conductometric	Easy to fabricate No reference electrode required Low frequency/amplitude source	Non-selective unless used in array format

set of chemicals. The first biosensor was the one that combined an electrochemical transducer (Clark amperometric oxygen electrode) with enzyme (glucose oxidase) as the sensing element for glucose detection.[2] The Clark electrode is a polarographic electrode used for measuring the concentration of oxygen in blood and gases. The sample is brought into contact with a membrane (usually polypropylene or PTFE - Teflon) through which oxygen diffuses into a measurement chamber containing potassium chloride solution. In the chamber there are two electrodes: one is a reference silver/silver chloride electrode and another is a platinum electrode coated with glass to expose only a tiny area of platinum (e.g., 20 μm diameter). The electric current flow between the two electrode s when polarized with a voltage of 600-800 mV determines the oxygen concentration in the solution.[1]

Among the biological recognition elements enzymes are by far the most important. The reason for this lies in the fact that these molecules provide not only the recognition of analyte-substrate, but also have the catalytic function important for the amplification of the signal.[3] Enzymes are quite flexible molecules and have various complex conformations with sometimes different catalytic activity.[4] The biorecognition molecules can be integrated in biosensors with a variety of electrochemical transducers (Table 1).

Among the most popular biosensor devices are amperometric sensors, which usually monitor the change in current at a fixed voltage induced by a redox reaction.[5] The popularity of amperometric sensors is explained largely due to their simplicity, ease of production and the low cost of the devices and instruments. A model of an amperometric sensor with chemical transduction by enzymatic reaction is showed in Figure 1. The signal in amperometric devices depends on the rate of the mass transfer to the electrode surface. In order to minimize the diffusion path of the detectable product of the reaction the enzyme should have close contact with the transducer.[6] Although in some cases it might be possible to monitor a direct electron transfer between enzyme and electrode,[6] normally redox centers of enzymes are located deep in the insulated protein shells, which makes a direct electron transfer unfeasible. The application of such enzymes in biosensors will require the presence of redox mediators. The role and the type of mediators vary according to their usage.

Hydrogen Peroxide - Substrate and Analyte

Hydrogen peroxide (H_2O_2) is one of the most important products or substrate of enzyme catalyzed oxidation reactions.[7,8] Most common enzymes used in biosensors are oxidases, which catalyze the model oxidation reactions:

$$\text{Substrate S} + O_2 \xrightarrow{oxidase} \text{Product P} + H_2O_2 \qquad (1)$$

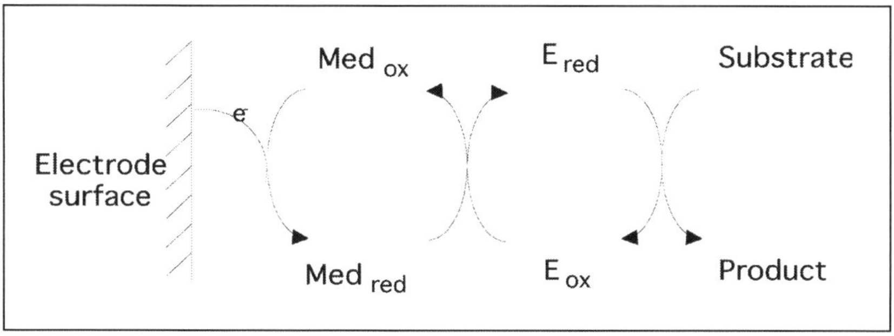

Figure 1. Mediated electron transfer, where Med_{ox} and Med_{red} are the oxidised and reduced forms of the mediator, and E_{red} and E_{ox} the reduced and oxidised forms of the enzyme.

The function of enzyme is to selectively oxidize analyte by the reduction of O_2 to H_2O_2.[9] Oxygen is the natural electron acceptor that oxidizes and is used in order to regenerate the enzyme during the reaction. Out of variety of enzymatic reactions that produce H_2O_2 (see Table 2), perhaps the most important in practical terms is the oxidation of glucose catalyzed by glucose oxidase (GOx). This well-studied reaction, which proceeds according to Equation 2 and results in production of H_2O_2, is used extensively in the development of glucose biosensors and assays:

$$\text{Glucose} + H_2O + O_2 \rightarrow \text{Gluconic acid} + H_2O_2 \quad (2)$$

The production of hydrogen peroxide is detected electrochemically and is then related to the concentration of glucose.

In addition to being a product/substrate of enzymatic reaction, hydrogen peroxide is by itself an important analyte. It plays an important role in natural oxidation processes as it is found in air, solids and water. Under different conditions very low concentrations of hydrogen peroxide could be determined, i.e., nano-molar range, in marine waters,[10] air,[11] and drinking water.[12] Furthermore, hydrogen peroxide could be determined at the level of a single cell during oxidative stress in food samples and it is used as a substrate in many immunoassays.[13]

The important areas of H_2O_2 application include industry (pharmaceutical, food, clinical), and environmental analyses (Table 2). Furthermore, its use as an antibacterial agent added to milk, demanding an established protocol for H_2O_2 detection in the food industry. So far the techniques

Table 2. List of enzymatic reactions, which produce or consume hydrogen peroxide

Enzyme	Production/Consumption of H_2O_2	Reference
Glucose oxidase	Production	14
Uricase	Production	14
Zinc oxide	Production	14
Horseradish peroxidase	Consumption	15,16
Tyrosinase (polyphenoloxidase)	Production	17
Glycolate oxidase	Production	18
Sarcosine oxidase or bovine albumin	Consumption	19
Lactate dehydrogenase and lactate oxidase	Indirect production	20
L-amino acid oxidase	Production	21
Catalase	Consumption	22
NADPH oxidase	Consumption	23

that have been used for the detection of hydrogen peroxide are enzymatic, spectrophotometric, thermo-optic and chemiluminescent assays.[24] Most of these techniques are time-consuming and/or suffer from various interferences. Due to this the development of fast, reliable and inexpensive biosensors for hydrogen peroxide detection are of prime importance.

The goal of the present review is an analysis of the electrochemical sensor approaches used for the detection of hydrogen peroxide. Due to similarity of the subjects the present review discusses both, the development of the biosensors where hydrogen peroxide is a product/substrate of enzymatic reaction and also the development of chemical sensors/biosensors specifically designed for the detection of this analyte. Several technical aspects of the development of sensors for hydrogen peroxide are reviewed in the following chapters: the choice of physical transducer, the choice of enzyme and immobilisation method, and the performance of sensors with respect to response time, sensitivity, linear range, detection limit and operational stability.

Sensors for Hydrogen Peroxide

Electrochemical Detection of Hydrogen Peroxide

The concentration of hydrogen peroxide can be measured directly using amperometric detection. A change in H_2O_2 concentration in the medium appears as a variation in the output current. The quantified parameters are magnitude of the sensor response, response time, and current response. It is desirable to measure signals in conditions when the linear relationship exists between the current value and the analyte concentration. At that point, the reactions are considered to be in steady state when "pseudoequilibrium" occurs between the species close to the sensor and their consumption at the indicative electrode. One of the serious problems associated with measurement of complex analytes is the possible interference of the redox species present in the sample. Several methods have been reported which aimed at reducing level of interference. These methods include use of perm-selective coatings,[24,25] use of artificial mediators,[2] or selective electrocatalysis.[26,27] The use of mediators or selective electrocatalysis helps to lower the detection potential to the level when the majority of interfering species are electroinactive.[28]

Erlenkotter et al have used platinum as a working electrode for direct detection of hydrogen peroxide.[19] The reaction that occurs is described below (Eqs. 3, 4).

$$Me(OH)_2 + H_2O_2 \rightarrow Me + 2H_2O + O_2 \tag{3}$$

$$Me + 2H_2O_2 \rightarrow Me(OH)_2 + 2e^- + 2H^+ \tag{4}$$

Hydrogen peroxide reduces the platinum oxides to metal and the reoxidation of the metal is performed electrochemically. Detection of hydrogen peroxide in this case depends on temperature, pH and the oxidation status of the platinum electrodes. The research has proved that the oxidation of H_2O_2 does not depend on the surface of the oxide films; however, it requires a stable oxidised surface for the reproducibility of the detection.[19] The difficulty with the application of these electrodes lies in relatively high price required for their manufacturing and in the high potential required for oxidation. The potential required to dismute (simultaneous oxidation and reduction) hydrogen peroxide on electrode is + 600 - +1200 mV versus a saturated calomel electrode (SCE).[29] The potential depends greatly on the nature of the working electrode (platinum, gold, graphite, graphite-polymer composite, etc.). On platinum the oxidation potential is +400 mV,[30] which is quite high and needs to be reduced in order to avoid any interference coming from real samples.[31] On carbon electrodes, which are much cheaper than platinum electrodes, the oxidation potential for hydrogen peroxide is even higher (>700 mV). The good compromise on price and performance was achieved for carbon (graphite) electrodes modified with rhodium,[32] mercury or platinum.[33] The mechanism of action of these additives lies in the ability of metals such as platinum and rhodium to increase the rate of electrons transfer between the enzymes and the conducting sites of the electrodes and speed up the response time of the measurements.[31]

The further reduction of the oxidation potential and enhancement of sensor signal can be achieved by using an enzyme, such as horseradish peroxidase.[34] The simplest electrode type is the one that consists of a layer of peroxidase molecules adsorbed on the electrode surface.[35] The sensor response, measured at the lowering overpotential of 0.6V vs SCE consists of the change in reduction current which is proportional to peroxide concentration. This type of sensors can be based on carbon black,[36] graphite,[4,37] gold,[36] and platinum electrodes.[38,39]

A variety of mediators—small organic molecules capable of lowering the redox potential—can be used for facilitating electron transfer between the enzyme catalytic centre and electrode. Mediated amperometric biosensors also have an advantage over nonmediated enzyme electrodes, since the mediator could replace oxygen as an electron acceptor. Hydrogen peroxide detection is known to be affected by oxygen concentration, since the gas is a cosubstrate of oxidase-catalyzed reactions (Eq. 5).

$$H_2O_2 \rightarrow O_2 + 2H + 2e^- \quad (5)$$

In this respect, mediated enzyme electrodes generate more reliable signals, less affected by possible interference from the oxygen in the sample. Furthermore, the employment of mediators with low redox potentials is advantageous in order to operate an enzyme electrode at potentials lower than the ones required for the dismutation (spontaneous oxidation/reduction) of hydrogen peroxide.[29] This is important for avoiding interference of other electroactive species that could be present in the sample solution. The details on the development and application of enzyme electrodes in the detection of hydrogen peroxide are discussed in the following chapter.

Enzyme Electrodes, Design and Performance

Choice of Enzyme

The enzymes that are used intensively in development of hydrogen peroxide sensors are horseradish peroxidase and catalase. Horseradish peroxidase (HRP) along with catalase are both hemic enzymes that contain Fe(III)-protoporphyrin as the prosthetic group. Both enzymes work with hydrogen peroxide as the substrate.[40] Detection schemes vary according to the method of the enzyme immobilisation, mediator, and type of enzyme (see Table 2). HRP catalyses the oxidation of numerous substrates (noted as SH) by hydrogen peroxide, following the general reaction:

$$2SH_2 + H_2O_2 \xrightarrow{HRP} 2SH^0 + 2H_2O \quad (6)$$

Direct reduction of hydrogen peroxide at a peroxidase modified electrode is demonstrated in Figure 2.

Reduction of HRP involves compounds I and II, E_1 and E_2 respectively as has been previously described, but its inactivation with excess of H_2O_2 involves the formation of compound-III or E_3 (oxyperoxidase) or the irreversible set of reactions which ends at the formation of a verdohemoprotein.[41,42]

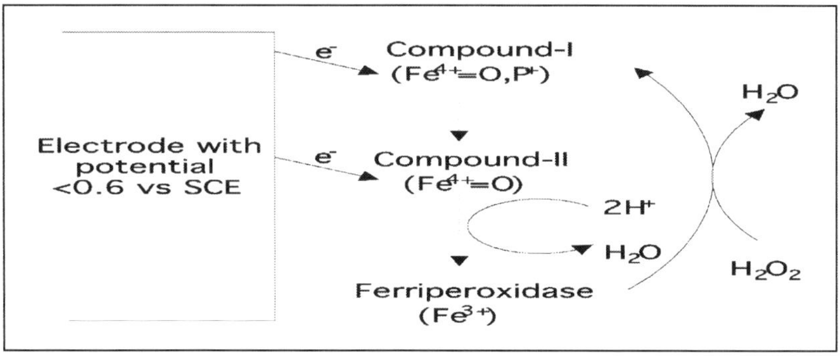

Figure 2. Direct bioreduction of hydrogen peroxide using a peroxidase modified enzyme at a potential below than 0.6 vs SCE. Compound-I is the oxyferryl π-cation radical heme intermediate 1 and compound-II is the oxyferryl intermediate 2, where both of them are the oxidised forms of the native ferriperoxidase. P^+ is the cation radical, which is localised on porphyrin ring or polypeptide chain.

Another enzyme which has been applied comparable times with horseradish peroxidase is catalase. Catalase is a heme protein belonging in the class of oxidoreductases with ferriprotoporphyrin-IX that acts as the redox centre for the enzyme. As mentioned before, this enzyme catalyses the degradation of hydrogen peroxide by the following reaction:

$$H_2O_2 + \underset{\text{Catalase}}{Fe(III)-E} \rightarrow H_2O + \underset{\text{(Compound-1)}}{O = Fe(IV)-E} \qquad (7)$$

$$H_2O + O = Fe(III) - E \rightarrow H_2O + O_2 + Fe(III) - E \qquad (8)$$

where hydrogen peroxide acts both as an electron acceptor and electron donor.[22]

Immobilisation

The factors that affect the performance of H_2O_2 biosensor are (i) the type of enzyme, (ii) the immobilisation method, and (iii) the thickness of the created enzyme layer.[6] Immobilisation processes are quite an important factor for the development of biosensors. This step has been studied extensively, in order to achieve easy operation, quick measurement and reduce the cost of the analysis.[43] The methods of enzymes immobilisation used for the development of hydrogen peroxide sensors can be divided into five major groups (Fig. 3):

- Physical adsorption;
- Covalent binding to an activated insoluble support;
- Encapsulation—capturing biological components with various forms of semi-permeable membranes;
- Entrapment in gels;
- Cross-linking by means of bifunctional reagents (this design could be also used together with adsorption and entrapment).

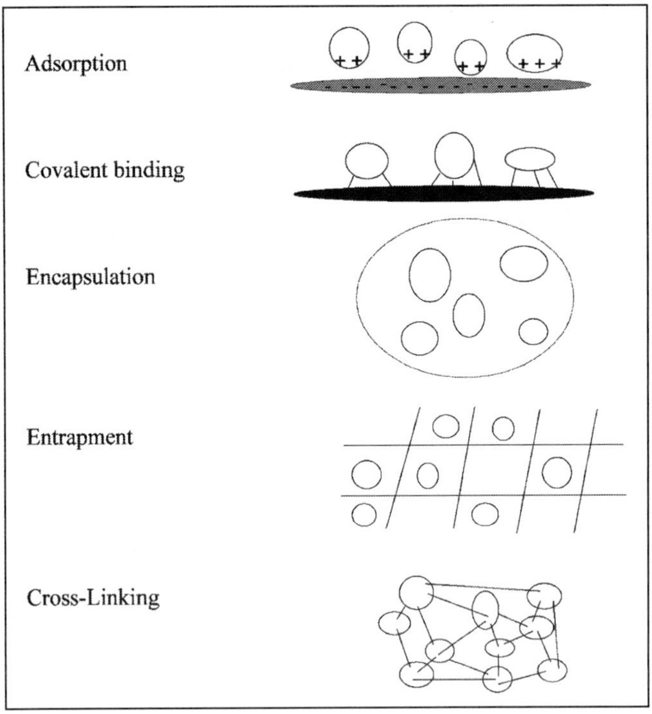

Figure 3. Schematic diagram of the principal immobilisation techniques (from Bardelett et al).[6]

Adsorption of enzymes has been used extensively for creating chemical transduction layers in a variety of biosensors. Enzymes in general are well adsorbed onto metal, metal oxide, carbon and glass surfaces that are used in transducers. The advantages of physical adsorption over other immobilisation techniques are the simplicity of the process along with low cost, little or no damage to enzymes, absence of chemical changes in enzyme structure, which might affect sensitivity and specificity, reversible character, and the possibility of regeneration of the catalytic surface.[44] Sometimes adsorption could be used along with cross-linking.[45] This procedure has several steps such as the mixing of the biological and chemical components under optimised conditions of pH, temperature and ionic strength followed by a period of incubation during which the chemical reaction proceeds. Immobilised materials are washed thoroughly for the removal of the nonbound biological components that would easily interfere in the solution and distort the data.

Covalent binding involves the formation of a covalent bond between the functional groups of enzyme and the support material. Proteins are bound through the involvement of amino, carboxyl, sulfhydryl, or aromatic parts of the chains of the amino acid in the macromolecule.[5] Some examples of amino acids group suitable for covalent binding are the amino group (NH_2) of lysine or arginine, as well as the carboxyl group (COOH) of aspartic acid or glutamic acid, the hydroxyl group (OH) of serine or threonine, and the sulfydryl group (SH) of cysteine.[43] The popular process for activation of functional groups and covalent enzyme immobilisation is the formation of a Schiff base, catalysed by carbodiimide. The hydrophilic factor of the supporting material is an important factor, which determines the efficiency of immobilisation. Polysaccharide polymers are very hydrophilic and some of them such as cellulose, dextran (Sephadex), starch, agarose, (Sepharose) have been extensively used for enzyme immobilisation.[43]

Encapsulation achieves the confinement of biological components by using various semi-permeable membranes.[46] Encapsulation allows for the enzymes to exist freely in solution, which is confined within the small area surrounded by the membrane. Macromolecules cannot cross the membrane barrier, which is permeable for small molecules only (substrates or products). Nylon and cellulose nitrate are the most popular materials used for the production of microcapsules that need to have a diameter between 10 and 100 μm diameters. Furthermore, biological cells could be used as capsules as it shown in erythrocytes based sensor.[43] Alternatively enzyme solution can be encapsulated in a thin layer, which covers the electrode and confined between the electrode and semi-permeable membrane surface.[47]

Entrapment differs from the above-mentioned techniques due to the fact that the movement of immobilised molecules is restricted by the presence of a gel. The porosity of this gel controls the enzyme mobility and the gel should be tight enough to avoid any leakage of the enzyme. Although the technique sets a barrier to the mass transfer and this would affect the reaction kinetics, the advantage is that the enzyme does not interact with the immobilized biocatalyst.[48] In some cases such as hydrogels, their swelling property would allow the enzymes to go through and leave the gel which could limit their application.[49] Apart from gels, entrapment could use a variety of organic and inorganic polymers produced on the transducer surface by electrochemical, photochemical, or plasma polymerisation. One such example involves enzyme entrapment into solidified materials such as paraffin. Petit et al studied the hydrodynamic conditions of paraffin and how it improves the stability of the enzyme-immobilised electrode and the reproducibility and repeatability of the results.[17] Many researchers have used entrapment of the enzyme into silica polymer that contains weak hydrogen bonds hybridised into sol-gels.[5,50] The fabrication of a biosensor with sol-gel is quite easy and could be done by syringing a mixture of the sol-gel and the enzyme solution on the electrode and allowing water to evaporate.[51] These organic and inorganic materials are able to incorporate both, mediators and enzymes for the fabrication of biosensors.[52] They could be prepared under ambient conditions and they have good porosity, high thermal and chemical stability and insignificant swelling in aqueous and nonaqueous solutions. The main limitation of sol-gel immobilisation techniques lies in the relatively low stability of inorganic membranes, which have a tendency to cracking. In order to overcome this problem Wang et al and Zhang et al have used hybrid inorganic-organic materials.[49,52] Polyvinyl alcohol (PVA) and hydrophobic poly(vinylpyridine) (PVP) have been used to compose the sol-gel composite material for developing a hydrogen peroxide biosensor.[51]

The final immobilisation procedure that have been used is cross-linking and involves joining the enzymes to each other forming a large, three-dimensional complex structure, achieved by physical or chemical methods. The chemical methods of cross-linking involve covalent bond formation between the protein molecules by means of a bi- multifunctional reagent such as glutaraldehyde and toluene diisocyanate.[43,53]

Apart from being a framework, the immobilisation matrix could have additional functions such as selective ion permeability, enhanced electrochemical conductivity, or mediation of electron transfer processes.[5]

Mediators

Using mediators in combination with enzyme is a practical alternative for the direct measurement of hydrogen peroxide and for the direct monitoring of enzymatic reaction. Another advantage is that the use of mediator could help to overcome the dependence of the amperometric biosensors on dissolved oxygen since the mediator can provide the essential electrons.[9]

In general, the mediated signal coming from an enzyme electrode is monitored through the measurement of the quantity of reduced redox mediator. In the case of horseradish peroxidase (HRP) based electrode the catalytic reaction that produces that mediator is described below:

$$H_2O_2 + 2H^+ + HRP_{red} \rightarrow 2H_2O + HRP_{ox} \tag{9}$$

$$MED_{ox} \xrightarrow{E} MED_{red} \tag{10}$$

The HRP_{red}, HRP_{ox}, MED_{red}, MED_{ox} are the reduced and oxidised forms of the enzyme and the mediator, respectively.

A variety of compounds have been used as mediators in enzyme biosensors (Table 3). The most common mediators used for hydrogen peroxide and glucose detection are ferrocenes and its derivatives.[2,54,55]

In reaction with HRP, H_2O_2 is oxidised to water by enzyme and ferrocene acts as an electron donor to the oxidised iron in the haem protein. The charge that is involved for the reduction of the ferricium ion on an electrode is proportional to the hydrogen peroxide concentration. Tatsuma et al[63] used soluble ferrocenemonocarboxylic acid as mediator, which operated at a potential of +150 mV vs. Ag/AgCl.[63] Ferrocene derivatives have been highly utilised since they exhibit excellent properties such as maintaining a one-electron redox couple even when their cyclopentadienyl rings are modified with several other substituents. In this way they can be used for modifying protein molecules.

An another group of mediators with similar properties are π-cyclopentadiennyl-π-dicarbollyliron complex ($CpFeC_2B_9H_{11}$), and pentaamminepyridineruthenium(II) complex $[Ru(NH_3)_5py](PF_6)]$ used by Frew et al.[64]

Zhang et al[10] have prepared a H_2O_2 sensor using Meldola's Blue (MDB) as mediator. MDB was incorporated into hydrogels in order to give a faster reaction rate.[52] Other mediators used for the detection of hydrogen peroxide are phenothiazine molecules such as methylene blue,[65,66] methylene green,[67] thionine,[68,69] $[Os(bpy)_2pyCl]^+$,[70] and toluidine blue.[71] Mediators are able to shuttle electrodes between electrodes and enzymes in several configurations such as soluble, associated in monolayer or multilayer, or incorporated in porous matrices. In all the cases it is necessary for the mediator to be regenerated having the appropriate medium and for the enzyme to play the role of a biocatalyst.

Since the mediators should be mobile in order to provide the electron flow between the enzyme catalytic centre and electrode, they are usually soluble in the electrolytic medium, and therefore could be lost during the repetitive measurements. This, in combination with inactivation of enzyme, due to its denaturing leads to the loss of sensitivity of the biosensor with time. The possible solution to this problem is using an excess of enzyme and mediators in the measurement.[31] Recently, a new technique has been developed where redox polymers were used in dual function: as immobilisation matrix and as materials facilitating electron-transfer.[72] In materials such as these, the mediator redox

Table 3. Classes of mediators that have been used for amperometric detection of hydrogen peroxide detection

Mediators	Ox/Red Potential (mV)	Ref.
Anthraquinones	-40	56
Methylene Blue	-156	56
p-Benzoquinones	+200	57
Ferrocene carboxylic acid	+295	55
Ferrocene dicarboxylic acid	+420	55
Aminoethyl ferrocene	+158	55
N,N,N-Trimethyl-N-ferrocenomethylammonium iodide	+137	55
1,1′-Bis(hydroxymethyl) ferrocene	+220	55
Hydroxymethyl ferrocene	+180	55
1,1′-Dimethyl-3-ethanolaminoferrocene	+490	55
Hexacyanoferrate (III)	-100	58
Phenothiazines	+405/+350	15
Ru, Os complexes	+500	59
Methylviologen MV	-660	60
Charge transfer complexes (e.g., tetracyano-p-chinone dimethane TCNQ- tetrathiafulvalene TTF)	+200	61
ABTS2,2′-azino-bis(3-ethylbenzothiazoline-6-sulfonate)	-200	62
Meldola's Blue (MD) within inorganic/organic hybrid material	-116	52
Ferrocenemonocarboxylic acid	+150	63
Ferrocenedicarboxylic acid	+150	63
Ferrocenecarboxaldehyde	+100	63
Pottasium ferrocyanide	+100	63

couple is not diffusing and therefore the electron transfer occurs by "hopping". Such mediator is not leaching out of the membrane of the sensor and therefore the use of a containment membrane is unnecessary. An example of this redox polymer material has been used by Hale et al[18] for a glycolate sensor.[18] The polymeric electron transfer mediators that were used were siloxane polymers with covalently attached ferrocene and 1,1-dimethylferrocene. With this procedure, electron transfer was efficiently transmitted from the reduced glycolate oxidase to the carbon-paste electrode. Garguilo et al[72] have developed amperometric sensors for hydrogen peroxide by immobilizing horseradish peroxidase (HRP) in a cross-linked redox polymer deposited on glassy carbon electrodes.[72] The redox polymer contained Os derivatives such as (PVP-$O_s(bpy)_2Cl$; bpy- 2,2'-bipyridine, PVP-poly(4-vinylpyridine) and $O_s(bpy)_2pyCl$ (PVP replaced with pyridine) covalently attached to the HRP enzyme.

Examples of Enzyme Biosensors Used for the Detection of Hydrogen Peroxide

Tatsuma et al[63] have immobilized HRP in a monolayer covalently attached to a tin oxide electrode.[63] The sensor used soluble ferrocenemonocarboxylic acid as mediator, which operated at a potential of +150 mV vs. Ag/AgCl. The electrode was chosen for its chemical and electrochemical stability as well as for the simplicity of modifying its surface with functional groups. The sensor had enhanced kinetic rate and as result increased sensitivity, with detection limit 10^{-8} M H_2O_2. The immobilisation of the enzyme as a monolayer could reduce also fabrication cost (which is especially important for the case when the enzyme is expensive).

Frew et al[64] developed H_2O_2 sensors using π-cyclopentadiennyl-π-dicarbollyliron and $[Ru(NH_3)_5py(PF_6)]$ complexes as mediators. Their biosensors exhibited sensitivity in the range of 5×10^{-8} - 6×10^{-6} M to hydrogen peroxide using a gold or pyrolytic graphite electrode.[75]

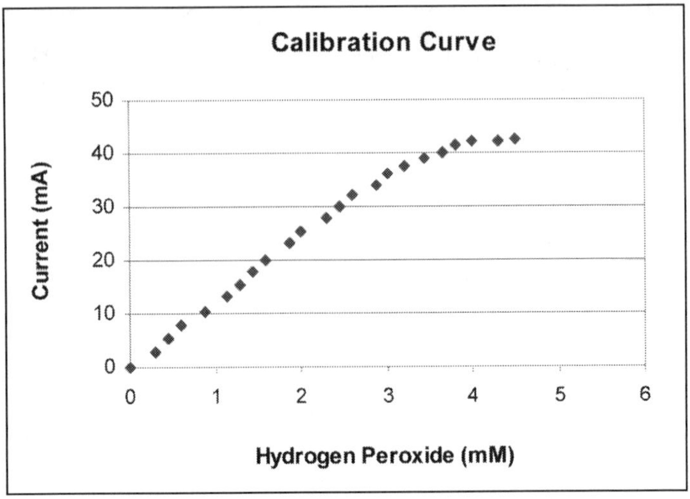

Figure 4. Calibration curve for the enzyme electrode for H_2O_2. The potential for the electrode is -50mV and the electrolyte is 50 mM sodium phosphate buffer, pH 7.5.

Zhang et al[10] have prepared a H_2O_2 sensor using MDB as mediator and electrode functionalised with a horseradish peroxidase-containing membrane.[52] MDB was incorporated into hydrogels in order to give a faster reaction rate. Their biosensor demonstrated a linear range in 1-0.6 mM concentration and good stability. The response time of the sensor was less than 25 seconds with a sensitivity of 75 nA μM cm^{-2}.

Li et al[73] have prepared an amperometric biosensor for hydrogen peroxide using a sol-gel immobilisation of horseradish peroxidase.[73] HRP was entrapped in a thin silica sol-gel matrix made from tetramethoxysilane polymerised on carbon paste electrodes. The hydrogen peroxide detection was performed in the presence of hexacyanoferrate (II). Under optimised conditions, the linear range for detection of hydrogen peroxide was 2×10^{-5} - 2.6×10^{-3} M. The enzyme electrode maintained up to 65% of its activity for the period of 35 days.

Wang et al[49] used HRP entrapped into hybrid PVA/silica sol for the measurement of hydrogen peroxide.[49] Their biosensor demonstrated a linear range in the concentrations 0.2-3.4 mM (curve at 3.4 mM in Fig. 4), with detection limit of 5×10^{-7} M. Furthermore, this biosensor exhibited high sensitivity (15 μA mM^{-1}) and fast response time (10 sec.). The results that were calculated from the calibration curve were in close agreement with the ones measured by a standard spectrophotometric method (Table 4).

Sergeyeva et al[74] have also used HRP along with ttb-CuPc thin films for the detection of H_2O_2.[74] Their biosensor achieved the detection in the range of 5 - 300 mM and an operational stability of 7 hours, which was improved to 90 days by storing the sensor at 4°C. To minimise any interference from the aqueous media the ttb-CuPc layer was covered by a hydrophobic gas-permeable membrane. Due to this, the ionic strength and the buffer capacity did not play a significant role on the measurements. Thus, the iodine that resulted from the peroxidase reaction could be monitored with the aid of the designed transducer (Fig. 4).

Miao and Tan[58] have fabricated an amperometric H_2O_2 biosensor using HRP and hexacyanoferrate (II) entrapped into silica sol-gel/chitosan film on the surface of carbon paste electrode.[58] The sensor operated at low potential (-100 mV) and had a similar sensitivity ($2.5 \times 10^{-4} \div 3.4 \times 10^{-3}$ M) as the one prepared by Wang et al[49] Their biosensor exhibited quite good reproducibility and stability. 85% of the original enzyme activity was preserved after 30 days of storage in a phosphate buffer solution at 4°C. The authors claimed that the high stability originated from the amino containing sol-gel/chitosan hybrid films, which provided a hydrophilic environment compatible with the biomolecules.

Table 4. Comparison of hydrogen peroxide concentrations in real samples between Wang's biosensor and the spectrophotometric method (Frew et al)[49,75]

	Concentration of Hydrogen Peroxide (μM)	
Sample No	**Measured by Biosensor**	**Determined by Spectrophotometric Method**
Milk		
1	21.16+0.37	21.94+0.29
2	12.05+0.16	12.35+0.13
3	18.31+0.26	18.71+0.21
4	20.09+0.33	20.62+0.27
Acid milk		
1	25.96+0.42	26.33+0.35
2	19.34+0.21	19.56+0.26
3	21.89+0.28	22.16+0.27
Juice		
1	8.46+0.18	8.75+0.12
2	9.39+0.20	9.56+0.11
3	5.22+0.16	5.54+0.08

Razola et al[15] have managed to detect hydrogen peroxide at a subnanomolar range.[15] Their biosensor consists of HRP immobilised on solid carbon paste along with phenothiazine as the redox mediator. The linear response of the biosensor from Razola et al, ranged from 2 nM-10 microM with a detection limit of 1 nM.[15] The immobilisation of HRP was achieved by cross-linking with GA and BSA. An incorporation of the electrode with Nafion film gave a H_2O_2 detection of 0.1 nM.

Wang and Dong[51] developed amperometric biosensor using a Nafion-methylene green modified electrode.[51] Nafion is a very popular electrode anion-exchange modifier and have been used in many applications.[4] Methylene green (soluble mediator) was incorporated into the Nafion thin layer and it was coated on the surface sol-gel-enzyme thin film immobilized on a glassy carbon electrode. The electrode revealed a sensitivity of 13.5 μa mM^{-1} with a detection of 1.0×10^{-7} M H_2O_2. The sensor response was fast reaching maximum in 20 s. The sensor response was influenced by the pH of the buffer with maximum observed at a pH of 6.5. Sensor had a good long-term stability.

Yabuki et al[76] have prepared a H_2O_2 sensor using ferrocene as mediator.[76] Their biosensor was fabricated with a poly-ion complex membrane containing physically entrapped microperoxidase (MP) and ferrocene and immobilized on glassy carbon electrode. The polyion membrane consisted of poly-L-lysine and poly(4-styrene sulfonate). Microperoxidase (heme peptide; heme octapeptide; nonapeptide and undecapeptide) has smaller molar mass than peroxidase (POD) and therefore a higher specific activity. The potential that was applied for this biosensor was 0.0V versus Ag/AgCl and the response current showed linear proportionality up to 20 μM H_2O_2 concentration. The electrode could be used for the period of 10 days and with a low detection limit of 0.5 μM (S/N = 5). The calibration curve for hydrogen peroxide is shown in Figure 5.

Karyakin and Karyakina[77] have developed a hydrogen peroxide sensor, based on Prussian Blue deposited on glassy carbon electrodes.[77] Prussian Blue was considered an "artificial peroxidase" due to its high catalytic activity and selectivity, which could be compared with biocatalysis. The application of Prussian Blue modified electrodes enabled the sensing of H_2O_2 at around 0 V vs. SCE. The electrocatalytic reduction of H_2O_2 in the presence of O_2 was found to be better for Prussian Blue deposited on glassy carbon electrodes than for platinum covered electrodes. Furthermore, these electrodes were more stable and active and less expensive than the platinum and peroxidase modified electrodes. The response was linear up to 0.1-100 μM and the detection limit found to be 10^{-7} M.

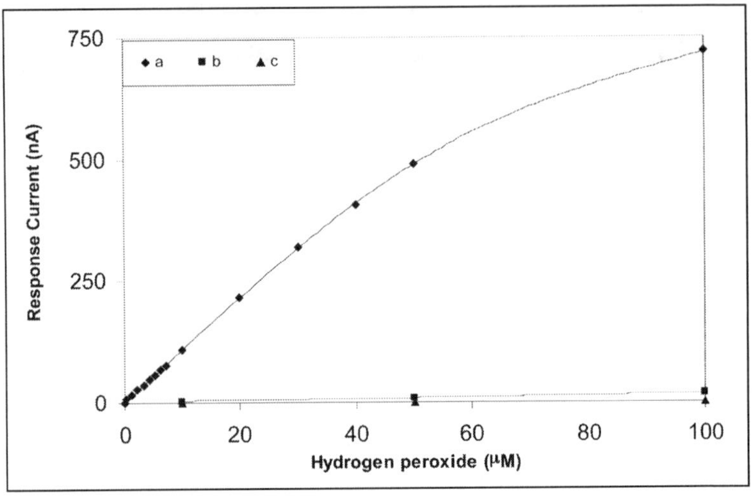

Figure 5. Hydrogen peroxide calibration curve for three different electrode systems: MP and ferrocene-immobilized polyion complex membrane (a), ferrocene-immobilized poly-ion complex membrane (b), and bare glassy electrodes (c).

It is interesting that the enzyme electrodes could be used for the detection of hydrogen peroxide also in organic solutions. Vijayakumar et al have used carbon paste electrodes modified with HRP, chemically derivatised with mPEG.[16] Covalent attachment of mPEG to the enzyme helps in preventing enzyme denaturing in organic media, which makes sensor operational in organic solvents such as benzene, toluene and chloroform.[78] In general, measurement in organic phase has advantages such as the possibility to monitor hydrophobic substrates, to remove microbial contamination, to avoid the possible side reactions, and to enhance the thermostability. However, the drawback for these sensors was that the electrode response for H_2O_2, based on direct electron transfer between the electrode and peroxidase is reduced. Additionally, enzymes could not maintain catalytic activity in an absolute nonaqueous environment; therefore, enzymes on organic phase enzyme electrodes (OPEEs) must retain a thin aqueous film in an organic media, which is difficult to control.[79]

The use of polymers in conjunction with metals and enzymes has been proved to be an appropriate alternative for the detection of hydrogen peroxide in both organic and aqueous media. Daly et al have used three polymers (polypyrrole, polyaniline, and 1,3-diaminobenzene) along with ruthenium (Ru), rhodium (Rh) and palladium (Pt) in several combinations, to provide a suitable H_2O_2 biosensor.[80] Both conductive and nonconductive polymers could be used in electrochemical sensors but the latter are mostly used for blocking interferon's or immobilising the biocomponents since they are not electroactive.[24-87] Daly et al proved that by the correct choice of potential and polymer, an enhanced electrochemical sensor could be obtained.[80] In conjunction with the appropriate metal, the sensor would keep the signal of the interferon's low and would enhance the response of the signal of interest being hydrogen peroxide.

Transition metal phthalocyanines on carbon paste electrodes have been also demonstrated for the determination of hydrogen peroxide using differential pulse cyclic voltammetry.[88] This type of compounds has been successfully used for antioxidants, substances that contain sulphur, reduced glutathione, organic peroxides, thiocyanate and selenocyanate. The versatility of metal phthalocyanines explains why Santamaria et al[88] used them for observing the electrochemical behaviour of hydrogen peroxide.[88] Carbon paste electrodes modified with nickel (II) phthalocyanine and copper (II) phthalocyanine showed a greater sensitivity and response than the non modified carbon paste electrodes, at pH 11.

One can conclude that the use of amperometric and potentiometric biosensors has achieved a considerable progress in the determination of hydrogen peroxide. The redox-enzyme and electrode structure achieved by many provides the basis for electrochemical biosensors. Enzymes need to be modified either by mutagenesis, or site-specific reactions that would provide structures with readily accessible sites. Many of the latter could be accomplished with the aid of a mediator.

Examples of PS-II Based Biosensors with Hydrogen Peroxide Production

Hydrogen peroxide has being known do be the product of the dismutation of O_2^- (free oxygen radical). This is the result of the oxidisable electron acceptors of photosystem I in a reaction that includes ferredoxin and NADP reduction to NADPH. However, several biosensors have being previously reported the production of hydrogen peroxide from photosystem II. The latter involves treatment of chloroplasts with hydroxylamine (NH_2OH) in order to oxidise H_2O_2 to O_2 through photosystem II.

Two examples have been reported involving the production of hydrogen through photosystem II. Both of them are based on the same principle of chloroplasts treatment. Nitrite and hydrogen peroxide are produced from the presence of hydroxylamine and the absence of artificial electron acceptors in the following reaction:

$$NH_2OH + 2O_2^- + H^+ \rightarrow NO_2^- + H_2O_2 + H_2O \quad (11)$$

Furthermore, the products of the equation 11 depend strongly on the concentrations of dibromothymoquinone and 2,3-dimethyl, 5,6-methylenedioxy p-benzoquinone used separately.[89] In the past, hydrogen peroxide has being reported to be a product of oxygen dismutation (O_2^-) through auto oxidation of reduced dyes via photosystem I.[90,91]

Pan and Izawa have used a Clark-type electrode in order to measure oxygen evolution or consumption through photosystem II. This was based again on treatment of chloroplasts with hydroxylamine that would inactive the O_2 evolving enzyme of the protein. An electron acceptor has also being used such as 2,5- dimethylquinone in the presence of dibromothymoquinone which is an antagonist of the plastoquinone and therefore it could eliminate the presence of photosystem I.[92]

Conclusion

Since hydrogen peroxide is the product of reactions catalysed by a large number of oxidase enzymes and is essential in food, pharmaceutical, and environmental analysis, its detection was and remains a necessity. Many attempts have been made in order to develop a biosensor that would be sensitive, stable, inexpensive and easy to handle. The most popular and efficient of them are amperometric enzyme biosensors, which utilised different types of mediators and enzymes, mostly peroxidase and catalase. Unfortunately many of the sensors developed do not meet the requirements for a practical device, which has a balance of technological characteristics (sensitivity, reliability, stability) and commercial adaptability (easy of mass production and low price). Thus a window of opportunity still remains open for future development. We hope that the present work will inspire other researches for further advances in the area of biosensors, in particular sensors for detection of such an important analyte as hydrogen peroxide.

References

1. Bannister JV, Higgins IJ, Turner APF. Development of amperometric biosensors for enzyme immunoassay. In: Blum LJ, Coulet PR, eds. Biosensor Principles and Applications. New York: Marcel Dekker Inc., 2001:7.
2. Cass AEG, Davis G, Francis GD et al. Ferrocene-mediated enzyme electrode for the amperometric detection of glucose. Anal Chem 1984; 56:667-671.
3. Manahan SE. Environmental Chemistry. 6th ed. Edited by CRC Press Inc., 1994.
4. Gorton L, Csoregi E, Dominguez Emneus J et al. Selective detection in flow analysis based on the combination of immobilized enzymes and chemically modified electrodes. Analytica Chimica Acta 1991; 250:203-248.
5. Cunningham AJ. Introduction to Bioanalytical Sensors. New York: Wiley, 1998..

6. Bardeletti G, Sechaud F, Coulet PR. Amperometric enzyme electrodes for substrate and enzyme activity determinations. In: Blum LJ, Coulet PR, eds. Biosensor Principles and Applications. New York: Marcel Dekker Inc., 1991:7.
7. Tang J, Wang B, Wu Z et al. Lipid membrane immobilized horseradish peroxidase biosensor for amperometric determination of hydrogen peroxide. Biosens Bioelectron 2003; 18:867-872.
8. Davis G. Electrochemical techniques for the development of amperometric biosensors. Biosensors 1985; 1:161.
9. Vreeke MS, Maidan R, Heller A. Hydrogen peroxide and b-Nicotinamide adenine dinucleotide sensing amperometric electrodes based on electrical connection of horseradish peroxidase redox centers to electrodes through a three-dimensional electron relaying polymer network. Anal Chem 1992; 64:3084-3090.
10. Zhang Ling-Su, Wong GTF. Optimal conditions and sample storage for the determination of H_2O_2 in marine waters by the scopoletin-horseradish peroxidase fluorometric method. Talanta 1999; 48:1031-1038.
11. Price D, Fauzi R, Mantoura C et al. Shipboard determination of hydrogen peroxide in the western Mediterranean Sea using flow injection with chemiluminescence detection. Analytica Chimica Acta 1998; 371:205-215.
12. Lin Jin-Ming, Arakawa H, Yamada M. Flow injection chemiluminescent determination of trace amounts of hydrogen peroxide in snow-water using KIO_4-K_2CO_3 system. Analytica Chimica Acta 1998; 371:171-176.
13. Jianzhong L, Zhang Z, Ling L. A simplified enzyme-based fiber optic sensor for hydrogen peroxide and oxidase substrates. Talanta 1994; 41:1999-2002.
14. Schumb WC, Satterfield CN, Wentworth RL. Isotope exchange in reactions. Hydrogen Peroxide. New York, London: Reinhold publishing corporation, Chapman and Hall, Ltd., 1955.
15. Razola SS, Aktas E, Vire JC et al. Reagentless enzyme electrode based on phenothiazine mediation of horseradish peroxidase for subnanomolar hydrogen peroxide determination. Analyst 1999; 125:79-85.
16. Vijayakumar A, Csoregi E, Ruzgas T et al. Comparison of carbon paste electrodes modified with native and polyethylene glycol derivatized horseradish peroxidases for the amperometric monitoring of H_2O_2. Sens Actuators B Chem 1996; 37:97-102.
17. Petit C, Gonzalez-Cortes A, Kauffmann JM. Preparation and characterization of a new enzyme electrode based on solid paraffin and activated graphite particles. Talanta 1995; 42:1783-1789.
18. Hale PD, Inagaki T, Sui LH et al. Amperometric glucose sensors based on glycolate oxidase and polymeric electron transfer mediators. Analytica Chimica Acta 1990; 228:31-37.
19. Erlenkotter A, Kottbus M, Chemnitius GC. Flexible amperometric transducers for biosensors based on a screen printed three electrode system. J Electroanal Chem 2000; 481:82-94.
20. Young SJ, Hart JP, Dowman AA et al. The nonspecific inhibition of enzymes by environmental pollutants: A study of a model system towards the development of electrochemical biosensor arrays. Biosens Bioelectron 2001; 16:887-894.
21. Setford SJ, White SF, Bolbot JA. Measurement of protein using an electrochemical bi-enzyme sensor. Biosens Bioelectron 2002; 17:79-86.
22. Varma S, Mitra CK. Bioelectrochemical studies on catalase modified glassy carbon paste electrodes. Electrochemistry Communications 2002; 4:151-157.
23. Zhou M, Diwu Z, Panchuk-Voloshina N et al. A stable nonfluorescent derivative of resorufin for the fluorometric determination of trace hydrogen peroxide: Applications in detecting the activity of phagocyte NADPH oxidase and other oxidases. Anal Biochem 1997; 253:162-168.
24. Sasso SV, Pierce RJ, Walla R et al. Electropolymerized 1,2-diaminobenzine as a meant to prevent interferences and fouling and to stabilize immobilized enzymes in electrochemical biosensors. Anal Chem 1990; 62:1111-1117.
25. Zhang Y, Hu Y, Wilson GS et al. Elimination of the acetaminophen interference in an implantable glucose sensor. Anal Chem 1994; 66:1183-1188.
26. Johnson DA, Cardosi MF, Vaughan DH. The electrochemistry of hydrogen peroxide on evaporated gold/palladium composite electrodes. Manufacture and Electrochemical Characterisation. Electroanalysis 1995; 7:520-526.
27. Newman JD, White SF, Tothill IE et al. Catalytic materials, membranes, and fabrication technologies suitable for the construction of amperometric biosensors. Anal Chem 1995; 67:4594-4599.
28. Arjsiriwat S, Tanticharoen M, Kirtikara Aoki K et al. Metal-dispersed conducting polymer-coated electrode used for oxidase based biosensor. Electrochemistry Communications 2000; 2:441-444.
29. Hendry SP, Higgins IJ, Bannister JV. Amperometric biosensors. Journal of Biotechnology 1990; 15:229-238.

30. Bennetto HP, Dekeyzer DR, Delaney GM et al. An amperometric biosensor for laboratory determination of glucose. Int Analyst 1987; 8:22-27.
31. Morales A, Céspedes F, Munoz Martínez-Fabregas E et al. Hydrogen peroxide amperometric biosensor based on a peroxidase-graphite-epoxy biocomposite. Analytica Chimica Acta 1996; 332:131-138.
32. Janata J, Josowick M, Devaney M. Chemical sensors. Anal Chem 1994; 66:207R.
33. Cardosi ME, Turner APF. Mediated electrochemistry: A practical approach to biosensing. In: Turner APF, ed. Advances in Biosensor. Vol. 1. Jai Press Ltd, 1991.
34. Schubert F, Wollenberger U, Pfeiffer D et al. Enzymatic Amplification and elimination in biosensors. In: Turner APF, ed. Advances in Biosensor. Vol. 1. Jai Press Ltd, 1991.
35. Ruzgas T, Csoregi E, Emneus J et al. Peroxidase-modified electrodes: Fundamentals and application. Analytica Chimica Acta 1996; 330:123-138.
36. Yaropolov AI, Malovick V, Varfolomeev SD et al. Electroreduction of hydrogen peroxide on an electrode with immobilised peroxidase. Dokl Akad Nauk SSSR 1979; 249:1399-1402.
37. Razumas VJ, Jasaitis JJ, Kulys JJ. 700-Electrocatalysis on enzyme-modified carbon materials. Bioelectrochem Bioenerg 1984; 12:297-322.
38. Diurliat H, Courteix A, Comtat M. Reactions of horseradish peroxidase on a platinum cathode. Bioelectrochem Bioenerg 1989; 22:197-209.
39. Comtat M, Durliat H. Some examples of the use of thin layer spectroelectrochemistry in the study of electron transfer between metals and enzymes. Biosens Bioelectron 1994; 9:663-668.
40. Lai ME, Bergel A. Electrochemical reduction of oxygen on glassy carbon: Catalysis by catalase. Electroanalytical Chemistry 2000; 494:30-40.
41. Nakajima R, Yamazaki I. The conversion of horseradish peroxidase C to a verdohemoprotein by a hydroperoxide derived enzymatically from indole-3-acetic acid and by m-nitroperoxybenzoic acid. J Biol Chem 1980; 255:2067-2071.
42. Rodriquez-Lopez JN, Hernandez-Ruiz J, Garcia-Canovas F et al. The inactivation and catalytic pathways of horseradish peroxidase with m-Chloroperoxybenzoic Acid. J Biol Chem 1997; 272:5469-5467.
43. Bickerstaff GF. Immobilisation of enzymes and cells. Scotland: University of Paisley, 1997.
44. Woodward J. Immobilised enzymes: Adsorption and covalent coupling. In: Woodward J, ed. Immobilized Cells and Enzymes: A Practical Approach. Oxford: IRL, 1985:3-17.
45. Blum LJ. Bio- and Chemi-Luminescence Sensors. France: World Scientific Publising CO Pte Ltd, CNRS, Univ Claude Bernard Lyon, 1997.
46. Bilitewski U, Turner APF. Biosensors for Environmental Monitoring. Harwood Academic Publishers, 2000.
47. Cabral JMS, Kennedy JF. Covalent and coordination immobilisation of proteins. In: Taylor RF, ed. Protein Immobilisation. New York: Marcel Dekker, 1991:73-138.
48. Brodelius P. Immobilized plant cells. In: Laskin AI, ed. Enzymes and Immobilized Cells in Biotechnology. London: Benjamin Cummings, 1985:109-148.
49. Wang B, Zhang J, Cheng G et al. Amperometric enzyme electrode for the determination of hydrogen peroxide based on sol-gel/hydrogel composite film. Analytica Chimica Acta 2000; 407:111-118.
50. Pankratov I, Lev O. Sol-gel derived renewable-surface biosensors. J Electroanal Chem 1995; 393:35-41.
51. Wang B, Dong S. Sol-gel-derived amperometric biosensor for hydrogen peroxide based on methylene green incorporated in Nafion film. Talanta 2000; 51:565-572.
52. Zhang J, Li B, Wang Z et al. Functionalized inorganic-organic composite material derivated by sol-gel for construction of mediated amperometric hydrogen peroxide biosensor. Analytica Chimica Acta 1999; 388:71-78.
53. Sarkar P, Tothill IE, Setford SJ et al. Screen-printed amperometric biosensors for the rapid measurement of L- and D- amino acids. The Analyst 1999; 124:865-870.
54. Vidal JC, Yague MA, Castillo JR. A chronoamperometric sensor for hydrogen peroxide based on electron transfer between immobilized. Sens Actuators B Chem 1994; 21:135-141.
55. Sadeghi SJ, Gilardi G, Cass AEG. Mediated electrochemistry of peroxidases-effects of variations in protein and mediator structures. Biosens Bioelectron 1997; 12:1191-1198.
56. Saby C, Mizutani F, Yabuki S. Glucose sensor based on carbon paste electrode incorporating poly(ethylene glycol)-modified glucose oxidase and various mediators. Analytica Chimica Acta 1995; 304:33-39.
57. Casero E, Darder M, Pariente F et al. Peroxidase enzyme electrodes as nitric oxide biosensors. Analytica Chimica Acta 2000; 403:1-9.
58. Miao Y, Tan SN. Amperometric hydrogen peroxide biosensor with silica sol-gel/chitosan film as immobilisation matrix. Analitica Chimica Acta 2001; 437:87-93.

59. Reiter S, Habermuller K, Schuhmann W. A reagentless glucose biosensor based on glucose oxidase entrapped into osmium-complex modified polypyrrole films. Sens Actuators B Chem 2001; 79:150-156.
60. Brunetti B, Ugo P, Moretto LM et al. Electrochemistry of phenothiazine and methylviologen biosensor electron-transfer mediators at nanoelectrode ensembles. J Electroanal Chem 2000; 491:166-174.
61. Sekine Y, Hall EAH. A lactulose sensor based on coupled enzyme reactions with a ring electrode fabricated from tetrathiafulvalen-tetracyanoquinodimetane. Biosens Bioelectron 1998; 13:995-1005.
62. Kadnikova EN, Kostic NM. Oxidation of ABTS by hydrogen peroxide catalyzed by horseradish peroxidase encapsulated into sol-gel glass. Effects of glass matrix on reactivity. J Mol Catal B Enzym 2002; 18:39-48.
63. Tatsuma T, Okawa Y, Watanabe T. Enzyme monolayer- and bilayer-modified tin oxide electrodes for the determination of hydrogen peroxide and glucose. Anal Chem 1989; 61:2352-2355.
64. Frew JE, Jones J, Scholes G. Spectrophotometric determination of hydrogen peroxide and organic hydroperoxides at low concentrations in aqueous solution. Analytica Chimica Acta 1983; 155:139-150.
65. Lei C, Deng J. Hydrogen peroxide sensor based on coimmobilized methylene green and horseradish peroxidase in the same montmorillonite-modified bovine serum albumin-glutaraldehyde matrix on a glassy carbon electrode surface. Anal Chem 1996; 68:3344-3349.
66. Qian J, Liu Y, Liu H et al. Characterization of regenerated silk fibroin membrane for immobilisation of peroxidase and construction of an amperometric hydrogen peroxide sensor employing phenazine methosulphate as electron shuttle. J Electroanal Chem 1995; 397:157-162.
67. Liu H, Ying T, Sun K et al. A reagentless biosensor highly sensitive to hydrogen peroxide based on new methylene blue N dispersed in Nafion gel as the electron shuttle. J Electroanal Chem 1996; 417:59-64.
68. Ruan C, Yang F, Lei C et al. Thionine covalently tethered to multilayer horseradish peroxidase in a self-assembled monolayer as an electron-transfer mediator. Anal Chem 1998; 70:1721-1725.
69. Xu Jing-Juan, Zhou Dong-Mei, Chen Hong-Yuan. A reagentless hydrogen peroxide biosensor based on the coimmobilisation of thionine and horseradish peroxidase by their cross-linking with glutaraldehyde on glassy carbon electrode. Electroanalysis 1998; 10:713-716.
70. Dequaire M, Limoges B, Moiroux J et al. Mediated electrochemistry of horseradish peroxidase. Catalysis and inhibition. J Am Chem Soc 2002; 124:240-253.
71. Rajendran V, Csöregi E, Okamoto Y et al. Amperometric peroxide sensor based on horseradish peroxidase and toluidine blue O-acrylamide polymer in carbon paste. Analytica Chimica Acta 1998; 373:241-251.
72. Garguilo MG, Huynh N, Proctor A et al. Amperometric sensors for peroxide, choline, and acetylcholine based on electron transfer between horseradish peroxidase and a redox polymer. Anal Chem 1993; 65:523-528.
73. Li Juan, Ngin Tan Swee, Ge Hailin. Silica sol-gel immobilized amperometric biosensor for hydrogen peroxide. Analytica Chimica Acta 1996; 335:137-145.
74. Sergeyeva TA, Lavrik NV, Rachkov AE et al. Hydrogen peroxide-sensitive enzyme sensor based on phthalocyanine film. Analytica Chimica Acta 1999; 391:289-297.
75. Frew JE, Harmer MA, O'Hill HA et al. A method for estimation of hydrogen peroxide based on mediated electron transfer reactions of peroxidases at electrodes. J Electroanal Chem 1986; 201:1-10.
76. Yabuki S, Mizutani F, Hirata Y. Preparation of a microperoxidase and ferrocene-immobilized polyion complex membrane for the detection of hydrogen peroxide. J Electroanal Chem 1999; 468:117-120.
77. Karyakin AA, Karyakina EE. Prussian Blue-based 'artificial peroxidase' as a transducer for hydrogen peroxide detection. Application to biosensors. Sens Actuators B Chem 1999; 57:268-273.
78. Inada Y, Takahashi K, Yoshimoto T et al. Application of polyehtylene glycol-modified enzymes in biotechnological processes: Organic solvent-soluble enzymes. Trends Biotechnol 1986; 4:190-194.
79. Dong S, Guo Y. Organic phase enzyme electrodes operated in water-free solvents. Anal Chem 1994; 66:3895-3899.
80. Daly DJ, O'Sullivan CK, Guilbault GG. The use of polymers coupled with metallised electrodes to allow H_2O_2 detection in the presence of electrochemical interferences. Talanta 1999; 49:667-678.
81. Umana M, Waller J. Protein-modified electrodes. The glucose oxidase/polypyrrole system. Anal Chem 1986; 58:2979-2983.
82. Trojanowicz M, Krawczynski vel krawczyk T. Enzyme entrapped polypyrrole modified electrode for flow-injection determination of glucose. Biosens Bioelectron 1990; 5:149-156.
83. Hoa DT, Kumar Suresh TN, Punekar NS et al. A biosensor based on conducting polymers. Anal Chem 1992; 64:2645-2646.

84. Van Os PJHJ, Bult A, Koopal CGJ et al. Glucose detection at bare and sputtered platinum electrodes coated with polypyrrole and glucose oxidase. Analytica Chimica Acta 1996; 335:209-216.
85. Foulds NC, Lowe CR. Immobilisation of glucose oxidase in ferrocene-modified pyrrole polymers. Anal Chem 1988; 60:2473-2478.
86. Madaras MP, Popescu IC, Ufer S et al. Microfabricated amperometric creatine and creatinine biosensors'. Biosens Bioelectron 1996; 11:vi.
87. Manowitz P, Stoecker PW, Yacynych AM. Galactose biosensors using composite polymers to prevent interferences. Biosens Bioelectron 1995; 10:359-370.
88. Santamaria M del Campo, Vazquez Barbado MD, Tascon Garcia ML et al. Determination of hydrogen peroxide by voltammetric techniques at carbon paste electrodes modified with transition metal phthalocyanines. Quimica Analitica 1998; 17:147-152.
89. Elstner EF, Frommeyer D. Production of hydrogen peroxide by photosystem II of spinach chloroplast lamellae. FEBS Letters 1978; 86:143-146.
90. Elstner EF, Kramer R. Role of superoxide free radical ion in photosynthetic ascorbate oxidation and ascorbate-mediated photophosphorylation. Biochimica et Biophysica Acta (BBA) - Bioenergetics 1973; 314:340-353.
91. Epel BL, Neumann J. The mechanism of the oxidation of ascorbate and Mn^{2+} by chloroplasts: The role of the radical superoxide. Biochimica et Biophysica Acta (BBA). Bioenergetics 1973; 325:520-529.
92. Pan RL, Izawa S. Photosystem II energy coupling in chloroplasts with H_2O_2 as electron donor. Biochimica et Biophysica Acta 1979; 547:311-319.

Chapter 17

Biodevices for Space Research

Dania Esposito,* Cecilia Faraloni, Floriana Fasolo, Andrea Margonelli, Giuseppe Torzillo, Alba Zanini and Maria Teresa Giardi

Abstract

This review focuses on the realisation of optical sensors able to monitor the effect of complex space radiation on biological components, based on the biosensor concept. A biosensor is a device that can reveal a biochemical variable using a biological component interfaced with a transducer. It issues an electric signal which is easy to process, depending on the analysed variable. Biosensors are useful to study the effect of stress conditions on living organisms. One of the goals of this research was to develop two types of biosensors able to monitor directly the response of oxygenic photosynthetic organisms to radiation present in space in view of their importance for future space colonization. In ground experiments and in balloon stratosphere flights, the photosynthetic process has been analysed at the level of photosystem II (PSII), the supramolecular pigment-protein complex in the chloroplast which catalyses the light-induced transfer of electrons from water to plastoquinone; PSII splits water into molecular oxygen, protons and electrons, thereby sustaining an aerobic atmosphere on Earth and providing the reducing equivalents necessary to fix carbon dioxide to organic molecules, creating biomass, food and fuel. The results indicated that presence of space radiation in the dark has a synergistic effect on photosystem II activity, suggesting that PSII D1 protein turnover may be involved in resistance to space stress. The resistance of the tested microorganisms to space stress seems to be related to their position on the evolutive scale of photosynthesis. The present studies allow to establish a regular and reliable correlation between measured physical characteristics of space radiation and biological radiation effect.

Introduction

In view of future space colonization research on the response of the biological life to space conditions is of great interest. In fact, after 20 years of experiments in Mir station and the results of after the Russian and NASA activities, it is now clear that space conditions are stressing and may affect living organisms both for the presence of cosmic and solar radiation and for the absence of gravity.[1]

Space radiation may be classified according to origin as: (i) galactic cosmic radiation (87% protons, 12% alfa, 1% HZE), with energies between 1 and 103 GeV; (ii) solar particle radiation, consisting of charged particles in large clouds, mainly protons with an energy of about 1GeV; (iii) geomagnetically trapped particle radiation, generated from the interaction of the radiation with the geomagnetic field comprising electrons with energies up to 7 MeV, protons with energies up to 600 MeV, and low energy heavy ions. Moreover, inside the spacecraft the radiation field is modified by interaction processes with the shielding material, originating secondary radiation composed by neutrons and heavy ions (HZE).[2]

*Corresponding Author: Dania Esposito—Institute of Cristallography, CNR, Via Salaria Km 29.300 – 00016, Monterotondo Scalo, Rome, Italy. Email: dania.esposito@mlib.cnr.it

Biotechnological Applications of Photosynthetic Proteins: Biochips, Biosensors and Biodevices, edited by Maria Teresa Giardi and Elena V. Piletska.

Interaction mechanism of ionizing radiation with the living organisms and matter depends on the type and energy of the radiation, and the characteristics of penetrated matter. The effects of the different types of radiation on biological matter can be various: neutrons, electrically uncharged, impact the nuclei of charged hydrogen atoms. The result is an elastic scattering process, ejecting protons. Beta rays are fast electrons that transfer energy, disrupting chemical bonds; this results in formation of radical or ionisation X and gamma rays release high-speed electrons from atoms first. Positively charged particles transfer energy to molecules in cells by essentially the same mechanisms. Radiation-induced ionisation may act directly on the cellular component molecules or indirectly on water molecules, causing water-derived radicals. Radicals react with nearby molecules in a short time, resulting in breakage of chemical bonds or oxidation of the affected molecules. The major effect in cells are either a single strand or both strands DNA breaks. However, the latter is believed to be much more important biologically, since its repair is more difficult and erroneous rejoining of broken ends may occur. These so-called mis-repairs result in induction of mutation, chromosome aberration, or cell death.[3]

It is clear that exposure of biological material to ionising radiation leads to a loss of function due to the modification of critical structures.[4] Results of former space experiments suggest that the biological effect of space radiation could be enhanced under microgravity.

One of the most important and discussed problems in radiation protection concerns the determination of the biological effectiveness of radiation on living organisms.[5,6] In order to assess the radiation risk for living beings during long-term spaceflights, it is very important to clarify whether combination of cosmic radiation and other factors have a synergistic effect on the cell functions.

Unicellular organisms were the first object in biological studies in space. Rapid multiplication, tiny size and simple procedure to keep growth conditions are the advantages in using these simplest biological systems to estimate cosmic radiation effect.[1,7] It seems that microgravity causes first damage modifying the vitality of the organisms with effect on cell division, cell components and their distribution. Many studies focus on the effects of space radiation and/or microgravity on the number, viability, kinetics of germination, growth rate and mutation frequency of spores formed in space.[8] A series of biological experiments were performed during space missions by European Space Agency (ESA) between 1987 and 1997, studying the effects of microgravity and cosmic radiation on microorganisms, on the embryogenesis of the stick insect, the radiation damage in plant seeds, bacterial spores and in plant tissues (http://www.estec.esa.nl/spaceflight/foton/fformer.htm).

Space environment affects almost all biological processes, in particular germination and flowering. Embryo lethality and lethal mutation frequency were observed in *Arabidopsis thaliana* seeds[9] and in other organisms such as *Escherichia coli* and *Bacillus subtilis*[10,11] while very little is known about the effects of ionising and non-ionising radiation on photosynthetic apparatus.

One of the goals of this research was to develop two types of biosensors able to monitor directly the response of oxygenic photosynthetic microorganisms to the space conditions.[12,13]

To measure radiation, physical sensors have been most widely utilised for biology studies so far.[14] Recently, studies on chromosome aberrations in the peripheral lymphocytes by Fluorescence in situ hybridisation and assays of micronuclei have been developed.[15]

The development of space research devices focuses on the concept of biosensor technology, based on the properties of the photosystem II (PSII) of distinct photosynthetic organisms. PSII is the supramolecular pigment-protein complex in the chloroplast which catalyses the light-induced transfer of electrons from water to plastoquinone. PSII splits water into molecular oxygen, protons and electrons, thereby sustaining an aerobic atmosphere on Earth and providing the reducing equivalents necessary to fix carbon dioxide, creating biomass, food and fuel. It consists of over 25 polypeptides, which make up the oxygen-evolving complex, light-harvesting chlorophyll protein complexes that capture light and a reaction centre involved in primary charge separation.[16] The photosynthetic apparatus represents a crucial contact point between the organism and its environment; its flexibility and stability are instrumental to survival under stress condition. D1 is the most important protein present in the PSII reaction centre. It is involved in the raising of an electron from the ground state to the excited state, after the absorption of a quantum of light by the primary donor chlorophyll dimer P680, from which it can pass to the primary acceptor plastoquinone Q_A.[16-18] If Q_A is already in the reduced state when the photon is absorbed, no electron transfer can take place and the chlorophyll fluorescence level increases by a factor of five. Analysis of the kinetics of fluorescence induction can thus be used to detect the activity of the photosystem II correlated to photosynthetic efficiency and oxygen evolution[19,20] useful also during radiation stress experiments.[21] Light energy absorbed by chlorophyll molecules, i.e., Photosynthetically Active Radiation(PAR), has different fates in the photosynthetic organisms: it can be used for the photochemistry of the photosynthesis process, while energy in excess can be dissipated as heat or it can be re-emitted as longer wavelength red/far-red light energy; this re-emission of light is termed chlorophyll Fluorescence. These three events occur in competition, so that any increase in the yield of one will result in a correspondent decrease in the yield of the other two. Hence, by measuring the yield of chlorophyll fluorescence, information about changes in the efficiency of photochemistry and heat dissipation can be obtained.[19,20]

The most widely used technique is fluorescence induction measuring changes in fluorescence yield when a light is switched on after a dark period. Under light, the fluorescence yield rapidly increase and then slowly decreases. This time-dependent intensity variation reflects photosynthetic electron transport and was first observed as early as 1960 by Kautsky. Most chlorophyll fluorescence (about 90%) is emitted by photosystem II. When a light is switched on, photosystem II transiently reduces its electron acceptor quinone Q_A, and the reduction is reflected by a transient increase in the fluorescence yield. Once PSII absorbs light and Q_A has accepted an electron, it is not able to accept another until it has passed the first one to a subsequent electron carrier (Q_B). During this period, the reaction centre is 'closed'. At any point in time, the presence of a proportion of closed reaction centres leads to an overall reduction in the efficiency of photochemistry and to a corresponding increase in the yield of fluorescence. The fluorescence induction kinetics can be used to study photosynthetic activity. The numerical parameters often considered are the ratio of the variable fluorescence (maximum minus initial fluorescence) to maximum fluorescence $(F_{max} - F_0)/F_{max}$, and the area above the fluorescence curve between F_0 and F_{max}, which is proportional to the pool size of the electron acceptors Q_A on the reducing side of photosystem II. If the electron transfer from the reaction centres to the quinone pool is blocked, such as during the binding of the photosynthetically active herbicides, this area will be dramatically reduced. Fluorescence has been successfully used as a fool to monitor the physiological state of the photosynthetic apparatus exposed to stress condition such as photoinhibition, air pollution, herbicides.[22-25]

D1 protein has a high turnover rate.[16] Turnover is a physiological process in photosynthetic organisms: the degradation and synthesis of new D1 protein genetically regulated is an important modulated mechanism during stress recovery. For this reason many biochemical analyses are focused on this important protein.

We have built an automatic fluorimeter able to monitor physiological state of photosynthetic organisms at the level of PSII directly under radiation exposure. This study is useful to follow the behaviour and the response of the organisms in space conditions during long-term missions and manned missions to Mars or inside International Space Station.

One limitation in simulation irradiation studies is represented by the fact that it is possible to obtain only fluxes of single, mono-energetic source often produced in a narrow unidirectional beam, while in space an organism is exposed to a flux of radiation with a wide spectrum of atomic mass and energy and from various directions.[22] This prompted to test the biosensor in flights to the Earth stratosphere.[23]

Present studies allow to establish a regular and reliable correlation between measured physical characteristics of radiation and biological radiation effects.

Experimental Methods

Intact Cells

Cultures of *Chlorella sorokiniana, Chlorella zofingiensis, Chlorococcum sp., Chlamydomonas reinhardtii* (Chlorophyta), *Arthrospira platensis* (Cyanobacteria), *Monodus subterraneus*

(Eustigmatophyta) were collected for the experiments performed utilising intact organisms. The organisms were grown in their media under an light intensity of 50 μmol photons/m^2s.

The cells were harvested by centrifugation and layered on agar-medium into two plastic devices, each provided with five cell chambers. The samples were covered with transparent polycarbonate plexiglas. The containers were sealed to avoid material loss during space flight.

Photosystem II as a Biomediator

For the experiments with extracted material, PSII particles were isolated from various photosynthetic higher plants (*Spinacea oleracea, Vicia faba, Medicago sativa)* with the method of Berthold.[24] Then samples of 2 microg/microL were immobilised inside a container (5 mm both in height and diameter) kept in the dark and frozen at –13 ± 2 °C utilising a suitable cooling system (10x10x10 cm, Fig. 1). The sample container was made of stainless steel. In the central hole there was the Peltier cell inserted on a PVC black support. In order to cool the sample, an additional external thermal insulation is needed. It is a box filled with Aspen Aerogel sheets appropriately cut. The fluorescence sample signal was recorded by an automatic multifluorimeter as reported in Figure 1.

Vitality measurements of PSII fluorescence were performed by optical fluorimeters (BioLumi purchased from Biosensor, www.biosensor.it, info@biosensor.it, Italy). They are multisite fluorimeters that provide fluorescence parameters (F_o, F_m, F_v, the ratio F_v/F_m and the area over the fluorescence curve). F_0 was calculated by using an algorithm that determines the line of best fit for the initial data points recorded at the onset of illumination; this line of best fit is extrapolated to time zero to determine F_o. Fm is the maximum value achieved while recording. F_v is the variable component of fluorescence. The F_v/F_m fluorescence ratio is proportional to the maximum photochemical quantum yield of the PSII.[25] The biodevice is 13x10x18 cm in size and 3 Kg in weight. It can measure fluorescence parameters on 10 samples, continuously for 30 days. The fluorimeter gives 10 inputs of red light at 650 nm with one second of interval between inputs on each sample. The red light covers a surface of 1 cm^2 and in such a way, a high reproducibility of the measurement is obtained, mediated over thousand points. The actinic light is provided by white LED's emitting in a range from 20 to 200 μmol photons/m^2s.

Radiation Source Facilities

Three main radiation source facilities were used for this study: CERN-EU high-energy reference field (CERF) facility, Joint Research Centre Dosimetry Division (Joint Research Centre at Ispra-Italy) and the S. Giovanni hospital (Turin, Italy).

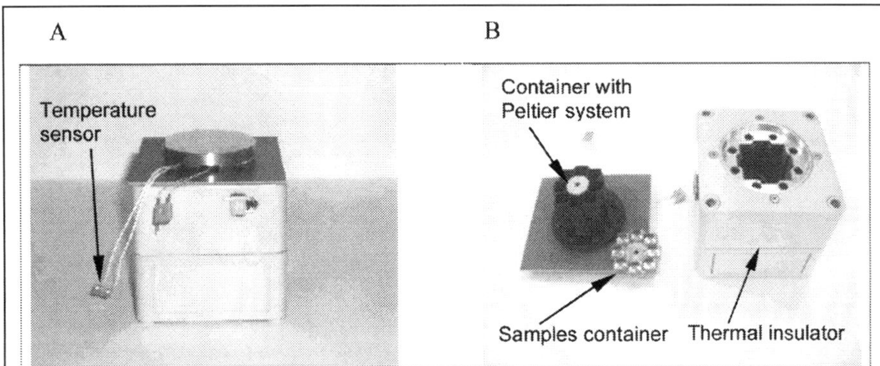

Figure 1. Refrigeration system to keep samples at –13 °C. A) external view with the temperature sensor; B) samples container and Peltier system.

The first source is installed in the secondary beam lines (H6) from the Super Proton Synchrotron (SPS). A proton beam is stopped in a copper target, 7 cm in diameter and 50 cm in length. These roof-shields produce almost uniform radiation fields over two areas of 2 x 2 m^2, each divided into 16 squares of 50x50 cm^2. Each element of these "grids" represents a reference exposure location. The intensity of the primary beam is monitored by an air-filled, precision ionisation chamber (PIC) at atmospheric pressure. One PIC-count corresponds to 2.2 x 10^4 particles (error ± 10%) impinging on the target. Typical values of dose equivalent rates are 1-2 nSv per PIC-count on top of the 40 cm iron roof-shield and 0.3 nSv per PIC-count outside the 80 cm concrete shields (roof and side). Behind the 80 cm concrete shield, the neutron spectrum has a second pronounced maximum at about 70 MeV and resembles the high-energy component of the radiation field created by cosmic rays at commercial flight altitude.[26]

The second radiation facility was at the Dosimetry Division of Joint Research Centre at Ispra-Italy, with the neutron source consisting of americium–beryllium with the following features: oxide of metallic Am-Be contained inside a steel capsule 1mm thick (activity of 5,994x10^{11} Bq; rate neutrons emission of 3.6 x $10^7 s^{-1}$).

The third radiation facility is a linear accelerator Linac Saturne producing fast electrons at 18 MeV present at the Radiotherapy Department at S. Giovanni Hospital in Turin (Italy).

Balloon Stratospheric Flights

The experiments were performed in a biodevice carried to an altitude of 38 Km by a balloon (see Fig. 2) 200 m long and made of polymeric material. The balloon, inflated with helium gas, was launched from the Milo base of Italian Space Agency (Trapani, Italy). The payload approximately followed 38° in latitude at an altitude that varied daily between 32 and 38 km. It was recovered at about 1400 km away from the launch site, in Spain, 24 hours later.[27]

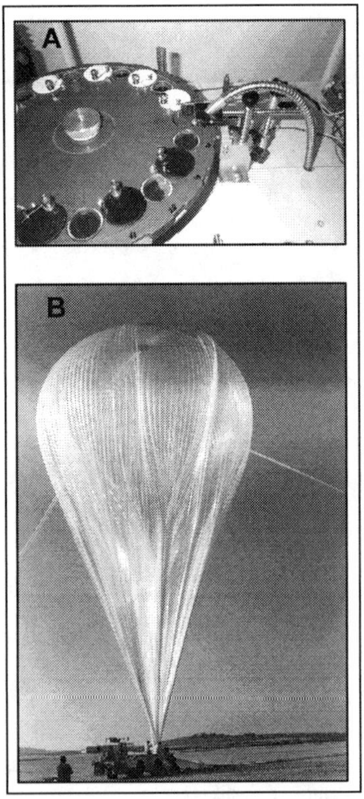

Figure 2. A) Automatic biodevice to measure photochemical efficiency, and B) the ASI stratospheric balloon utilised for the flight.[27]

During the flight, the organisms were kept immobilised on agar medium under various experimental conditions: (i) partially shielded from cosmic radiation, (ii) dark, (iii) light conditions (Fig. 2A). The biodevice was kept inside a thermostated and pressured container.

In order to avoid an excess of light, the external solar light was filtered by a black cover reducing to 10% of the external illumination.

Monte Carlo Simulation

In order to obtain a simulation of radiation environment outside and inside the stratospheric balloon, and to evaluate the absorbed dose on biological samples, the Monte Carlo method was applied.

The Monte Carlo algorithm was first introduced by Metropolis[28] to calculate properties of substances that may be considered as composed of interacting individual molecules. Today, the name 'Monte Carlo' is related to numerical algorithms using random numbers, and it is utilised to find an approximate solution of a complex system. It is often used for stochastic problems that are too complex to be solved analytically. The Monte Carlo algorithm has been extended to a broad range of physical and biological systems.[29]

The adopted GEANT3 program simulates the passage of elementary particles through the matter. The principal applications of GEANT in High Energy Physics are the transport of particles through an experimental set up for the simulation of detector response and the graphical representation of the set up and of the particle trajectories. The adopted GEANT3 system allows (i) the description of an experimental set-up by a structure of geometrical volumes; (ii) the transport of particles through the various regions of the set-up, taking into account geometrical volume boundaries and physical effects according to the nature of the particles themselves, their interactions with matter and the magnetic field; (iii) the record of particle trajectories and the response of the sensitive detectors; (iv) the visualisation of the detectors and the particle trajectories.

Results

For the development of methods able to estimate biological effects of radiation, two approaches were pursued in this study. The first approach concerned the utilisation of intact photosynthetic microorganisms, monitored by fluorescence emission as signal transduction.[19] The second approach utilised extracted and frozen biological material, also monitored by fluorescence of PSII. For its particular feature, after isolation and suitable immobilisation, the multiprotein complex PSII maintained the activity.

Intact Cells as a Biological Component

These experiments were essential to determine the behaviour of algae strains to cosmic radiation stress, focusing on the response of photosynthetic activity in simulation studies with radiation exposure, under different light conditions. The automatic multifluorimeter (BioLumi, www.biosensor.it) recorded the temperature as well as the fluorescence parameters (F_o, F_m, F_v, the ratio F_v/F_m, Area). After 10 minute dark adaptation, cells were exposed to a 7h light and 12 h dark cycle. Table 1 shows the growth light provided by white LEDs used during the experiments.

In an attempt to mimic space conditions, some radiation experiments were carried out by applying cycles of rapid irradiation, followed by a recovery period and again a period of irradiation.

Figure 3 shows the photochemical efficiency (F_v/F_m) measured in *Chlamydomonas reinhardtii* and *Chlorella sorokiniana* exposed to fast neutrons at CERN laboratory at a dose rate of 0.2 mSv/h (0-800MeV). This experiment can be broken up into three phases: a first phase during which algae, immobilized on agar medium, were irradiated for 7 hours (dose rate: 0.20 mSv/h), a second phase during which the irradiated samples were allowed to recover for about 30 hours, and a third phase during which samples were again irradiated for 7 hours (dose rate: 0.26 mSv/h). The dose rates during the two radiation phases were different because the intensity of the primary beam obtained in the Synchrotron is never constant along the exposure time. During the experiments the algae were illuminated by white LEDs at various growth light intensities (e.g., 20, 70, 120 μmol photons/m^2s).

Table 1. Scheme of various beams utilized in the source facilities for exposure to the ionizing radiation of the photosynthetic material under different light conditions

Radiation Source	Dose Rate	MeV	Exposure Time	Light Condition (μmol photons/ m^{2}s)	Radiation Facility
Fast neutrons Am-Be (19 Ci)	0.126 mGy/h 0.11 mSv/h 3.9 mSv/h	3 (mean)	24 h 5 h 42 h 45 h	0-150-300 0-20-70-120	JRC-ISPRA-IT
Fast neutrons H6 SPS	0.23 mSv/h	0-800	7 h 13 h 16 h 24 h	0-150-300 0 0-20-70-120 5	CERN (Geneva-SW)
Gamma LINAC-SATURNE	3 Gy/min	18	40 min	0-150-300	S. Giovanni Hospital (Turin-IT)
Gamma ^{60}Co	1.540 Gy/h 4.2 Gy/h	1	2 h	5	JRC-ISPRA
Gamma ^{137}Cs	0.11 Gy/h 1.2 mGy/h	0,66	24 h 10 h	0-20-70-120 0	JRC-ISPRA CNR
Electrons LINAC-SIEMENS	3 Gy/min	6	9 min	5	S. Giovanni Hospital (Turin-IT)

The F_v/F_m ratios of *Chlamydomonas reinhardtii* (Fig. 3A) and *Chlorella sorokiniana* (Fig. 3B) during the exposure to fast neutrons are expressed as percentage of the initial values for each sample. In The control sample (not irradiated) kept at 120 μmol photons/m^{2}s the F_v/F_m ratio showed almost constant during the experiment.

As it can be seen from Figure 3, the effect of irradiation of cells to fast neutrons was dependent on the light intensity at which cultures were exposed. The highest inhibition was found when algal samples were irradiated in the dark conditions, while in illuminated cells the inhibition increased with increasing light intensity.

The results indicated that dark conditions together with irradiation have a synergistic effect. This suggests that D1 protein turnover may be involved in resistance to radiation stress. This explanation is supported by the fact that in the absence of light no recovery from stressed D1 can take place. On the other hand, it seems that light condition (20, 70 μmol photons/m^{2}s) allows an adaptation to radiation during the second phase exposure; as a matter of fact there was a recovery in the F_v/F_m ratio, in particular in *Chlamydomonas reinhardtii.*

From Tables 2 and 3, it was clear that there was a relationship between the light intensity and the effect or radiation on photochemical efficiency. There was an enhancement of F_v/F_m values of samples under radiation exposure, as compared to the not irradiated sample. High light, however, causes a decrease of photochemical efficiency.

Gamma rays are different from neutrons for their action mechanism on the matter. The damage process due to radiation is present in all organisms tested (Table 3), but with slight differences among the species. In particular the cyanobacterium *Arthrospira platensis* and the Eustigmatophyta

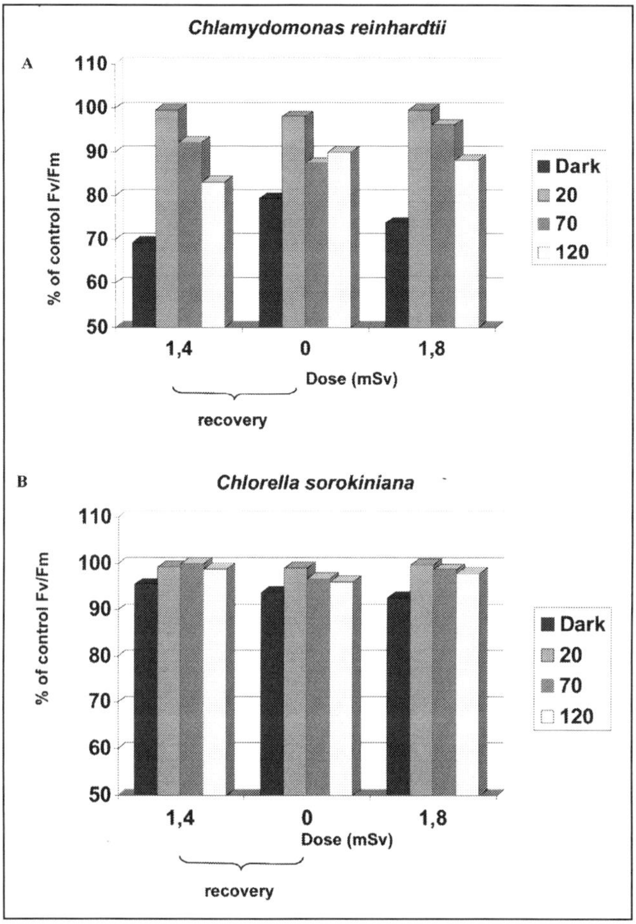

Figure 3. A) *Chlamydomonas reinhardtii* and B) *Chlorella sorokiniana* cells immobilyzed in agar exposed to fast neutrons (dose rate 0.2 mSv/h. Light conditions: dark, 20, 70, 120 μmol photons/m^{2}s (CERN source facility). The variation of fluorescence rate F_v/F_m during the radiation exposure was found to be significant over ± 0.4% (SD).

Table 2. Photosynthetic efficiency change

Chlamydomonas reinhardtii	
Light Conditions(μmol photons/m^{2}s)	**% Fv/Fm**
20	5.3%
70	16.5%
120	-21.4%

Chlamydomonas reinhardtii exposed to fast neutrons under various light conditions (Time 42 h; dose-rate 3.9 mSv/h; energy 3 MeV). The values of the right column are obtained from the difference between photochemical efficiency of irradiated and not irradiated samples multiplied by 10^2. $Fv/Fm_{t=0}$ = 0.753. SD, ± 0.4%. % Fv/Fm: variation of the Fv/Fm (final value minus initial value) due to irradiation.

Table 3. Photosynthetic efficiency decrease

Organisms	% Fv/Fm Decrease of Initial Value
Chlorella zofingiensis	6.9%
Chlorella sorokiniana	6.6%
Chlorococcum sp.	6.1%
Arthrospira platensis	5.1%
Monodus subterraneus	2.3%

Gamma ray exposure at light condition of 150 μmol photons/m^2s. Time: 20 min; dose-rate: 3 Gy/min; energy: 18 MeV. The values are obtained from the difference between photochemical efficiency of irradiated and not irradiated samples multiplied by 100. $Fv/Fm_{t=0}$ = 0.709. SD, ± 0.4%.% Fv/Fm: variation of the Fv/Fm (final minus initial value) due to irradiation.

Monodus subterraneus are clearly less affected than the green algae *Chlorella sorokiniana* and *Chlorella zofingiensis.*

Stratospheric Flights

When the balloon had reached the stratosphere (Fig. 4), during the first ASI BIRBA flight (Launch 10/07/2000–h.06:16UTC. Landing 11/07/2000–5:14UTC, visit internet web site http://www.asint.tp.asi.it/birba/birba.htm) a number of unexpected measurement failures of the fluorescence sensor were recorded, revealing an electric disturbance probably due to radiation beams. This is in accordance with the particular high level of solar activity recorded in the year 2000, the 11th year of the solar cycle (see internet site on Solar Chart: http://hiraiso.crl.go.jp/). Because the sunspot numbers are related to the solar activity, it is clear from the graph (Fig. 5) that during the balloon flight an extraordinary solar activity occurred. This activity was especially high on the day 10th of July 2000, compared to the same period of the year 2001, causing an increase of the electromagnetic radiation in the stratosphere.

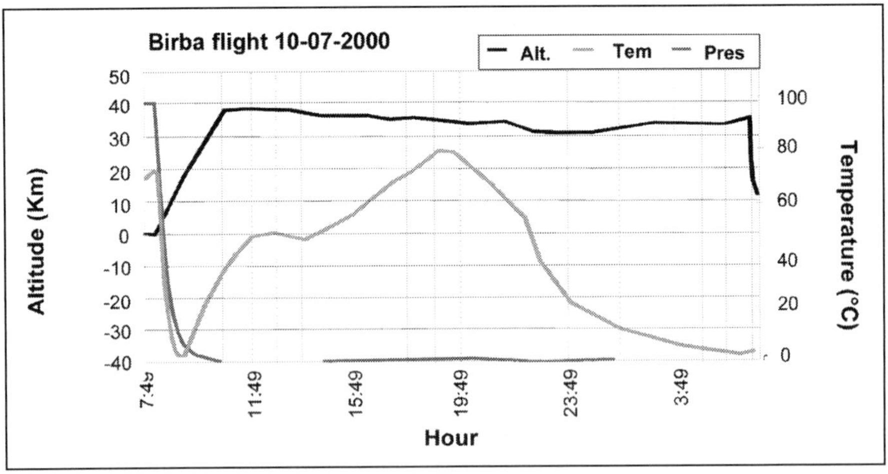

Figure 4. Data of temperature, altitude and pressure taken inside the box during stratospheric ASI flight in 2000.

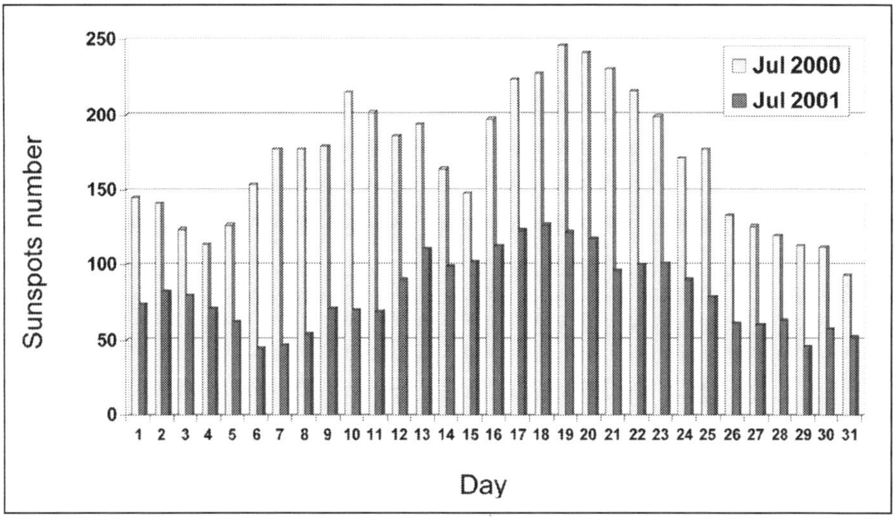

Figure 5. Comparison between the daily sunspot number of July 2000 and July 2001 (results are elaborated from the data base of National Oceanic and Atmospheric Administration, ftp://ftp.ngdc.noaa.gov/STP/SOLAR_DATA)

Meanwhile the electromagnetic disturbance, although the temperature and light was observed by the enclosed instruments and sensors to be constant inside the flying box, the photosynthetic organisms showed a strong inhibition in the photochemical activity (about 80% reduction of activity) which resulted in the death of the cells.

During the flight, an average of 15 PAR and a maximum of 40 PAR was detected, indicating that the external light was not sufficiently high to photoinhibit the photosystem II. During the flight, data from physic passive dosimeters collected the fast neutron components (10 keV ÷ 20 MeV) at the various altitudes.[30] A dose equivalent of 4.15 mSv/h ± 0.83 was measured, that was only relatively high. Therefore the main cause for death of the microorganisms was probably due to the particular high electromagnetic solar activity; indeed, local heat, impairment of polarity and changes of electric field are induced in the membrane of living cells by electromagnetic field.[31,32] In ground tests utilising single beam radiation, the combined effect of single radiation (gamma, fast neutrons) and PAR was synergistic and increased inhibition of photosynthetic activity but it was never so inhibitory of photosynthesis as observed in flight.

One hypothesis is that complex space radiation may particularly interfere with redox signalling of the photosynthetic electron transfer.

During the second balloon flight into the stratosphere, (Launch: 02/07/2001 h 5:47 am UTC. Landing: 03/07/2001 h 2:32 am UTC) various microalgal species, that in simulation studies had shown good tolerance to high dose of gamma radiation (a thousand mSv), were tested. The various organisms kept on agar media under various experimental conditions showed differential behaviour during the flight (Table 4).

Table 4 shows that the Eustigmatophyta *Monodus subterraneus* seems to maintain the highest photochemical efficiency after 10 hours flight to stratosphere in the dark, while *Chlorella sorokiniana* shows the highest inhibition. In control experiments dark conditions did not affect photosynthetic activity. The passive physic dosemeter in the box detected a dose equivalent to 0.62 mSv/h ± 0.124, a value lower than that detected during the first flight (4.15 mSv/h). It is important to point out that for technical reasons the second balloon flight reached only an average altitude of 29.4 Km, while the first flight reached 38 km.

The GEANT code (Fig. 6) allowed to estimate the cosmic cascade.

Table 4. Photosynthetic efficiency decrease

	% Fv/Fm Decrease of the Initial Value		
	Shielded	**Dark**	**Light**
Chlorella sorokiniana	7.3	19.6	8.9
Chlorella zofingiensis	6.3	7.8	3.7
Chlorococcum sp.	2.2	5.4	4.2
Monodus subterraneus	5.2	2.3	3.8

Percentage decrease of photochemical efficiency after a 10 hour flight to stratosphere with a balloon (38 Km maximum height). SD, ± 0.3%

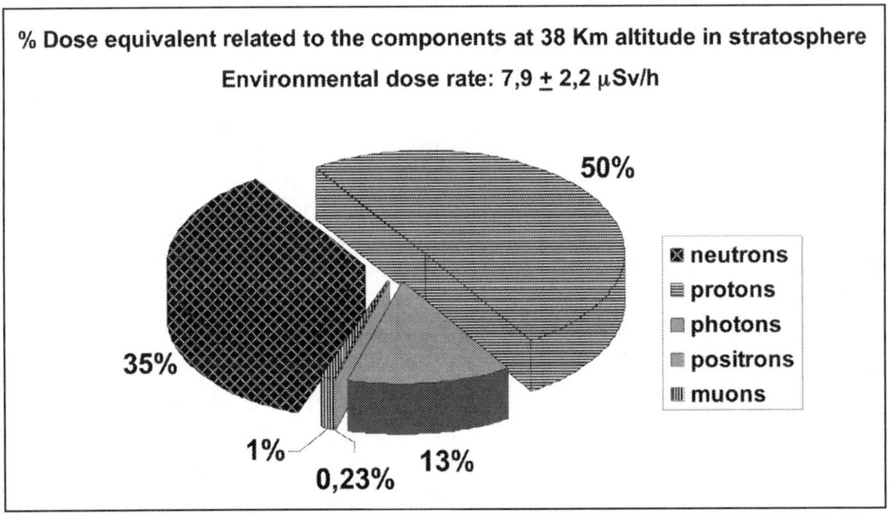

Figure 6. Percentage of environmental dose equivalent for various particles at the altitude of 38 Km (stratosphere) as determined GEANT 3 simulation. Environmental dose rate: 7.9 ± 2.2 μSv/h

Photosystem II as a Biomediator

A second approach of this study was to develop a biosensor based on the Radiation Target Analysis (RTA).[33] According to the RTA theory there is an exponential decrease in biochemical activity of an enzyme exposed to irradiation that is correlated to the absorbed energy and proportional to the minimum mass of the molecules possessing this activity. The empirical relationship Log Mr = 5.89-log D_{37}-0.0028T, determined by Beauregard and Potier, correlates the minimum active mass of an enzyme to the dose of radiation.[4] The relation considers the temperature dependence on the radiation inactivation. D_{37} is the dose of radiation in MRad that inhibits the enzyme activity to 37% of the initial value and T is the temperature in °C of the sample during irradiation. To avoid the diffusion of free radicals the irradiation must be performed at low temperatures (-13 ± 2 °C), so that a single hit causes damage to the enzyme.[34,35]

High Dose Exposure

In order to understand the effects of radiation on photosystem II, we first analysed isolated PSII membranes with polyclonal antibodies against the major proteins, exposing samples to high-energy

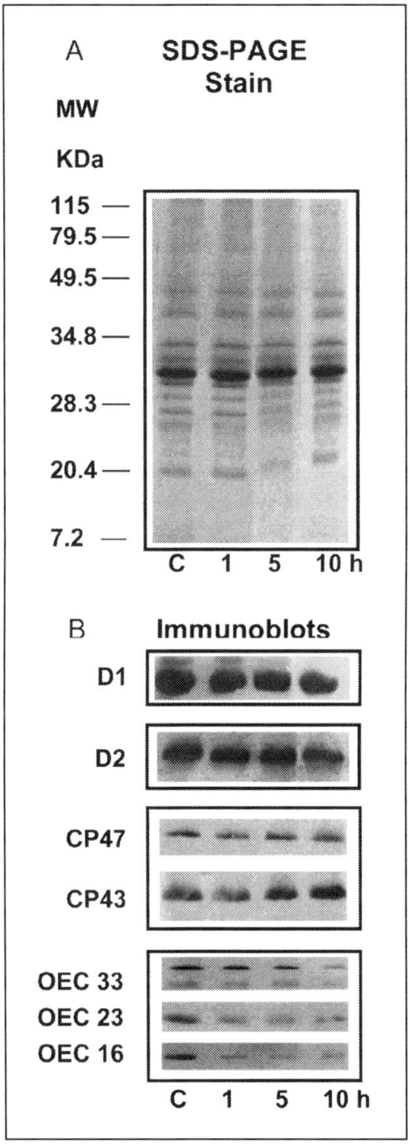

Figure 7. Biochemical analyses of gamma-irradiated PSII particles with a ^{60}Co source of 8.03 kGy/h. A) SDS-PAGE. B) Immunoblot analyses with polyclonal antibodies against the main PSII proteins. The gels and blots were replicated.

γ-radiation at a dose rate of 8.03 kGy/h. The analyses showed that irradiation at a high dose rate, initially results in the release of the proteins of the oxygen evolution complex and later affects the content of the D1 protein and slightly of D2 protein, while the content of the internal antennae CP43 and CP47 proteins appears to remain constant (Fig. 7).

As reported above, changes in the activity of frozen PSII thylakoids and PSII complexes were monitored by the fluorescence technique.[36] The loss of activity reported as the log of the activity versus the dose of high dose radiation (Fig. 8) shows a quite linear correlation.

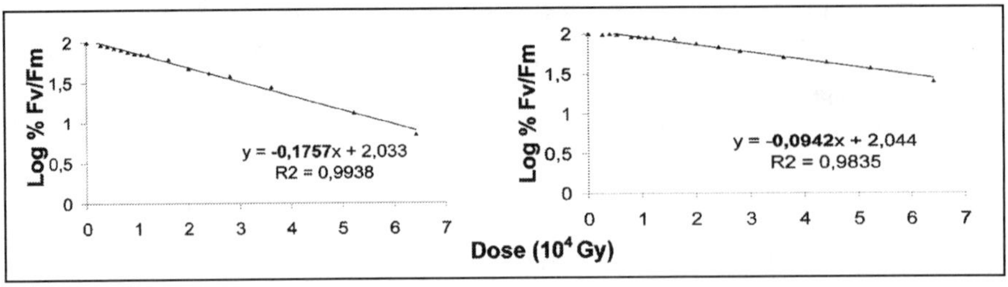

Figure 8. Radiation target analyses of thylakoids. High energy γ-radiation was provided by a ^{60}Co-source of a dose rate at the sample position of 0.803 Mrad/h. The membrane temperature was maintained at -18 ± 1°C. Initial PSII activity of the control Fv/Fm averaged 0.856 ± 0.005.

Low Dose Exposure

We tested the response of various fluorescence parameters to radiation. Among all tested values (F_o, F_m, F_v/F_m, area, F_j, F_P etc.) the modification of the area over the fluorescence curve was more relevant at the low radiation dose exposure (Fig. 9).

A good linear correlation was found between the dose rate and the signal decrease. It was interesting to note that exposure of PSII particles to both gamma rays (A) and fast neutrons (B) gave calibration curves with similar slope at a similar dose rate. These findings are important to focalise the studies on the testing PSII isolated from various photosynthetic organisms (Giardi et al in preparation).

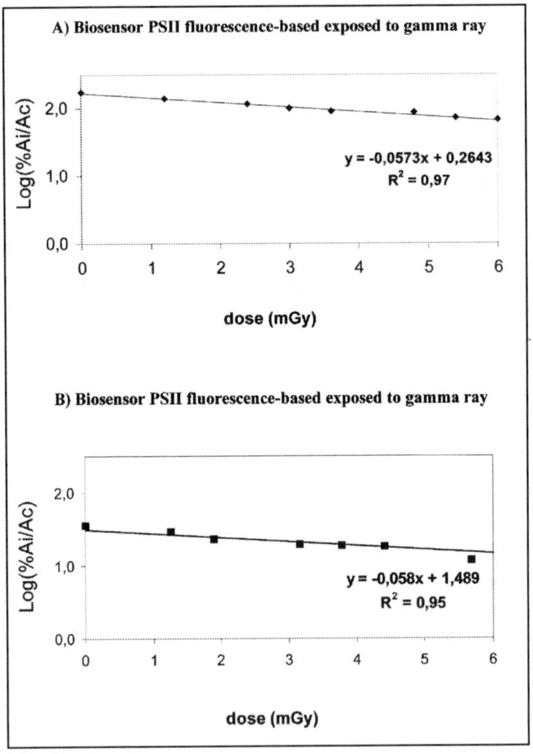

Figure 9. Calibration curve (A) of PSII particles from *Spinacea oleracea* exposed to gamma rays (^{137}Cs dose rate 1.2 mGy/h) and (B) fast neutrons (dose rate 1.26 mGy/h)- JRC, Ispra, (IT). The temperature and pH were highly controlled. Ai/Ac: ratio of area irradiated /control

Conclusion

It is worthwhile to point out the advantages of biosensors in space (inside spacecraft or ISS) and Mars missions; they are: (i) highly sensitive (ii) miniature size, (iii) portable, and (iv) allow continuous monitoring of the radiation effects during both space missions and in International Space Station. They are also suitable for photosynthetic organisms in long-term missions to increase oxygen production and biomass yield in spacecrafts. Our efforts are focused on the design and construction of automatic biodevices, the production of several PSII protein mutants, either resistant or oversensitive to radiation, and the identification of the PSII target to radiation in view of future space colonization. In this way we wish to improve our understanding on the effects of CSR on the oxygen-evolving activity of photosynthetic organisms, to identify strategies and mechanisms for protection from space stress, and to determine growth light conditions that maximise the oxygen evolution of microorganism cultures in space.

Another important point is the development of methods for radiation and biology dosimetry, and research estimating the biologically significant radiation loads on human body in space. A biological indicator of radiation dose could be useful in those cases where no other physical dose estimates is available or sufficiently reliable, giving an important contribution in radioprotection: the importance of utilising such sensors in space is related to the need to check the radiation exposure of humans. There is a considerable interest about radiation production of free radicals and their effects on the enzymatic activity.[37] Radiation inactivation is a method used successfully to analyse a wide range of biologically active proteins.[38] During the interaction with gamma rays and high-energy electrons, the energy deposited in protein molecules results in breakage of many covalent bounds and, as a direct effect, a decrease of biological activity which is exponentially related to the radiation dose, because of the random nature of radiation interactions. However, when dealing with protein in a frozen state, only direct effects are considered since damage to individual molecules has no effect elsewhere. Bearing in mind this theory, we have developed a biosensor for radiation using the frozen PSII particles as target able to correlate the activity with the radiation dose.

Understanding the mechanism and response to space stress among the various photosynthetic species is made difficult by both the complex nature of space radiation, and by the complexity of damaging process to several cell targets. A typical solar flare is accompanied by high energy phenomena such as non-thermal emissions of gamma rays, hard and soft-rays and radio waves of wide frequency band. Actually flares differ in their structure, time evolution, and the relative importance of various channels of energy release. As a result any experiment in space unfortunately is not repeatable.

However, laboratory studies using radiation from several sources available in Europe and experiments in balloon flights to stratosphere as well, gave strong support the observation that radiation may interfere with the process of photosynthesis at the PSII level. Therefore, most likely high resistance of microalgae to space radiation is the product of multiple protective strategies, among which replacement of D1-protein by high turnover may play an important role. In sensitive species, repair may be either ineffective or cannot keep up with the rate of damage. It was observed that light seems to play an important role in the radiation stress response. Both dark and high illumination enhance the effect of space radiation, while low light protects against radiation damage. Dark damage is probably because the possibility of recovery is precluded by the reduced protein turnover rate. The findings support the involvement of D1 protein turnover in damage and recovery from space radiation. This is in accordance with previous observations on the important role of D1 protein turnover in counteracting PSII stress in general.[39] The photosynthetic apparatus represents a crucial contact point between the organism and its environment. Its flexibility on the one hand and stability on the other are essential to cope with environmental stress.[40] Of central importance is the turnover ability of D1 protein. The D1 protein is the product of the *psbA* gene, and its synthesis and degradation are known to be regulated by light through redox state modification of PQ quinone. Environmental extremes that negatively impact photosynthesis seem to act primarily by damaging PSII at the level of D1 protein metabolism. Previous experiments have shown that degradation of the D1 protein is involved in the functional damage.[39,41-44] Thus, in an intact system with active protein synthesis, the activity of PSII not only depends on physiological light intensity but also on the presence of interfering radiation and the capacity of the organism to increase D1 protein turnover. Organisms under any type of stress are always characterized by increased turnover of the D1-reaction center polypeptide of PSII. Thus, the D1 protein is a major factor of RCII instability and its replacement after degradation is a primary component of the PSII repair cycle.[16,45-48]

The unstable character of the D1-RC protein has been conserved throughout evolution among oxygen-evolving species. This fact is physiologically significant to the survival of photosynthetic oxygenic organisms under the extreme space conditions.

Gamma rays are completely different from neutrons for their action on the matter. Electromagnetic radiation as gamma rays knock electrons off atoms or molecules causing primary ionisation. This damage process was assessed in all tested organisms (Table 4).

It is interesting to note that the effect of radiation seems to inversely correlated to the cell dimensions, being the *Monodus subterraneus* less affected than the *Chlorella sorokiniana* and *Chlorella zofingiensis.* In a screening of microalgal species exposed to UV-B,[41] the sensitivity was correlated to cell dimensions, that is, the lower the cell diameter the higher the damage recorded. This relationship seems to be valid in our space experiments. As a matter of fact, among the organisms tested *Chlorella sorokiniana,* (Chlorophyta), which showed the highest sensitivity, presents the smallest cell diameter (average cell diameter 3.7 μm). On the other hand, *Monodus subterraneus* (Eustigmatophyta) which showed the lowest inhibition and has the bigger diameter (more than 9 μm). Perhaps large cell cross-sections may provide effective shading of internal structures.

Therefore the observed response to cosmic radiation stress seems to be related to the evolution of the photosynthetic organisms. Cyanobacteria (as *Arthrospira platensis,* prokaryote) and Eustigmatophyte (as *Monodus subterraneus*) show a lower damage than green algae (as *Chlamydomonas reinhardtii*). Photosynthesis was well-established on the Earth by approximately 3.5 billion years ago, and it is widely believed that the ancient photosynthetic organisms had metabolic capabilities similar to those of modern cyanobacteria. This implies that the development of the two photosystems and the ability to evolve oxygen occurred very early in the Earth's history and that a presumed phase of evolution involving non-oxygen evolving photosynthetic organisms took place even earlier.[42,43] Evolution of oxygenic photosynthesis with its CO_2 fixation and oxygen release enabled life on Earth as we experience it today. The capacity to perform oxygenic photosynthesis was passed on to the algae and higher plants by an endosymbiotic event that may have turned a cyanobacterium into a cell organelle, the chloroplast. Within this constraint, the individual characteristics of the donor can be modified to adjust to changes in the photosynthetic pathways and to different light conditions and the presence of ionising radiation as well. For example, in order to gain the functional capability to oxidize water, it was necessary for the primary donor to both become highly oxidizing and coordinate electron and proton transfer with a metal complex.[44]

It could be useful to develop a more sensitive and stable monitoring model with this biological matter (from photosynthetic organisms) to add a new and effective approach to the traditional dosimetry systems. This approach utilising *Chlamydomonas reinhardtii* mutated at the level of D1 protein (Johanningmeier, Esposito, Torzillo, Faraloni, Zanini and Giardi, in preparation) will be undertaken in the next ESA flights called Biopan-Foton towards deep space.

Acknowledgements

This work was supported by the Italian Space Agency. The authors thank Dr. V. Cotronei for his assistance during ASI balloon flights to stratosphere.

The authors thank Dr. G. Angelini for the use of his laboratory facilities for the balloon ASI flights, Dr. U. Nastasi (S. Giovanni Hospital) for irradiation facilities; P. Turkosky, D. Walker, for the assistance at JRC-Ispra, and L. Visca (INFN) for the assistance in elaboration of physical data.

References

1. Nechitailo GS, Mashinsky AL. In: Space Biology; Studies at Orbital Station. Moscow: Mir Publisher, 1993.
2. Spurny F. Radiation doses at high altitudes and during space flights. Radiat Phys Chem 2001; 61:301-307.
3. Ballarini F, Ottolenghi A. Models of chromosome aberration induction: an example based on radiation track structure. Cytogenet Genome Res 2004; 104(1-4):149-56.
4. Beauregard G, Potier M. Temperature dependence of the radiation inactivation of proteins. Anal Biochem 1985; 150:117-120.
5. Le Maire M, Thauvette L, De Foresta B et al. Effects of ionizing radiations on proteins. Biochem J 1990; 267:431-439.
6. Cui FZ, Lin YB, Zhang DM et al. Irradiation effects on secondary structure of protein induced by KeV ions. Radiat Phys Chem 2001; 60:35-38.
7. Ohnishi T, Takahashi A, Ohnishi K. Biological effects of space radiation. Biol Sci Space 2001; 15 (Suppl S):203-10.
8. Takahashi A, Ohnishi K, Takahashi S et al. Differentiation of Dictyostelium discoideum vegetative cells into spores during Earth orbit in space. Adv Space Res 2001; 28(4):549-53.
9. Kranz AR, Bork U, Bucker H et al. Biological damage by ionizing cosmic rays in dry Arabidopsis seeds. Int J Rad Appl Instrum D 1990; 17(2):155-65.
10. Horneck G, Eschweiler U, Reitz G et al. Biological responses to space: results of the experiment "Exobiological Unit" of ERA on EURECA I. Adv Space Res 1995; 16(8):105-18.
11. Horneck G, Rettberg P, Reitz G et al. Protection of bacterial spores in space, a contribution to the discussion on Panspermia. Orig Life Evol Biosph 2001; 31(6):527-47.
12. Koblizek M, Masojidek J, Komenda J et al. An ultrasensitive PSII-based biosensor for monitoring a class of photosynthetic herbicides. Biotechnol Bioeng 1998; 60:664-669.
13. Giardi MT, Koblízek M, Masojídek J. Photosystem II-based biosensors for detection of pollutants. Biosens Bioelectron 2001; 16:1027-1033.
14. Bonting SL. Utilization of biosensors and chemical sensors for space application. Biosens Bioelectron 1992; 7:535-548.
15. Vaijapurkar SG, Deepshikha A, Chaudhuri SK et al. Gamma-irradiated onions as a biological indicator of radiation dose. Radiat Meas 2001; 33:833-836.
16. Mattoo A, Giardi MT, Raskind A et al. Dynamic metabolism of photosystem II reaction center proteins and pigments. a review. Physiol Plant 1999; 107:454-461.
17. Giardi MT. Phosphorylation and disassembly of photosystem II as an early stage of photoinhibition. Planta 1993; 190:107-113.
18. Giardi MT, Komenda J, Masojidek J. Role of protein phosphorylation on the sensitivity of photosystem II to strong illumination. Physiol Plant 1994; 92:181-187.
19. Krause GH. Chlorophyll Fluorescence and Photosynthesis: The Basics. Annu Rev Plant Mol Biol 1991; 42:313-349.
20. Lazàr D. Chlorophyll fluorescence induction. Biochim Biophys Acta 1999; 1412:1-28.
21. Saakov V, Lang M, Schindlef C et al. Changes in chlorophyll fluorescence and photosynthetic activity of French bean leaves induced by gamma radiation. Photosynthetica 1992; 27(3):369-383.
22. Horneck G. Radiobiological Experiment in space: a review. Nicl Tracks Radiat Meas 1992; 20:85-205.
23. Giardi MT, Esposito D, Saviano R et al. Interactive experiments in the Stratosphere: an automatic System to monitor the effect of space radiation on oxygenic organisms in BIRBA flight. In: Monti R, Bonifazi C, eds. Microgravity and Space Station Utilization (MSSU). Vol 2. Napoli: Liguori Editore, 2002:100-105.
24. Berthold DA, Babcock GT, Yocum CF. A highly resolved oxygen-evolving photosystem II preparation from spinach thylakoid membranes. FEBS Lett 1981; 134:231-234.
25. Govindjee. Sixty-three years since Kautsky: Chlorophyll a Fluorescence. Aust J Plant Physiol 1995; 22:131-160.
26. Mitaroff A, Cern MS. The CERN-EU high-energy reference field (CERF) facility for dosimetry at commercial flight altitudes and in space. Radiat Prot Dosimetry 2002; 102(1):7-22.
27. Angelini G, Ragni P, Esposito D et al. A device to study the effect of space radiation on photosynthetic organisms. Phys Med 2001; 17(suppl 1):267-8
28. Metropolis N, Rosenbluth AW, Rosenbluth MN et al. Equation of state calculations by fast computing machines. J Chem Phys 1953; 21(6):1087–1092.

29. Tremmel IG, Kirchhoff H, Weis E et al. Dependence of plastoquinol diffusion on the shape, size, and density of integral thylakoid proteins. Biochim Biophys Acta 2003; 1607:97–109.
30. Zanini A, Ongaro C, Manfredotti C et al. Neutron spectrometry at various altitudes in atmosphere by passive detector technique. Il Nuovo Cimento C 2001; 024:471-784.
31. Dubinin NA, Vaulina EN. Evolution and Gravitation, Nauka Moscow 1979;17-38.
32. Neichitailo G, Gordeev A. Effect of artificial electric firelds on plants grown under microgravity condition. Adv Space Res 2001; 28:629-631.
33. Kempner ES. Novel predictions from radiation target analysis. Trends Biochem Sci 1993; 18(7):236-9.
34. Angelini G, Ragni P, Esposito D et al. A biodosimeter to reveal gamma radiation. Upica CNR 2000; 17.
35. Angelini G, Alessandrelli S, Pace E et al. Effect of γ-radiation on Photosystem II activities for the realisation of Biosensors. Proceedings of the II Workshop on chemical sensors and Biosensors 1999; 358-367.
36. Esposito D, Pace E, Margonelli A et al. A biodosimeter that utilises isolated enzymes to detect of ionising radiation. Radiat Prot Dosimetry 2002; 99(1-4):303-305.
37. Miller JH, Bolger G, Kempner E. Radiation target analysis of enzymes with stable free radicals. Radiat Phys Chem 2001; 62:33-38.
38. Kempner ES. Molecular size determination of enzymes by radiation inactivation. Adv Enzymol 1988; 61:107-147.
39. Giardi MT, Rigoni F, Barbato R et al. Relationships between heterogeneity of the PSII core complex from grana particles and phosphorylation. Biochem Biophys Res Commun 1991; 176(3):1298-305.
40. Torzillo G, Bernardini P, Masojidek J. On line monitoring of chlorophyll fluorescence to assess the extent of photoinhibition of photosynthesis induced by high oxygenic and low temperature and its effect of production of outdoor cultures of Spirulina platensis (cyanobacteria). J Phycol 1998; 34:504-510.
41. Xiong F , Lederer F, Lukavsky J et al. Screening of freshwater algae (Chlorophyta, Chromophyta) for ultraviolet-B sensitivity of the photosynthetic apparatus. J Plant Physiol. 1996; 148:42-48.
42. Turner PE, Chao L. Sex and the evolution of intrahost competition in RNA virus phi6. Genetics 1998; 50:523-532.
43. Woese C. The universal ancestor. Proc Natl Acad Sci USA 1998; 95:6854–6859.
44. Blankenship R, Hartman H. The origin and evolution of oxygenic photosynthesis. Trends Biochem Sci 1998; 23:94–97.
45. Franco E, Alessandrelli S, Masojidek J et al. Modulation of D1 protein turnover under cadmium and heat stresses monitored by (35S) methionine incorporation. Plant Sci 1999; 144:53-61.
46. Tsiotis G, Psylinakis M, Woplensinger B et al. Investigation of the structure of spinach photosystem II reaction center complex. Eur J Biochem 1999; 259(1-2):320-4.
47. Giardi MT, Masojidek J, Godde D. Effects of stresses on the turnover of D1 reaction centre II protein: review, Physiol Plantarum 1997; 101:635-642.
48. Geiken B, Masojídek J, Rizzuto M et al. Incorporation of (35S) methionine in higher plants reveals that stimulation of the D1 reaction centre II protein turnover accompanies tolerance to heavy metal stress. Plant Cell Environ 1998; 21:1265-1273.

CHAPTER 18

Successes in the Development and Application of Innovative Techniques

Eleftherios Touloupakis,* Giovanni Basile, Emanuela Pace, Maria Teresa Giardi and Flavia di Costa

Introduction

Up to now research centres and companies have developed their scientific disciplines and technologies within disparate sectors. Today, the trend is to combine these individual disciplines to meet a common goal. Converging technology represents the application and integration of complementary disciplines towards new fields. The technologies applied in various scientific fields often overlap; however, their union can result in much more than the sum of the single component, leading to unexpected and novel solutions.

A good example of the so called "technological revolution" is the convergence of cognitive computing, intelligent and smart materials and genomics applied to create the emerging field of biosensor technology. From these convergences, great changes and new challenges in the fields of environmental science, medicine and safety are expected. Recently, research institutions and companies have recognized the scientific importance of biosensors as reliable miniaturized devices able to perform monitoring, that can be applied to several areas of analysis. In fact biosensors, as discrete analytical devices, are able to measure analytes selectively, often in a natural matrix, without prior separation of multi component samples, producing quantitative data within minutes. In recent years, micro- and nanofabrication techniques, advances in biochemistry and genetic engineering and the finding and preparation of new materials has enabled the development of novel biosensors, multianalyte biosensors, and microfabricated integrated biosensing systems.

History

The origins of the use of electrodes as sensors dates back to 1956, when L.C. Clark developed an electrode for oxygen detection.[1] Only later, in 1969, G.G. Guilbault and J. Montalvo described the first enzyme-electrode.[2] It was a urea specific enzyme electrode for the determination of urea in blood fluids. In 1973 the Yellow Springs Instrument Company launched a glucose analyser based on the amperometric detection of hydrogen peroxide. In 1982 the first needle-type enzyme electrode for subcutaneous implantation was described.[3] In 1987 MediSense (Cambridge, USA) presented a pen-sized meter for home blood-glucose monitoring based onto screen-printed enzyme electrodes. The BIAcore, produced in 1990, was based on surface plasma resonance technology to monitor affinity reactions in real time. In 1996 the first biosensor for DNA detection, called Affymetrix, was introduced. In recent years, interest in biosensor research has continued to increase, as testified by the many scientific publications and patents developed.

*Corresponding Author: Eleftherios Touloupakis—Department of Chemistry, University of Crete, 71409 Iraklion, Crete, Greece. Email: toulou_e@chemistry.uoc.gr

Biotechnological Applications of Photosynthetic Proteins: Biochips, Biosensors and Biodevices, edited by Maria Teresa Giardi and Elena V. Piletska. ©2006 Landes Bioscience.

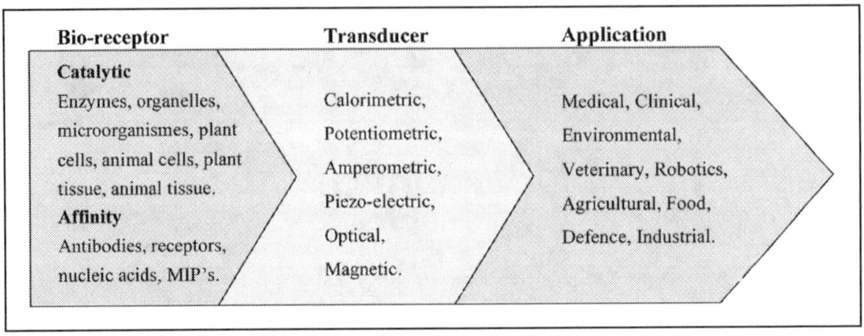

Figure 1. Potential bio-receptors and their applications.

Biosensor

Previously, quantitative and qualitative analyses were performed by analytic chemistry techniques which assure good reliability and precision. However, these traditional techniques are complicated and the measurements must be carried out by skilled operators. Moreover, these techniques often involve laborious pretreatment of samples and a combination of various apparatus and techniques. One of the main features of the biosensor is the selectivity of the bio-receptor for the specific chemical target, which is also conserved in presence of other, potentially interfering, species. The bio-receptor is generally an enzyme or a molecular receptor, such an antibody, immobilized onto the surface of a transducer. Specific interactions between the chemical target and the bio-receptor produce a physicochemical change which is detected and measured by the transducer. The latter is an energy converter that provides quantitative or semi-quantitative analytical information.[4] So the selectivity allows in situ real time measurements, which, together with the miniature size of the biosensor, its low cost, and its wide range of potential applications, has generated great commercial interest. However, to this date a set of problems associated with bio-receptors prevent their widespread commercialisation. They are: low stability and high cost, the need for additional compounds (cofactors), difficulties with the immobilisation technology, poor performance in nonaqueous media and, finally, poor compatibility with micro technology.

Concerning immobilization technology, many methods have been tested to improve bio-receptor stability and to conserve its bioactivity. The most widely used methods of immobilization are: (i) physical or chemical adsorption on a solid surface; (ii) covalent binding; (iii) entrapment within a membrane, surfactant matrix, polymer or microcapsule; (iv) cross-linking between molecules.

The choice of the most suitable immobilization method depends either on the kind of biological element or the transduction system.

The transducer transforms the chemical information into a measurable signal; it can take many forms depending upon the parameters being measured. The most common are electrochemical, optical, mass and thermal. The measured parameters may be converted into an electrical signal, depending on the type and application of the biosensor. Figure 1 indicates potential bio-receptors and their applications.

Advantages

The principal advantages to using biosensors are: high specificity and sensitivity, simple to use, low-cost instrumentation, fast response times, minimum sample pretreatment, small dimensions and ease of transport for in situ measurements.

The immobilized biological recognition element can be regenerated and reused for repeated assays, allowing continuous or multiple assays. Antibody-based biosensors, for example, have been shown to reversibly respond to chemical compounds, within seconds or minutes. Table 1 indicates the most important features for a good quality biosensor.

Table 1. The most important features for a good quality biosensor

Analytical Features	Chemical	Physical
High sensitivity stability high selectivity reversibility short time of analysis	Detection of low analyte concentrations non pretreated samples	Small portable automatic cheap easy-to-use

Table 2. Focus of research on biosensor development

Research	Scientific Research Centers, Universities
Agrofood	Food industries
Environment	Agencies for the environmental protection, industries
Medicine	Research centers, hospitals
Defence	Armed forces, public forces of public safety, institutions
Space	Space research agency

Applications

Biosensor technology has great potential as a revolutionary method for the analysis and control of biological systems.

Some potential applications of biosensors are: agricultural and veterinary analysis through monitoring of food and drink production and the fermentation process; waste management by environmental monitoring of pollution and microbial contamination; clinical diagnosis through pharmaceutical and drug analyses; mining and toxic gas monitoring; military and defensive applications; and aerospace personnel safety. Table 2 illustrates the focus of research on biosensor development.

Technical Challenges in Biosensor Tech

The miniaturization of the devices represents a goal of biosensor technology as it cuts production costs and decreases sample amount. Nanotechnology is applied to the development of microelectromechanical systems and nanobiosensing devices as well. These new analytical systems will be provided with miniature components, such as pumps, detectors, valves, reactors, and thermostats.

The development of an efficient sensing apparatus is based on the knowledge of the bio-reactivity of the biological mean. A deep understanding of the biochemistry, the electron flow, the cofactor participation, as well as of the interfering conditions, is of great importance in the building of an efficient system. In particular, the planning and development of the electric circuit is based on the electrical properties of the biological material.

The close relationship between chemical interaction, triggered biochemical response and signal detected, assures a real-time safety monitoring system.

The pretreatment of samples for the analyses should be minimal or absent to avoid time-consuming and costly procedures. Moreover, the bio-receptor interaction with chemicals should be made independent of physical parameters such as temperature and pH. It is extremely important to obtain a reproducible biochemical response even under different sample conditions. Accordingly, the interface between the bio-receptor and the inorganic transducer should be stable, while the transducer response should be accurate, reproducible and linear over the useful analytical range.

Market Potential

There is an increasing demand for cheap and reliable sensors that allow not only routine monitoring in the laboratory, but also analyses in the hospital ward, emergency and operating rooms. Moreover, the demand by patients for personal self-checking biosensors for use in the monitoring and control of some treatable condition, such as diabetes, is increasing.

One of the most exploitable markets for biosensors is that for immediate assay. Up to now glucose sensing for the control of diabetes represents the most widespread commercial product. However, many other new markets are still available for exploitation, such as medical diagnostics, environmental control, pharmaceuticals, food science and defence. The balance between market opportunities and technical/financial obstacles will control the future expansion of biosensor technology.

Commercial Requirements for Biosensors

The biosensor must satisfy some important general requirements to be commercially successful. These features are summarized in Table 3.

Moreover, in order to convert a prototype into a commercial product it is necessary to satisfy some rules of market demand. Is the proposed biosensor patented? Does it respect the concerning laws and regulations? Is it manufacturable? What are the unfulfilled needs in the market? What is the target market? Who are the competitors? What is the dimension of the market? When will the final goal be achieved? Are there positive feedbacks from the market?

The answers to these questions can give an idea of the commercial potentialities of a new biosensor. Table 4 shows some companies that have already commercialised biosensors.

A commercialisation plan should promote revenue. The product is often over-engineered, too expensive and delayed with respect to the market demand. To prevent this and to manage the stakeholder need, it is best to define the borders within the development plan and to proceed step by step. Hence, each milestone must be verified for its validity, with respect to the aim and the duration of the progressive evolution. No definite rules are established but any successful company must carry out a clear processing program.[5]

Table 3. Some important general requirements for successful marketing methods of biosensor

Instrument Features	Measurement Features	Market Features
Physical robustness	Reliable immobilization procedure	Size/weight appropriate for use
Accuracy and reproducibility	Fully automated	Acceptability by user
Sensitivity and resolution	Minimal logistical burden	Service requirements
Dynamic range	Speed of response	Large scale manufacturing
Stability of the biological material in the bio-receptor	Highly specific bio-receptor for the purpose of the analyses	Biocompatibility of in vivo biosensor
Insensitivity to environmental conditions	No false positives	Running costs and life
Ease of testing and calibration	Reliability and self-checking capability	The bio-receptor should be non contaminating
Insensitivity to electrical and other environmental interferences	Reproducible placement of small volume of bio-receptor	Capital cost

Table 4. Companies that have already commercialised biosensors

Biosensor.srl	It offers basic instruments for realization of biosensors and biosensors upon request. www.biosensor.it
Medtronic	The CGMS (Continuous Glucose Monitoring System), a device for subcutaneous implantation. www.medtronic.com
Cygnus	The GlucoWatch Biographer is a biosensor device for continuous glucose monitoring.[6] www.cygnus.com
Biosensor Applications	This company focuses on biosensors in general, including sensors for bioprocesses. www.biosensor.se
Biacore AB	It offers several models of biosensors based on SPR (surface plasmon resonance) technology.[7] www.biacore.com
Research International	The RAPTOR Plus was developed by the US Naval Research Laboratory (NRL).[8] www.resrchintl.com
Nomadics Inc.	It develops advanced sensors, instrumentation, and mobile technology products. www.nomadics.com
Affinity Sensors	It focuses on evanescent wave biosensors. www.affinity-sensors.com
Bioanalytical Technologies	The company sets up the production and improvement of liquid-crystalline DNA biosensor technology. www.biosensor.ru
Advanced Biosensor Technology [ABTECH]	It produces chemical and biological sensor devices. www.abtechsci.com
Inventus BioTec	A company for bioanalytic, biosensors, and diagnostic. www.inventus-biotec.com
National Research Development Corporation	BOD biosensors. www.nrdcindia.com/index.htm
Inventus BioTec	Produces biological sensor devices (GlucoSens, LactoSens). www.inventusbiotec.com

Future Challenges

Application and commercialisation of biosensor technology has lagged behind the output of research laboratories. Although many biosensors and several patents are produced each year, very few of these play a prominent role in clinical diagnostics, the food industry, environmental, agricultural, or veterinary applications. There are many reasons for the slow rate of technology transfer from research laboratories to the marketplace: cost considerations, stability and sensitivity issues, quality assurance, and instrumentation design. Many of the main barriers are technical: methods of sensor calibration, methods of producing inexpensive and reliable sensors, stabilisation and storage of biosensors, and total integration of the sensor system. Another problem is the need for a multidisciplinary team, which is not always available in a factory.

Biosensor advancement in the commercial world could also be accelerated by the use of intelligent instrumentation, electronics, and multivariate signal processing methods. Increasing attention must be paid to the engineering of both the basic components and the device as a whole.

Conclusions

New companies can grow by exploiting the advantages of biosensor technology. The last two decades have seen the increment of technological investments and therefore the growth of small start-up enterprises whose intent was the development of the innovative technologies of commercial interest. However, the youth and small dimensions of these enterprises are responsible for their poor experience in the trend marketing. As a consequence, their innovative products do not get the commercial attention that they deserve. Frequently they produce one or more technologies that can be introduced in the market, in which case the key challenge is how to convert their potential into revenue for the company as well as the stakeholders. Actually, the main problem for a new enterprise lies less in the technology than on the commercialisation aspects of the innovative product. The commercialisation process defines the company strategy and its location in the business' marketing.

Regarding biosensor production, its marketing is in competition with many chemical analytical systems as well as conventional laboratory techniques. Most of the patents are presented by scientific institutions as biosensor prototypes without a real potential for commercialisation. The institutions should strictly collaborate with industries to define the properties of a real marketable product. Economical and technical problems should be overcome, together with the manufacture of the sensing and transductions components and their increased reliability and stability.

Biosensors sensing is generally less stable than conventional technologies, though the detection properties of a biosensor, its miniaturization, automation and easily use, make it advanced. Increasing progress within the biosensing field will lead to the ability to overcome technological obstacles and will improve the analytical detection. In the near future the development of biosensors miniature enough to be implanted into humans, animals and plants are expected.

Finally, the potential applications of biosensors based on the integration of scientific research and industrial technology seem to be unlimited.

Acknowledgement

This work was supported by the EU Contract No QLK3-CT-2001-01629.

References

1. Clark LC. Monitor and control of blood and oxygen in tissue. Trans Am Soc Artif Intern Organs 1956; 2:41-48.
2. Guilbault GG, Montalvo J. A urea specific enzyme electrode. J Am Chem Soc 1969; 91:2164.
3. Shichiri M, Kawamori R, Yamasaki T et al. Wearable-type artificial pancreas with needle-type glucose sensor. Lancet 1982; ii:1129-1131.
4. Thevenot DR, Toth K, Durst RA et al. Electrochemical biosensors: Recommended definitions and classification. Biosens Bioelectron 2001; 16:121-131.
5. Laird I, Sjoblom L. Commercializing technology: Why is it so difficult to be disciplined? Business Horizons 2004; 47(1):65-71.
6. Tierney MJ, Kim HL, Burns MD et al. Electroanalysis of glucose in transcutaneously extracted samples. Electroanalysis 2000; 12:666-671.
7. Elliot CT, Baxter GA, Crooks SRH et al. The development of a rapid immunobiosensor screening method for the detection of residues of sulphadiazine. Food Agric Immunol 1999; 11:19-27.
8. King KD, Vanniere JM, Leblanc JL et al. Automated fiber optic biosensor for multiplexed immunoassays. Environ Sci Technol 2000; 34(13):2845-2850.

Index

E

F

G

H

I

L

M

N

O

P

Q

R

S

T

W

X